DATE DUE

DEC 04 2014			

HIGHSMITH   # 45220

Contents

Chapter 1	Introduction and Overview *Carol M. Duffus and John H. Duffus*		1
Chapter 2	Toxic Amino Acids *J. P. Felix D'Mello*		22
Chapter 3	Lectins *George Grant*		49
Chapter 4	Proteinase Inhibitors *Grenville Norton*		68
Chapter 5	Antigenic Proteins *J. P. Felix D'Mello*		108
Chapter 6	Glucosinolates *Alan J. Duncan*		126
Chapter 7	Alkaloids *David S. Petterson, David J. Harris, and David G. Allen*		148
Chapter 8	Condensed Tannins *D. Wynne Griffiths*		180
Chapter 9	Cyanogens *Raymond H. Davis*		202
Chapter 10	Mycotoxins *Brian Flannigan*		226
Chapter 11	Fibrous Polysaccharides *Martin Eastwood and Christine A. Edwards*		258

Chapter 12 Saponins 285
G. Roger Fenwick, Keith R. Price, Chigen Tsukamoto, and Kazuyoshi Okubo

Subject Index 329

Contributors

David G. Allen, *Agricultural Chemistry Laboratory, Chemistry Centre (WA), 125 Hay Street, East Perth, WA 6004, Australia*
Raymond H. Davis, *University of London, Wye College, Wye, Ashford, Kent TN25 5AH, UK*
Alan J. Duncan, *Macaulay Land Use Research Institute, Pentlandfield, Roslin, Midlothian EH25 9RF, UK*
Carol M. Duffus, *Department of Crop Science and Technology, The Scottish Agricultural College — Edinburgh, West Mains Road, Edinburgh EH9 3JG, UK*
John H. Duffus, *Department of Biological Sciences, Heriot-Watt University, Riccarton, Edinburgh EH14 4AS, UK*
J. P. Felix D'Mello, *Department of Crop Science and Technology, The Scottish Agricultural College — Edinburgh, West Mains Road, Edinburgh EH9 3JG, UK*
Martin Eastwood, *Gastro Intestinal Laboratory, Western General Hospital, Crewe Road, Edinburgh, UK*
Christine A. Edwards, *Department of Human Nutrition, University of Glasgow, Yorkhill Hospitals, Glasgow, UK*
G. Roger Fenwick, *AFRC Institute of Food Research, Colney Lane, Norwich NR4 7UA, UK*
Brian Flannigan, *Department of Biological Sciences, Heriot-Watt University, Riccarton, Edinburgh EH14 4AS, UK*
George Grant, *Rowett Research Institute, Greenburn Road, Bucksburn, Aberdeen AB2 9SB, UK*
D. Wynne Griffiths, *Scottish Crop Research Institute, Invergowrie, Dundee DD2 5A, UK*
David J. Harris, *Agricultural Chemistry Laboratory, Chemistry Centre (WA), 125 Hay Street, East Perth, WA 6004, Australia*
Grenville Norton, *Department of Applied Biochemistry and Food Science, University of Nottingham, School of Agriculture, Sutton Bonington, Loughborough LE12 5RD, UK*
Kazuyoshi Okubo, *Department of Food Chemistry and Biotechnology, Faculty of Agriculture, Tohoku University, 1-1 Tsutsuitori, Amamiyacho, Sendai, Japan 980*
David S. Petterson, *Western Australia Department of Agriculture, Baron-Hay Court, South Perth, WA 6151, Australia*

Keith R. Price, *AFRC Institute of Food Research, Colney Lane, Norwich NR4 7UA, UK*

Chigen Tsukamoto, *Department of Food Chemistry and Biotechnology, Faculty of Agriculture, Tohoku University, 1-1 Tsutsuitori, Amamiyacho, Sendai, Japan 980*

CHAPTER 1

Introduction and Overview

CAROL M. DUFFUS AND JOHN H. DUFFUS

1 Introduction

The presence of naturally occurring anti-nutritional factors and other toxic substances is one of the main factors to be considered in the use of crops, both as food and as industrial raw materials. The purpose of this chapter is to provide a general background to the subject and, in addition, to deal briefly with some of the lesser known toxic substances whose importance does not seem at present to justify separate treatment.

2 Crop Plants

In the present context we will be using the term crop plant to mean any plant which is harvested for use in the diet of Man and other animals. The major food crops are the cereals which include wheat, maize, rice, and barley. Other food crops of importance include vegetables, fruits, pulses, nuts, and seeds.

Also contained in this definition of crop plants are the grasses and other forage crops, which are a major item in the diet of herbivores.

The parts of a crop plant which may be used as food sources include the foliage, buds, stems, roots, fruits, and tubers. Much is eaten following processing and cooking and these post-harvest treatments often result in an improvement in quality, both by improving digestibility and by destroying endogenous toxic substances.

3 Contribution of Plants to the Diet of Man and Other Animals

As far as the diet of Man is concerned, undernutrition is caused primarily by an inadequate intake of dietary energy. When energy intake is sufficient, specific nutrient deficiencies, such as protein, seldom pose a problem. For this reason, the relative contribution of the different plant food groups to the human diet is determined on the basis of *per caput* dietary energy supplies.[1] The changes over the years from 1961 to 1981 in the contribution of individual plant products to dietary energy supplies in Man are shown in Table 1. These data also provide an indication of food consumption. The figures show that in the developed market economies (*e.g.* Western Europe and USA) there was a drop in the shares of cereal grains, roots, and tubers, with this shortfall being made up largely from vegetable oils and fats, as well as alcoholic beverages. This suggests that the protein content of the plant-derived portion of the diet may have fallen, since neither fats, oils, nor alcoholic beverages have a significant protein content. There was at the same time however, a small increase in the amount of animal protein eaten, which would compensate for the reduced protein content of the plant-based dietary constituents. A much greater decrease in the contribution of cereals was observed in Eastern Europe and the USSR over the same period. It should be noted however, that the consumption of cereals in this economic group was initially some 16 per cent higher than in the developed market economies. The share of roots and tubers also fell. In this case, the difference was made up by a marked rise in the proportion of animal products with smaller rises in vegetables, fruits, vegetable oils, fats, and alcoholic beverages.

On a world basis, there has been very little change in the share of the major food groups to total dietary energy supplies over the period 1961–1981. The falls in the share of pulses, nuts, and seeds, and of roots and tubers, were compensated for by increases in sugar and in vegetable oils and fats. Overall the figures show that food from crop plants makes by far the greatest contribution to *per caput* daily energy supplied. In the far East (not shown) the contribution for 1979–1981 was over 94 per cent.

The share of animal products in total *per caput* energy supplies represents an additional but unseen contribution by crop plants to the human diet. The natural food of the domesticated herbivore is pasture herbage, which normally contains many species of grasses, legumes, and herbs. Cultivated grasslands, on the other hand, normally consist of one or a small group of species. Forage crops other than grasses include members of the family Leguminosae and the genus *Brassica*. The forage legumes are superior to grasses in protein and mineral content and commonly grown species include the clovers (*Trifolium* spp.) and lucerne (*Medicago sativa* L.). Since many tropical pastures are deficient in indigenous legumes there is a continuing search for new species suitable for introduction as forage legumes. Promising newcomers include *Leucaena leucocephala*, *Indigofera spicata*, and the grain legume *Canavalia ensiformis*. However, as we will see later (Chapter 2) the presence of toxic non-protein amino acids has restricted their full exploitation.

Cereal grains provide the major component of the concentrate ration of ruminant animals. Non-ruminant species, such as pigs and poultry depend upon them as a source of dietary energy. The residues remaining after extraction of oil from oilseeds are rich in protein (400–550 $g\ kg^{-1}$) and serve as

[1] FAO, 'The Fifth World Food Survey', Food and Agriculture Organization of the United Nations, Rome, 1987.

Table 1 Share of major food groups in total per caput dietary energy supplies 1961–63, 1969–71, and 1979–81 (per cent of total)[1]

Food Group	Developed market economies			Eastern Europe and USSR			World		
	1961–3	1969–71	1979–81	1961–3	1969–71	1979–81	1961–3	1969–71	1979–81
VEGETABLE PRODUCTS	69.2	68.4	68.3	77.6	75.0	72.9	83.5	83.5	83.7
Cereals	31.1	27.4	26.4	47.2	41.4	37.5	50.1	49.6	50.2
Pulses, nuts, and seeds	2.6	2.4	2.4	1.6	1.5	1.4	5.3	4.7	3.9
Roots and tubers	4.9	4.2	3.7	7.9	7.3	6.0	8.2	7.8	6.9
Sugar	12.6	13.5	13.0	10.0	12.0	13.0	8.5	9.0	9.1
Vegetables and fruits	4.6	5.0	4.8	2.7	3.0	3.6	3.8	3.7	4.0
Vegetables oils and fats	7.9	9.7	11.3	4.5	5.0	6.0	4.9	5.6	6.6
Stimulants and spices	0.7	0.8	0.8	0.2	0.3	0.4	0.4	0.5	0.4
Alcoholic beverages	4.8	5.4	5.9	3.5	4.5	5.0	2.3	2.6	2.6
ANIMAL PRODUCTS	30.8	31.6	3.17	22.4	25.0	27.1	16.5	16.5	16.3

Table 2 Components of some typical diets for farm animals in production

Dairy Cows

Weight (kg)	Milk yield (kg day^{-1})	Grass silage	(DM)	Barley	Soyabean meal
600	25	39	(9.8)	6.5	1.2
		Hay			
600	25	10.2		8.9	0.9

Beef Cattle

Weight (kg)	Liveweight gain (kg day^{-1})	Grass silage	(DM)	Barley	Soyabean meal
350	1.0	20	(5.1)	2.5	0
450	1.0	26	(6.4)	2.7	0
		Hay			
350	1.0	4.7		3.6	0.5
450	1.0	5.5		4.2	0.6

Growing Pigs

	Cereals	Soyabean meal	Fish meal
50 kg liveweight	1.51	0.35	0.1
Pregnant sow	2.02	0.16	0.06
Lactating sow	4.81	0.80	0.15

(data from Dr C. A. Morgan, The Scottish Agricultural College, Edinburgh)

supplementary protein sources for ruminant animals in intensive and semi-intensive production systems. They are also commonly used as dietary protein supplements for non-ruminant farm animals whose principal dietary component is cereal grains. The seeds from which these cakes or meals are derived, generally contain anti-nutritional substances. Some of these may be removed or inactivated during the process of oil extraction but where they are not, additional detoxification processes are required before they can be incorporated into animal dietary formulations. The main components of some typical animal diets are shown in Table 2.

4 Nutritive Value of Plants and Plant Products

The potential value of a food for supplying nutrients can be determined by chemical analysis. However its actual value can be arrived at only after making allowance for losses which occur during digestion, absorption, and metabolism. Furthermore, the conventional chemical analysis of food composition does not include estimates of the amounts of anti-nutritional factors present. Composition therefore is not a direct measure of nutritive value. Nevertheless, determinations of food composition have proved to be a useful indicator of nutritive value.

For man, the most widely used reference book of food composition is

McCance and Widdowson's 'The Composition of Foods'[2] together with its more recent supplements.[3] These cover over one thousand separate food items, most of which are commonly eaten in the UK and Western Europe. The values in the table do give some indication of the availability of carbohydrates and energy from a particular food. That is, values for carbohydrates are a measure of free sugars, dextrins, and starch, expressed as monosaccharides. These are the 'available' carbohydrates and are assumed to be digested and absorbed by Man, ultimately entering the system as glucose or glucogenic precursors. Dietary fibre, on the other hand, is now reported[4] both as 'total fibre' and 'total non-starch polysaccharides' together with five of the components or subdivisions of fibre which may have physiological activity. Such figures give an indication of the proportion of dietary carbohydrate (possibly including lignin) which is not digested. Energy values (in kJ) are calculated from the amounts of protein, fat, carbohydrate, and alcohol in the foods using energy conversion factors. The composition of some common foods is shown in Table 3. The science and practice of farm animal nutrition is rather more exact than that of human nutrition. As well as determining the quantities in which nutrients are required by the different classes of farm animal, a major part of this subject is devoted to assessing, as precisely as possible, the dietary availability of nutrients. As with the human diet, the presence of naturally-occurring anti-nutritional factors in animal foods is of major interest and much effort is expended in assessing their effect on animal metabolism, growth, and performance, in addition to developing detoxification protocols.

Foods for farm animals are evaluated in terms of their digestibility, energy content, energy value, and protein utilization. The digestibility of a food is most accurately defined as that proportion which is not excreted in the faeces and which is therefore assumed to be absorbed. For the determination of digestibility, the food is fed to the animals in known amounts and the output of faeces measured. For poultry, digestibility measurements are complicated by the fact that faeces and urine are voided together. If it is assumed that most urinary nitrogen is present as uric acid, or that most faecal nitrogen is present as true protein, then a chemical separation of faeces and urine may be achieved.

Digestibility determinations are difficult to carry out. Hence, *in vitro* methods, often quite successful, have been developed in which the food is treated with digestive enzymes isolated from the alimentary tract. For example, the digestibility of foods for ruminants can be measured by treating them first with rumen liquor and then with pepsin.

A number of factors influence digestibility. These include food composition, dietary composition, food preparation, processing, cooking, and quantity of food eaten (see also next section). In relation to food composition, the fibre fraction has probably the greatest influence on digestibility. The physiological

---

[2] A. A. Paul and D. A. T. Southgate, 'McCance and Widdowson's – The Composition of Foods', 4th edition, HMSO, London, 1978.
[3] B. Holland, I. D. Unwin, and D. H. Buss, 'Cereals and Cereal Products', Third supplement to 'McCance and Widdowson's The Composition of Foods', 4th edition, Royal Society of Chemistry, Ministry of Agriculture, Fisheries and Food, Nottingham, 1988.
[4] R. W. Wenlock, L. M. Sivell, and I. B. Agater, *J. Sci. Food Agric.*, 1985, **36**, 113.

Table 3 *Composition of some common foods (g per 100 g food)*[3]

Food	Water	Protein	Fat	Carbohydrate	Energy value (kJ)	Starch	Sugars	Dietary fibre[b]
Wheat flour								
White **plain**	14.0	9.4	1.3	77.7	1450	76.2	1.5[a]	3.6
Wholemeal	14.0	12.7	2.2	63.9	1318	61.8	2.1[b]	8.6
White bread								
average	37.3	8.4	1.9	49.3	1002	46.7	2.6	3.8
Wheat bran	8.3	14.1	5.5	26.8	872	23	3.8	39.6
Fruit cake								
Rich Dundee	20.6	4.9	12.5	50.7	1357	11.9	38.8	3.4
Naan bread	28.8	8.9	12.5	50.1	1415	44.6	5.5	2.2
Oatmeal								
Quick cook **raw**	8.2	11.2	9.2	66.0	1587	64.9	1.1	6.8
Soyabean flour								
low fat	7.0	45.3	7.2	28.2	1488	14.8	13.4	13.3

[a] includes the glucofructan levosin
[b] determined by the method of Southgate[4]

effects of dietary fibre have generated considerable interest in relation to digestion in non-ruminant farm animals and in Man. In general, increasing the fibre content of the diet increases faecal bulk and frequency of elimination. Fibre may also be associated with inhibitory effects on the absorption and retention of minerals such as zinc, calcium, and magnesium. Phytic acid (inositol hexaphosphate), a component of the outer endosperm of cereals, and therefore a constituent of bran fractions, can bind calcium, magnesium, zinc, and iron thus decreasing their absorption from the tract. It may be that some of the observed inhibitory effects of cereal fibre can be attributed to the presence of this molecule. In ruminants, the relative proportions of fibre to starch and other components of the concentrate ration, influence the numbers and populations of microbial species present in the rumen. Thus, roughage diets high in cellulose and other non-starch polysaccharides give rise to high concentrations of acetic acid in the rumen. As the proportion of concentrates (starch and protein) increases, the proportion of acetic acid falls and propionic acid rises, leading to a depression of milk fat content in lactating ruminants.

Energy Value

In the scientific rationing of farm animals, as with Man, the first consideration in the evaluation of foods is given to those nutrients supplying energy. Thus, if the energy supply is sufficient, then any deficiency, say of an essential mineral element or vitamin, can be made up by adding small amounts of the purified component to the final dietary formulation. The energy value of foods may be stated as net, metabolizable, digestible, (or even gross) energy.[5] Digestible energy is the gross energy less the energy contained in the faeces. Metabolizable energy is the digestible energy less the energy lost in urine and, if a ruminant, in combustible gases, such as methane. Net energy value of a food is the metabolizable energy less the losses of energy as heat (heat increment). In most European countries, energy systems for pigs are based on metabolizable energy. Energy systems for ruminants are more complex because of the wider variety of foods involved and the range of different digestive processes taking place in the ruminant digestive tract. In the present energy system in the UK, energy values of foods are expressed as metabolizable energy, and the metabolizable energy value of a ration is calculated by adding up the contributions of the individual feed constituents. In turn the energy requirements of the animals are expressed as net energy. The essential feature of the interface is a series of equations to predict efficiency of utilization of metabolizable energy for maintenance, growth, and lactation.

Protein Evaluation

Different criteria are required for the evaluation of food proteins for ruminant and non-ruminant animals. Both classes of animals require an adequate supply

[5] P. McDonald, R. A. Edwards, and J. F. D. Greenhalgh, 'Animal Nutrition', Longman, Harlow, UK, 1988.

of indispensable amino acids in the proper quantity plus sufficient dispensable amino acids to meet requirements for synthesis of protein and other nitrogenous substances. In non-ruminants, amino acids are absorbed in the proportions present in the original dietary protein. But in ruminants, some of the ingested protein and non-protein nitrogenous substances are incorporated into microbial protein in the rumen and eventually passed into the small intestine where they are absorbed, together with any amino acids derived from rumen undegradable protein. This means that the pattern and quantities of absorbed amino acids may differ considerably from those originally present in the food. Different approaches are therefore necessary in the evaluation of food protein sources for ruminant and non-ruminant animals.

The protein in foods destined for non-ruminants may be evaluated by a variety of procedures including: 'biological value' which measures the proportion of absorbed nitrogen which is retained by the animal; or chemical scoring techniques designed to establish the dietary pattern of amino acids most appropriate to the needs of the animal. Systems of evaluating protein foods for ruminants centre on the concept of 'metabolizable protein' developed by the Agricultural and Food Research Council of the United Kingdom.[6,7]

5 Nutritive Value of Plants and Plant Products—Post-harvest Events

The nutritive value of plants and plant products may be influenced considerably by post-harvest events. These include storage, processing, and cooking.

Storage

In temperate climates (including Europe) there is only one growing season for most crops. Consequently, the harvested material has to be stored if demand is to be satisfied throughout the year. During this period, changes may take place in nutritive value. For example, in potatoes stored at low temperatures there may be a marked increase in the sugar content. This will affect nutritive value very little but has a major effect on crisping and chipping quality. More serious perhaps, is the effect of storage on the content of toxic glycoalkaloids (Chapter 7). Both duration of storage and exposure to light during harvesting, handling, and marketing, can result in an increase in the content of tuber glycoalkaloids. The nutritive value of cereal grains in storage does not normally alter significantly but if put into store at too high a moisture content, mycotoxins can be produced as a result of fungal infections (Chapter 10).

---

[6] Agricultural Research Council, 'The Nutrient Requirements of Ruminant Livestock', Commonwealth Agricultural Bureaux, Farnham Royal, 1980, Chapter 4, p. 121.

[7] Interdepartmental Working Party Report on Protein Requirement of Ruminants, Commonwealth Agricultural Bureaux International, Wallingford, 1992 (in press).

Processing and Cooking

The nutritive value of a food can be altered markedly following processing and/or cooking. For example, the wheat grain undergoes considerable changes in composition during threshing, storage, milling, fermentation, and baking. Thus, during threshing, the lignified and fibrous outer glumes and paleae are removed, leaving a single-seeded wheat fruit or caryopsis. If stored at a moisture content between 16 and 30 per cent, fungal growth can take place with the production of mycotoxins as we saw previously (also Chapter 10). Above 30 per cent moisture content, bacterial growth may lead to a marked rise in temperature and even charring. During the milling of grain to produce white flour, the endosperm, which accounts for about 82 per cent of the intact grain, is separated from the outer layers (bran) and embryo (germ) and milled. Thus the theoretical maximum extraction rate for white flour should be around 82 per cent. At this level of extraction however, a proportion of bran and germ will end up in the milled product. In practice therefore, extraction rates for white flour are normally less than 75 per cent. Depending on the extraction rate then, flours will have different nutritive values (Table 3). Thus wholemeal flour (100 per cent extraction) has higher dietary fibre, phytic acid, protein, and lipid content than white flour (75 per cent extraction). Since it includes the outer layers it may also contain higher amounts of pesticides, mycotoxins, and aerial contaminants. The higher lipid content of wholemeal flour, derived from the aleurone and embryo, means that the shelf life is shorter than that of white flour. Interestingly, the phytic acid content of unleavened breads (made without yeast) is greater than that of breads made by baking a dough containing yeast. The reason appears to be that yeast contains a phytase which releases the bound mineral elements from the phytic acid and thus accounts for the greater availability of iron and other minerals from breads made with a fermented dough (Figure 1).[8]

The nutritive value of oilseed meals and cakes depends very much on the process used to extract the oil from the original oilseeds. If, for example, the oil is removed from the seed by low temperature solvent extraction, then protein quality is not adversely affected. However, since the temperature of extraction is low, many of the anti-nutritional substances present are not inactivated and may be present in the meals or cakes left after oil extraction. Soyabeans, which have a low original oil content of around 160 to 210 g kg^{-1} are normally solvent extracted, and the residual meals contain a number of toxic substances including allergenic, goitrogenic, and anticoagulant factors. For non-ruminant animals therefore, a heat treatment called toasting follows solvent extraction, which, if carefully controlled, inactivates the toxic factors while not reducing protein quality. In contrast, the oil from seeds with higher original oil content, such as groundnuts or cottonseeds, is removed in a process where the seeds are passed through a screw press. This involves higher temperatures than direct

---

[8] P. Le François, A. Verel, and Y. Audidier, in 'Nutrient Availability: Chemical and Biological Aspects', ed. D. Southgate, I. Johnson, and G. R. Fenwick, Royal Society of Chemistry, Cambridge, Special Publication No. 72, 1989, p. 161.

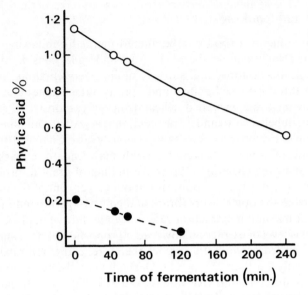

Figure 1 *Phytic acid content of bread doughs during fermentation:*[8] *Wholemeal* —○—; *White* —●—

solvent extraction. Increased pressures are also generated within the press. While these conditions may improve nutritive value by inactivating anti-nutritional factors, they may simultaneously result in protein denaturation and as a consequence lower the protein nutritive value.

6 The Nature of Toxicity

All substances are potentially toxic.[9] The only factor which determines whether they cause harm or not is the degree of exposure. In foodstuffs this will depend upon the concentration present in the food initially and on the amount consumed. The precise relationship between exposure and harm caused depends upon the chemical properties of a given substance and the way in which it is metabolized by the organism at risk. The details of this relationship have been worked out for a number of drugs and their actions on human beings and this knowledge forms the basis of the science of quantitative structure activity relationships (QSAR) which has been invaluable in the production of new drugs. However, the application of QSAR to wider areas of toxicological assessment will take time. Hence, we are still dependent on the information gained from toxicological tests for our current assessment of potential toxicity.

Toxicological testing has become increasingly codified in recent years with the coming of the Notification of New Substances Regulations as part of the effort of the Commission of the European Communities (CEC) to ensure chemical safety in the light of growing knowledge of the incidence of disease

[9] J. H. Duffus, 'Environmental Toxicology', Edward Arnold, London, 1980.

Table 4 *The minimum information required for introduction of a new substance under UK and CEC regulations*

Identity: Name/trade name; formula (empirical/structural); composition; methods of detection/determination.

Uses and Precautions: Proposed uses; estimated production/importation; handling/storage/transport methods and precautions; emergency measures.

Physicochemical properties: Melting point; relative density; vapour pressure; surface tension; water solubility; fat solubility; partition coefficient (octanol:water); flash point; flammability; explosive properties; auto-flammability; oxidizing properties.

Toxicological studies: Acute toxicity (oral/inhalation/cutaneous); skin and eye irritancy; skin sensitization; sub-acute toxicity (28 day); mutagenicity (bacterial and non-bacterial).

Ecotoxicological studies: Toxicity to fish; toxicity to *Daphnia*; degradation data (BOD/COD).

Possibility of rendering the substance harmless: For industry; for public; declaration concerning the possibility of unfavourable effects; proposed classification and labelling; proposals for any recommended precautions for safe use.

Note: For a detailed review of these requirements, see J. L. Vosser, 'The European Community Chemicals Notification Scheme and Environmental Hazard Assessment' in 'Toxic Hazard Assessment of Chemicals', ed. M. L. Richardson, Royal Society of Chemistry, London, 1986, p. 117 (see also reference 10).

caused by chemical exposures. Table 4 lists the minimum information (Technical Dossier Base Set) now required under CEC and UK regulations, including the toxicity tests that must be carried out before widespread use of a new substance can be considered.

Of course, the information now required for new chemicals is rarely available on the toxic substances present in crop plants and it will be many years before it has been compiled. In the meantime, an attempt to prepare such a dossier will still make a good start to any assessment of the hazards likely to be associated with any given potentially toxic substance.

Assessment of hazard is the first stage in establishing, for regulatory purposes, supposedly safe human consumption levels of potential toxicants in food. It is essential to know the levels of the toxicant in the foodstuff under consideration, the relationship between intake of toxicant and production of adverse effects, and the normal rate of consumption of the foodstuff. From this information, one can calculate whether any harm is likely to be occurring from current consumption patterns. If so, immediate action must be taken to reduce intake of that foodstuff to tolerable levels. Normally a tolerable level is one which is appreciably less than any known to cause harm. Safety factors are applied to allow for degrees of uncertainty. For example, if there are reliable human data relating exposure to harm (usually following epidemiological studies), the safe level is deemed to be 10 times less than the minimum level known to cause harm. If there are no good human data relating to mammals, the safe level is

[10] M. L. Richardson (ed.), 'Toxic Hazard Assessment of Chemicals', Royal Society of Chemistry, London, 1986.

taken to be 100 times less than that known to cause harm. If the available mammalian data are limited, the factor applied is 1000. In the absence of any data relating harmful effects to exposure, these must be obtained urgently for any substance giving cause for concern. However, many substances which are potentially toxic have not been subjected to scrutiny because they are found in traditional foods which have never been considered harmful and are therefore classified as 'Generally Regarded as Safe (GRAS)'.

7 The Principal Toxic Substances Present in Crop Plants

The principal toxic substances so far identified in crop plants come from a wide range of chemical types.[11-15] Not all of these substances will be covered in this book but only those which have so far attracted the most concern and attention. Briefly, the substances to be considered are the following:

Toxic Amino Acids

The potentially toxic amino acids are not normal components of proteins but occur free in many plants and particularly in the Leguminosae. This has limited the use of leguminous plants as food. The toxic amino acids are usually concentrated in seeds but can be found throughout the whole plant. Often their toxic action affects the nervous system but a wide range of other effects have been reported. Of particular interest in this group of substances are the lathyrogens.[16,17] Lathyrogens occur throughout the plant but are present in the highest concentrations in seeds. Lathyrogens include L-α,γ-diaminobutyric acid, β-N-oxalyl-α,β-diaminopropionic acid, and γ-N-oxalyl-α,γ-diaminobutyric acid which are found in several Lathyrus species including the chick pea (*Lathyrus sativus*) and β-N-(γ-glutamyl) aminopropionitrile found in other species including the sweet pea (*Lathyrus odoratus*), the caley pea or hairy vetchling (*Lathyrus hirsutus*), and the everlasting or wild pea (*Lathyrus sylvestris*).

There are two syndromes of lathyrism. Osteolathyrism is characterized by severe skeletal abnormalities and, in some animals, by haemorrhaging. It has been produced experimentally by feeding β-aminopropionitrile. Neurolathyrism is the other syndrome and is the condition usually referred to as lathyrism. It is characterized in animals by weakness and paralysis of the

---

[11] M. R. Cooper and A. W. Johnson, 'Poisonous Plants in Britain and Their Effects on Animals and Man', HMSO, London, 1984.
[12] J. B. Harris (ed.), 'Natural Toxins: Animal, Plant and Microbial', Clarendon Press, Oxford, 1986.
[13] R. F. Keeler and A. T. Tu, 'Handbook of Natural Toxins', Marcel Dekker, Basel, 1983.
[14] I. E. Liener (ed.), 'Toxic Constituents of Plant Foodstuffs', 2nd edition, Academic Press, New York, 1980.
[15] D. H. Watson (ed.), 'Natural Toxicants in Food', Ellis Horwood, Chichester, 1987.
[16] D. N. Roy, *Nutr. Abstr. Rev.*, 1981, **51**, 691.
[17] G. A. Rosenthal, 'Plant Nonprotein Amino and Imino Acids', Academic Press, New York, 1982.

hindlegs and difficulty in breathing. Horses are particularly susceptible but cattle and sheep are also affected. In humans, paralysis is usually confined to the legs and there may be muscle tremors. It should be emphasized that when only relatively small quantities of chickpeas are eaten as part of a mixed diet, they are quite harmless.

Lectins

Lectins (phytohaemagglutinins) are proteins or glycoproteins of non-immune origin which have multiple highly specific carbohydrate binding sites. They were originally identified in the castor bean but are now known to be widespread in the plant kingdom. They are particularly concentrated in legume seeds and have been shown to cause gastroenteritis, nausea, and diarrhoea in man and cattle.

Protease Inhibitors

Protease inhibitors and especially trypsin inhibitors are often considered the main cause of death in experimental animals fed raw soyabeans. They are found in most if not all species of beans and also in lupins.

Allergens

Certain individuals may show a more or less violent local or generalized reaction (allergy) after ingestion or contact with a particular plant protein that does not have any adverse effect on most other individuals.[18] Unlike the effects of other toxic principles, the intensity of the reaction does not depend on the quantity of the toxicant but on the sensitivity of the affected individual. Allergens are not toxic constituents in the strict sense. Their toxic action is the result of an altered immunological response in those animals or people who ingest them.

Perhaps the most important of the substances which can act as allergens is gluten. A condition called coeliac sprue or gluten induced enteropathy is seen in susceptible individuals who eat this protein. The disease is characterized by flattening of the gut mucosal surface and infiltration of the epithelial layer and lamina propria with inflammatory cells. This leads to malabsorption and diarrhoea, increased fat excretion, bloating, and weight loss. These responses involve both direct toxic effects of the gluten and immunological responses.

Glucosinolates

Glucosinolates (formerly called thioglucosides) are hydrolysed when wet raw plant material is crushed to give glucose, an acid sulphate ion, and goitrogens (thiocyanate, isothiocyanate, and goitrin). Many different glucosinolates have

---

[18] J. H. Dean, M. J. Murray, and E. C. Ward, in 'Casarett and Doull's Toxicology', 3rd edition, ed. C. D. Klaasen, M. O. Amdur, and J. Doull, Macmillan, New York, 1986, p. 245.

been isolated from members of the Cruciferae family cultivated as fodder plants, but only one or two glucosinolates are present in relatively large amounts in any given plant species. Glucosinolates are often at their highest concentrations in seeds. The goitrogenic effects of glucosinolates cannot be alleviated by giving iodine, as the mechanism of the goitrogenicity is different from that involved in goitre resulting from the iodine deficiency.

Thiocyanate ions may also be present in plants of families other than the Cruciferae as a result of the breakdown of cyanogenic glycosides. Goitrogens of undetermined chemical structure have been reported in other plants but little is known about them.

Alkaloids

Alkaloids are not a homogeneous group of compounds that can be defined solely on their chemical structure or pharmacological activity. The term 'alkaloid' (alkali-like) was proposed by the German pharmacist K. F. W. Meissner in 1819, after the first recorded isolation of a crystalline constituent (narcotine) from opium in 1803. Now over 4000 are known, although it is estimated that they are present in less than 10% of all plant species.

In general, it is recognized that an alkaloid must be a product of metabolism, and must contain at least one nitrogen atom that can act as a base. Two further qualifications are that it has a complex molecular structure with the nitrogen atom as part of a heterocyclic ring and that it shows pharmacological activity. Alkaloids have been isolated from the roots, seeds, leaves, or bark of some members of at least 40% of plant families, with a number of the families being particularly rich in alkaloids, *e.g.* Amaryllidaceae, Buxaceae, Compositae, Euphorbiaceae, Leguminosae, Liliaceae, Papaveraceae, Ranunculaceae, and Solanaceae.

It is thought that many toxic alkaloids produce their effects by mimicing or blocking the action of nerve transmitters. Some clinical features common to the acute poisoning produced by such alkaloids are excess salivation (or its absence), dilation or constriction of the pupil, vomiting, abdominal pain, diarrhoea, incoordination, convulsions, and coma. Poisoning by the pyrrolizidine alkaloids is very different; it is chronic and the principle organ affected is the liver. Some alkaloids (*e.g.* those of *Conium maculatum*) are teratogenic, causing defects in the foetus when plants containing them are eaten by the mother.

Polyphenolics

Polyphenolics (tannins) are complex phenolic polymers which vary in chemical structure and biological activity. They are found in a large number of agricultural crops including cereals, pulses, and forages. They produce an astringent reaction in the mouth and have the ability to tan leather. On the basis of their chemical structure they can be classified into two main groups: the hydrolysable tannins, which are glycosides, and the condensed tannins. The latter are the more widely distributed in plants, but pass through the digestive tract unchanged and are generally not toxic, although large quantities can give rise to gastroenteritis. The condensed tannins have also been reported to cause growth depression in chicks.

Cyanogens

Cyanogenic glycosides are not toxic as such, but only through their release of hydrocyanic acid (HCN) which occurs when they have been broken down either after ingestion or as a result of plant cell damage before ingestion. Trace amounts of cyanogenic glycosides are widespread in the plant kingdom, but relatively high concentrations are found only in certain plants. Most cases of cyanide poisoning are caused by consumption of plants of the families Rosaceae, Leguminosae, and Gramineae. In general the highest concentrations are found in the leaves.

In man and monogastric animals, such as horse, pig, dog, and cat, the acidic contents of the stomach inhibit the action of the plant enzymes, although these may be active during subsequent digestion in the duodenum. In ruminants (cattle, sheep, and goats) the cyanogenic material remains in the rumen at neutral pH, which favours plant enzyme action, while the material is physically broken down further by rumination and the action of rumen bacteria; bacterial enzymes may also play a part in hydrolysing the glycoside. This may explain the greater susceptibility to cyanogenic plants that has been reported for ruminants.

Cyanide absorption is very rapid, and if sufficient is available (the minimum oral lethal dose is 2–4 mg kg^{-1} body weight) its toxic action is also very rapid, with death occurring from a few minutes to an hour after ingestion.

Mycotoxins

Mycotoxins were recognized as a cause of diseases (mycotoxicoses) in animals only quite recently. Mycotoxins are produced by fungi contaminating crops. Aflatoxin, a mycotoxin, was shown to be the cause of death of thousands of poultry in Britain during 1960. It was found subsequently that this toxin also induced tumours of the liver in fish and mammals. Many other mycotoxins have been identified and some of them such as ochratoxin, citrinin, zearalenone, and trichothecenes have been isolated from animal feeds. All feedstuffs may become contaminated if improperly dried and stored.

Fibrous Polysaccharides

Fibrous polysaccharides have received much attention in recent years because of their possible role in prevention of a range of diseases prevalent in western communities, diseases such as coronary heart disease, diabetes, stroke, diverticular disease, and large bowel cancer. These polysaccharides are generally constituents of the plant cell wall and therefore to be found in all crops to varying extents. However, the chemistry of these molecules varies considerably depending on their origin and hence many of the broad generalizations that have been stated regarding dietary fibre and its properties must be regarded with some scepticism. The potential toxicity of these substances is largely associated with their ability to make essential nutrients unavailable. Of particular importance is the reduction in availability of minerals such as calcium, zinc, iron, phosphorus, magnesium, and copper.

Saponins

Saponins are water-soluble plant constituents that are distinguished by their capacity to form a soapy foam even at low concentrations, their bitter taste, and ability to haemolyse red blood cells. They are glycosides with a non-sugar aglycone portion which is termed a sapogenin. Saponins are classified according to the chemical nature of the sapogenin into two major groups: steroidal and triterpenoid saponins. Saponins are widely distributed in the plant kingdom and are found in various forage legumes such as lucerne (*Medicago sativa*) and clover (*Trifolium repens*). Saponins can occur in all parts of plants, although their concentration is affected by variety and stage of growth. Saponins are generally harmless to mammals when ingested, although large quantities can be irritant and cause vomiting and diarrhoea. They are, however, highly toxic to fish and snails. The saponins of lucerne can inhibit the growth of chicks and depress egg production.

8 Other Potentially Toxic Substances Which May Be Present in Plants

Amongst the other potentially toxic substances which may be present in plants are the following:

Pesticide Residues

Pesticides are substances specifically used to control agricultural and public health pests and to increase crop production. Because of their universal use, they are common contaminants of crops and their processed products.[19] Since they are designed to be biocides, they have to be treated with care and the permitted safe levels of residues present in foodstuffs carefully assessed and effectively enforced. However, consideration of pesticides is a vast subject in itself and does not come within the remit of this book.

---

[19] S. D. Murphy, in 'Casarett and Doull's Toxicology', 3rd edition, ed. C. D. Klaasen, M. O. Amdur, and J. Doull, Macmillan, New York, 1986, p. 519.

Metals

Metal contamination of plants[20] may arise naturally from the nature of the soil in which they grow or as a result of human activities such as the use of metal contaminated sewage sludge as a fertilizer or the use of a metal-based pesticide such as those based on methylmercury. Again consideration of metal contamination does not come within the remit of this book but it is important to consider the possibility when looking at a situation where poisoning may have occurred.

Nitrates and Nitrites

Plants absorb nitrates from the soil and usually convert them rapidly into other nitrogenous compounds.[21] Under certain conditions, however, some plants may accumulate quite high concentrations of nitrates. Nitrates are not very toxic, but they are readily converted by bacteria in the alimentary tract into the much more toxic nitrites. Several factors can increase the accumulation of nitrates by plants and these include drought, shade, the use of herbicides and, in particular, application of nitrogenous fertilizers. Plants most likely to accumulate nitrates are beet, mangels, turnips, swedes, rape, and kale. Nitrites pass easily from the gastrointestinal tract into the blood, where they combine with haemoglobin in the red blood cells to form methaemoglobin, a compound that is incapable of taking up and transporting oxygen. Consequently the clinical signs of nitrite poisoning are those associated with oxygen deficiency and include general weakness and a fall in blood pressure. Death may follow from asphyxia. Young animals and babies are particularly at risk because the small volume of blood that they contain requires only a small amount of nitrite to convert all the haemoglobin to methaemoglobin.

Acids, Alcohols, and Esters

Of the potentially toxic acids occurring in crop plants,[22] fluoroacetic acid is historically of great importance because of the work by Sir Rudolph Peters which led to the discovery of 'lethal synthesis'. This is the name he gave to the conversion of fluoroacetic acid to fluorocitric acid by the enzymes associated with the citric acid cycle.

In practical terms, perhaps the most important potentially toxic acid is oxalic acid. Oxalic acid and its salts, the oxalates, occur naturally in nearly all living organisms but some members of certain plant families (*e.g.* Chenopodiaceae, Geraniaceae, and Polygonaceae) can contain large amounts, mainly as the soluble sodium or potassium salts and the insoluble calcium salts. The oxalate

---

[20] R. A. Goyer, in 'Casarett and Doull's Toxicology', 3rd edition, ed. C. D. Klaasen, M. O. Amdur, and J. Doull, Macmillan, New York, 1986, p. 582.
[21] R. E. Menzer and J. O. Nelson, in 'Casarett and Doull's Toxicology', 3rd edition, ed. C. D. Klaasen, M. O. Amdur, and J. Doull, Macmillan, New York, 1986, p. 825.
[22] M. R. Cooper and A. W. Johnson, 'Poisonous Plants in Britain and Their Effects on Animals and Man', HMSO, London, 1984, p. 17.

content of plants is mainly a species characteristic but wide variations can occur within a species. Variation depends upon the age of the plant, the season, the climate, and the type of soil. Anatomical variations also occur. The highest concentrations commonly occur in the leaves and the lowest in the roots. The oxalate content of many plants tends to increase as the plants mature, an example being the leaves of rhubarb. Other plants show a rapid rise in oxalate content during the early stages of growth, followed by a decrease as the plants mature, e.g. *Atriplex*, mangels, sugar beet leaves, and sugar beet roots. To be potentially dangerous the plants must contain 10% or more oxalic acid on a dry weight basis.

Under natural conditions plants with a high oxalate content are readily eaten by livestock, but if large amounts of such plants are eaten over a short period, acute poisoning can occur. When oxalates are ingested by ruminants, they may be broken down by the rumen micro-organisms, they may combine with free calcium in the digestive tract to form insoluble calcium oxalate that is excreted in the faeces, or they can be absorbed into the blood stream. What happens to the oxalates depends upon a variety of factors, including the amount and the chemical form of the oxalates, their rate of ingestion, the previous grazing history of the animal, the amount of calcium in the diet, and the nutritional status of the animal.

Ruminants can become adapted to a high oxalate intake by developing a ruminal flora that can break down and utilize oxalates. Horses and pigs can also become adapted by developing a similar type of flora in their large intestines. Hungry or undernourished animals are less tolerant of oxalates, as they absorb a higher proportion of those present. Oxalate poisoning occurs principally when hungry cattle or sheep are allowed to graze heavy growths of plants with a high oxalate content.

The main adverse effect of oxalate ingestion is hypocalcaemia that results from the combination of the oxalates with calcium in the bloodstream. The calcium oxalate produced is deposited in various tissues, and especially in the kidneys. Signs of acute poisoning in cattle and sheep include rapid and laboured breathing, depression, weakness, staggering gait, recumbency, coma, and death.

Cardioactive Glycosides

On hydrolysis, cardiac glycosides[23] yield aglycones that have a steroid structure and a highly specific action on the heart, increasing contractility of heart muscle and slowing the heart rate. The sugar portion of the glycoside has no physiological activity of its own but often greatly enhances the cardiac activity of the aglycone. The best known members of this group are the digitalis glycosides present in the foxglove (*Digitalis purpurea*). These have long been used in medicine. Over 400 cardiac glycosides have been isolated. Besides the cardiac effects, the cardiac glycosides can also produce gastroenteritis and diarrhoea;

[23] W. C. Evans, 'Trease and Evans' Pharmacognosy', 13th edition, Balliere-Tindall, London 1989.

many of the other signs are a direct result of the inability of the heart to circulate blood. If lethal quantities are eaten, death occurs in 12–24 hours. With sublethal quantities the clinical signs may persist for 2–3 days.

Bracken Carcinogens

Bracken contains two carcinogens, quercetin and kaempferol.[24] Quercetin is a flavonol found in many crop plants as well as bracken. Kaempferol occurs in bracken in the glycoside complexes, astragalin and tyleroside. These substances have been shown to cause bladder cancer in the rat but it is still uncertain whether they cause the tumours associated with ingestion of bracken by cattle and sheep.

Bracken also contains a substance or group of substances that can cause acute haemorrhagic poisoning in cattle, the cyanogenic glycoside, prunasin, and thiaminase, which can induce thiamine deficiency in non-ruminant animals such as the horse and the pig.

Favism Factors

Favism (fabism, fabismus) is characterized by haemolytic anaemia of varying severity. It can occur after inhaling pollen or eating the seeds of the broad bean, *Vicia faba*.[25] It is prevalent in some islands and coastal regions of the Mediterranean. Susceptibility is dependent on the genetically transmitted, male sex-linked deficiency of the enzyme glucose-6-phosphate dehydrogenase. Thus, since this deficiency is relatively rare outside some islands and coastal regions of the Mediterranean, broad beans are safe to eat for most people. The precise chemical nature of the substances which cause favism is still a matter of debate.

Photosensitive Agents and Photosensitization

Some plants contain substances that can make non-pigmented or slightly pigmented skin hypersensitive to the ultraviolet radiation in sunlight.[26] After ingestion, these substances are absorbed unchanged into the circulating blood and thus reach the skin, where they may be excited by ultraviolet rays and then induce chemical changes that lead to cell damage. Some of these substances, the furocoumarins, can cause photosensitivity simply by skin contact.

The body responds to photosensitized cell damage by itchiness, redness, heat, oedema, and swelling of the affected skin. Blisters may develop and break, giving rise to scabs and secondary infections; skin necrosis may occur. Animals with photosensitive substances in their skin are said to be in a state of photosensitivity. The reaction that occurs in photosensitive animals after

---

[24] G. M. Williams and J. Weisburger, in 'Casarett and Doull's Toxicology', 3rd edition, ed. C. D. Klaasen, M. O. Amdur, and J. Doull, Macmillan, New York, 1986, p. 99.
[25] M. R. Cooper and A. W. Johnson, 'Poisonous Plants in Britain and Their Effects on Animals and Man', HMSO, London, 1984, p. 112.
[26] M. R. Cooper and A. W. Johnson, 'Poisonous Plants in Britain and Their Effects on Animals and Man', HMSO, London, 1984, p. 18.

exposure to sunlight is called photosensitization. The severity of the clinical signs of photosensitization is determined by the amount of photosensitive substance in the skin, the intensity of the ultraviolet radiation, and the duration of exposure.

In addition to the primary photosensitivity in which the photosensitive agent remains unchanged during its transfer from the plant to the skin, there is a secondary or hepatogenous photosensitivity in which the photosensitive substance is a normal breakdown product of digestion, usually eliminated by the liver. One such substance, phylloerythrin, is formed in the digestive tract of ruminants during the microbial degradation of chlorophyll, and, after absorption, is transported to the liver, where it is excreted in the bile. Any liver damage affecting bile excretion leads to the accumulation of phylloerythrin in the blood and thus in the skin, where it can be excited by ultraviolet rays. The liver damage leading to this secondary photosensitivity can result from disease, hepatoxic drugs, industrial or agricultural chemicals, mycotoxins (produced by fungi infecting plant material) as well as from plant toxins themselves.

Primary photosensitivity in animals has been produced by eating buckwheat (*Fagopyrum esculentum*) and St John's wort (*Hypericum perforatum*). Primary photosensitivity in man has also occurred in people handling vegetables of the Umbelliferae family, such as parsnips, carrots, and celery. In some of these cases, the furocoumarins responsible result from the presence of fungal pathogens on the plants (*e.g. Sclerotinia* spp. on celery). The photosensitization occurring with bog asphodel (*Narthecium ossifragum*) in sheep is of the hepatogenous type as is also that occurring with ingestion of mycotoxins in dead plant material.

9 Concluding Comments

Currently there is a great increase in concern about the food that we eat and the possibility of harm arising from contamination with micro-organisms or pesticide and other residues acquired at some point in the growth, harvesting, or processing of crop plants. Less attention has been paid to the naturally occurring substances which may cause or contribute to illness in animals or humans. However, a great deal of relevant research on these substances has been carried out in recent years and we are beginning to get an understanding of the scope of the potential problem. This chapter has attempted to give a preliminary overview of our current state of knowledge and the rest of this book will give a detailed consideration of most of the relevant substances for which this is possible. Inevitably, the coverage is selective but it should help to provide a basis from which future research may develop. Above all, it should help to put the problems posed by the naturally occurring potentially toxic substances in crop plants into perspective.

CHAPTER 2

Toxic Amino Acids

J. P. FELIX D'MELLO

1 Introduction

By virtue of their limited biosynthetic competence, animals are unable to manufacture the carbon skeletons, or keto acids, of ten amino acids.[1] These amino acids are designated as indispensable for the maintenance, growth, and reproduction of animals. A common core of ten amino acids appears to be indispensable for all vertebrate species (Table 1). It should be noted that arginine can be synthesized by mammals, although not at a sufficiently rapid rate to satisfy the demands for growth. Non-ruminant animals are totally dependent upon a dietary supply of the indispensable amino acids, but ruminant animals derive a substantial proportion of their requirements from microbial protein synthesized within the rumen. Those amino acids which the animal is able to synthesize within its own tissues are termed dispensable. However, it is widely recognized that young animals will not achieve their genetically determined growth potential if the dietary nitrogen is supplied entirely in the form of the indispensable amino acids. Thus, although an animal exhibits a specific requirement for the indispensable amino acids, an optimal combination of the dispensable amino acids should also be provided in order to elicit maximum efficiency of utilization of all amino acids. A third group comprising the non-protein amino acids is conventionally accorded with potent toxic properties.[2-4] These occur in unconjugated forms in a wide range of plants, but particularly in leguminous species. The ubiquitous occurrence of these amino acids has undermined efforts to fully exploit, as food sources, such promising

---

[1] J. P. F. D'Mello, in 'Recent Advances in Animal Nutrition', ed. W. Haresign and D. Lewis, Butterworths, London, 1979, p. 1.
[2] G. A. Rosenthal and E. A. Bell, in 'Herbivores, their Interaction with Secondary Plant Metabolites', ed. G. A. Rosenthal and D. H. Janzen, Academic Press, New York, 1979, p. 353.
[3] G. A. Rosenthal, 'Plant Non-protein Amino and Imino Acids', Academic Press, New York, 1982, Chapter 3, p. 57.
[4] J. P. F. D'Mello, in 'Anti-nutritional Factors, Potentially Toxic Substances in Plants', ed. J. P. F. D'Mello, C. M. Duffus, and J. H. Duffus, Association of Applied Biologists, Warwick, 1989, p. 29.

Table 1 *Nutritional classification of amino acids*

Indispensable	Dispensable	Toxic (non-protein)
Lysine	Glycine	Canavanine
Methionine	Cystine	Indospicine
Threonine	Serine	Homoarginine
Tryptophan	Proline	Mimosine
Isoleucine	Alanine	3,4-Dihydroxyphenylalanine
Leucine	Aspartic acid	β-Cyanoalanine
Valine	Glutamic acid	β-N-Oxalyl-α, β-diaminopropionic acid
Phenylalanine	Tyrosine	α,γ-Diaminobutyric acid
Arginine		β-Aminoproprionitrile
Histidine		Selenomethionine
		Se-Methylselenocysteine
		Selenocystathionine
		S-Methylcysteine sulphoxide
		Djenkolic acid

species as *Canavalia*, *Indigofera*, *Leucaena*, *Lathyrus*, and *Brassica*. However, the issue of toxicity transcends the boundaries imposed by the nutritional classification shown in Table 1. Consequently, all amino acids should be regarded as possessing the potential to precipitate deleterious effects in animals and man. In addition, the toxicity of many non-protein amino acids is determined to a significant extent by complex interactions with the nutritionally important amino acids. Thus, canavanine induces its adverse effects by competing with arginine in metabolic reactions, and lysine exacerbates this toxicity in avian species by virtue of its own antagonism with arginine.[5]

While the fundamental biochemical functions of the toxic non-protein amino acids still await elucidation, there is convincing evidence of their role in higher plants as the agencies of chemical defence against predation and disease.[2,3] In particular, the inhibitory effects of canavanine on insect development and survival are well established.

This chapter reviews the toxicity and mode of action of the non-protein amino acids in higher animals and also considers methods of detoxification. In addressing these issues, cognisance will be given to mounting evidence of differences between mammals and avian species and between ruminant and non-ruminant animals in their sensitivity to non-protein amino acids. The toxic nature of amino acids will be considered in its widest sense in view of structural similarities and interactions between the non-protein and the indispensable amino acids. Recognition will also be given to recent innovations in analytical methodology which highlight the diverse range and distribution of non-protein amino acids, particularly among the leguminous species of crop plants.

[5] J. P. F. D'Mello, T. Acamovic, and A. G. Walker, *Trop. Agric. (Trinidad)*, 1989, **66**, 201.

2 Distribution of Toxic Non-protein Amino Acids

Legumes contain higher concentrations and a more diverse array of toxic non-protein amino acids than any other plant species, and the seed is generally the most concentrated source of these compounds. As shown in Table 2, canavanine is ubiquitous among leguminous species[3,6-9] with concentrations of up to 127 g kg^{-1} dry weight in the seed of *Dioclea megacarpa*. Marked reductions in canavanine concentrations in the germinating seed of *Canavalia ensiformis* (jack bean) have been reported.[10,11] However, much of this data has arisen from non-specific colorimetric assays. Although interest in colorimetric methodology has persisted,[12] alternative methods based on chromatographic separation have recently been described. Using an ion-exchange chromatographic procedure D'Mello *et al.* showed that canavanine concentrations declined only marginally during germination of *C. ensiformis*, the lowest values occurring 24 hours after germination in the dark.[6] A more rapid and sensitive technique based on high performance liquid chromatography was subsequently described by Acamovic and D'Mello[13] which allowed the simultaneous analysis of both canavanine and its degradation product, canaline. The latter compound was not detected in the seed of *C. ensiformis*. The diverse distribution of toxic non-protein amino acids is further illustrated by the occurrence of canavanine with indospicine[8,14] in the seeds of *Indigofera spicata* (trailing or creeping indigo). Homoarginine (Table 2) is present in at least two *Lathyrus* species[15,16] in combination with β-N-oxalyl-α,β-diaminopropionic acid.

Leucaena leucocephala (ipil-ipil; koa haole) is an ubiquitous tropical legume, with some species yielding timber and others capable of providing prolific quantities of palatable forage. However, the presence of the toxic aromatic amino acid, mimosine, restricts the use of this legume in animal nutrition. Concentrations of mimosine in the leaf usually range between 10–25 g kg^{-1} dry matter[17] but even higher concentrations occur in the seed (Table 2). Wide discrepancies in mimosine content of *Leucaena* leaf have been recorded[17] with variations arising, principally, from deficiencies in analytical methodology. The traditional FeCl$_3$ colorimetric assay for mimosine[18] is influenced by factors such as pH, the presence of other phenolic compounds, and variability in recoveries caused by the use of charcoal in the decolourization procedure.[19,20] These

---

[6] J. P. F. D'Mello, A. G. Walker, and T. Acamovic, *Trop. Agric. (Trinidad)*, 1988, **65**, 376.
[7] B. Lucas, A. Guerrero, L. Sigales, and A. Sotelo, *Nutr. Rep. Int.*, 1988, **37**, 545.
[8] R. W. Miller and C. R. Smith, *J. Agric. Food Chem.*, 1973, **21**, 909.
[9] A. K. Shqueir, D. L. Brown, and K. C. Klasing, *Anim. Feed Sci. Technol.*, 1989, **25**, 137.
[10] J. H. Johnstone, *Biochem. J.*, 1956, **64**, 21.
[11] G. A. Rosenthal, *Plant Physiol.*, 1970, **46**, 273.
[12] J. Cacho, M. A. Garcia, and I. Ferrando, *Analyst (London)*, 1989, **114**, 965.
[13] T. Acamovic and J. P. F. D'Mello, *J. Sci. Food Agric.*, 1990, **50**, 63.
[14] M. P. Hegarty and A. W. Pound, *Aust. J. Biol. Sci.*, 1970, **23**, 831.
[15] J. P. F. D'Mello, T. Acamovic, and A. G. Walker, unpublished results.
[16] E. A. Bell, *Nature (London)*, 1964, **203**, 378.
[17] J. P. F. D'Mello and T. Acamovic, *Anim. Feed Sci. Technol.*, 1989, **26**, 1.
[18] H. Matsumoto and D. G. Sherman, *Arch. Biochem. Biophys.*, 1951, **33**, 195.
[19] M. P. Hegarty, R. D. Court, and M. P. Thorne, *Aust. J. Agric. Res.*, 1964, **15**, 168.
[20] R. G. Megarrity, *J. Sci. Food. Agric.*, 1978, **29**, 182.

Table 2 *Concentrations of some toxic, non-protein amino acids in seeds of leguminous plants*

Amino acid	Legume species	Concentration (g kg^{-1} dry weight)
Canavanine	Canavalia ensiformis	51[a]
	Gliricidia sepium	40[b]
	Robinia pseudocacia	98[c]
	Dioclea megacarpa	127[c]
	Indigofera spicata	8.9[d]
	Sesbania sesban	6.2[e]
Indospicine	Indigofera spicata	20[f]
Homoarginine	Lathyrus cicera	12[g]
Mimosine	Leucaena leucocephala	145[h]
	Pithecolobium ondulatum	8.4[h]
β-Cyanoalanine	Vicia sativa	1.5[i]
β-N-Oxalyl-α,β-diaminopropionic acid	Lathyrus sativus	25[c]
α,γ-Diaminobutyric acid	Lathyrus sylvestris	16[c]
Djenkolic acid	Leucaena esculenta	2.2[h]
	Pithecolobium lobatum	20[c]

[a] J. P. F. D'Mello, A. G. Walker, and T. Acamovic, *Trop. Agric. (Trinidad)*, 1988, **65**, 376.
[b] B. Lucas, A. Guerrero, L. Sigales, and A. Sotelo, *Nutr. Rep. Int.*, 1988, **37**, 545.
[c] G. A. Rosenthal, 'Plant Non-protein Amino and Imino Acids', Academic Press, New York, Chapter 3, p. 57.
[d] R. W. Miller and C. R. Smith, *J. Agric. Food Chem.*, 1973, **21**, 909.
[e] A. A. Shqueir, D. L. Brown, and K. C. Klasing, *Anim. Feed Sci. Technol.*, 1989, **25**, 137.
[f] M. P. Hegarty and A. W. Pound, *Aust. J. Biol. Sci.*, 1970, **23**, 831.
[g] J. P. F. D'Mello and A. G. Walker, unpublished results.
[h] J. P. F. D'Mello and T. Acamovic, *Anim. Feed Sci. Technol.*, 1989, **26**, 1.
[i] C. Ressler, S. N. Nigam, and Y. H. Giza, *J. Am. Chem. Soc.*, 1969, **91**, 2758.

disadvantages highlighted the need for alternative and specific methods of analysis. Acamovic and D'Mello[21] developed a specific ion-exchange chromatographic (IEC) technique for the determination of mimosine. However, this method was slow and, more importantly, could not be employed to determine the major metabolite of mimosine, 3-hydroxy-4(1H)-pyridone (3,4-DHP). These disadvantages were overcome when Acamovic *et al.* subsequently developed a method based on high performance liquid chromatography (HPLC) which allowed the simultaneous and rapid estimation of both mimosine and 3,4-DHP.[22]

The lathyrogenic group of non-protein amino acids comprises β-cyanoalanine (BCNA), β-N-oxalyl-α, β-diaminopropionic acid (ODAP), α,γ-diaminobutyric acid (DABA), and β-aminopropionitrile (BAPN). As shown in Figure 1, BCNA plays a central role in the synthesis of other lathyrogenic amino

[21] T. Acamovic and J. P. F. D'Mello, *J. Chromatogr.*, 1981, **206**, 416.
[22] T. Acamovic, J. P. F. D'Mello, and K. W. Fraser, *J. Chromatogr.*, 1982, **236**, 169.

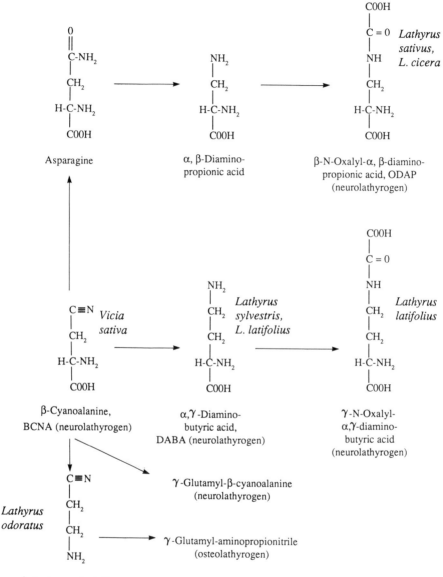

Figure 1 *The lathyrogenic amino acids: their structural relationships and distribution* (Adapted from 'Toxicants Occurring Naturally in Foods', 1966, p. 42; and from 'Herbivores, their Interaction with Secondary Plant Metabolites', 1979, p. 353, with permission from National Academy of Sciences, National Academy Press, Washington, DC and from Academic Press, New York.)[23,2]

acids,[23] its occurrence in *Vicia sativa* (Table 2) being augmented by the presence of a glutamyl dipeptide of BCNA. Concentrations of this dipeptide range from 6 g kg^{-1} in the seed to 26 g kg^{-1} in the seedling on a dry weight basis.[24] Although *Lathyrus sativus* is a rich source of ODAP (Table 2), this neurotoxic amino acid also occurs in *L. cicera*, *L. clymenum* as well as species of *Crotalaria* and *Acacia*.[3] Several other species of *Lathyrus* including *L. sylvestris* and *L. latifolius* contain another neurotoxin, DABA which also occurs as an oxalyl derivative (Figure 1), a compound which is more abundant than the parent amino acid in *L. latifolius*[3]; DABA is absent from the seeds of *L. sativus* and *L. cicera*. The principal source of BAPN appears to be the seed of *L. odoratus* (sweet pea)[3] in which it also occurs as the glutamyl dipeptide (Figure 1).

Several plants contain toxic structural analogues of the sulphur containing amino acids in which the sulphur atom is replaced by selenium. In *Astragalus* species, the predominant selenoamino acids are *Se*-methylselenocysteine and selenocystathionine.[25] A comprehensive array of selenoamino acids, including selenomethionine has been identified in other plants.[3]

Forage and root *Brassica* crops are fairly widely grown in temperate regions as fodder for ruminant animals. However, the presence of the non-protein amino acid, *S*-methylcysteine sulphoxide (SMCO)[26,27] is a significant deterrent in the exploitation of these crops. Typical concentrations of SMCO in *Brassica oleracea* (kale) range from 40–60 g kg^{-1} forage dry weight.[26] Highest concentrations are found in the young leaves and growing shoots and increases also occur as the crop matures. Djenkolic acid (Table 2) is another sulphur containing amino acid with toxic properties,[3] occurring principally in the seed of *Pithecolobium lobatum* (djenkol bean); lower concentrations have been detected in the seeds of *P. ondulatum* (2.8 g kg^{-1}) and of *Leucaena esculenta*[7] (Table 2).

3 Structural Features of Non-protein Amino Acids

The non-protein amino acids are generally classified on the basis of their structural relationships with the nutritionally important amino acids and their metabolites or on the basis of their physiological properties.[3] Thus canavanine, indospicine, and homoarginine are recognized by their structural similarity with arginine, an intermediate of the urea cycle (Figures 2 and 3). In addition, the indispensable amino acid, lysine acts as a potent antagonist of arginine and, consequently, is capable of modulating the toxicity of certain analogues of urea cycle substrates. Structural congeners also exist for other intermediates of the urea cycle and these are best illustrated by the unique set of reactions involving canavanine biosynthesis (Figure 3). It is now established that *Canavalia ensiformis* contains the full complement of the ornithine–urea cycle enzymes that are required for the synthesis of these analogues.[3]

---

[23] I. E. Liener, in 'Toxicants Occurring Naturally in Foods', National Academy of Sciences, Washington, 1966, pp. 42.
[24] C. Ressler, S. N. Nigam, and Y. H. Giza, *J. Am. Chem. Soc.*, 1969, **91**, 2758.
[25] A. Shrift, *Annu. Rev. Plant Physiol.*, 1969, **20**, 475.
[26] R. H. Smith, *Vet. Rec.*, 1980, **107**, 12.
[27] T. N. Barry and T. R. Manley, *Br. J. Nutr.*, 1985, **54**, 753.

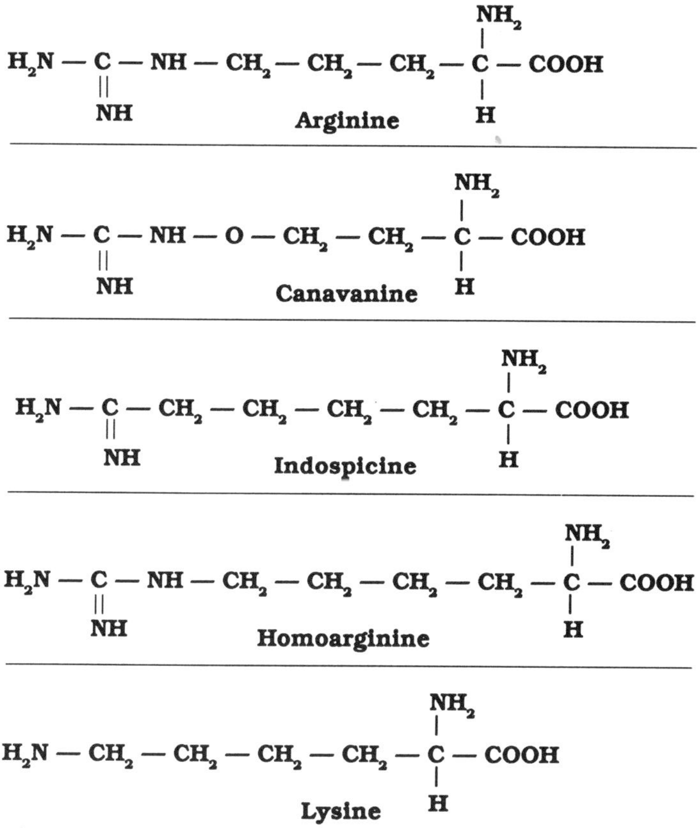

$$H_2N-\underset{\underset{NH}{\|}}{C}-NH-CH_2-CH_2-CH_2-\underset{H}{\overset{NH_2}{\underset{|}{C}}}-COOH$$
Arginine

$$H_2N-\underset{\underset{NH}{\|}}{C}-NH-O-CH_2-CH_2-\underset{H}{\overset{NH_2}{\underset{|}{C}}}-COOH$$
Canavanine

$$H_2N-\underset{\underset{NH}{\|}}{C}-CH_2-CH_2-CH_2-CH_2-\underset{H}{\overset{NH_2}{\underset{|}{C}}}-COOH$$
Indospicine

$$H_2N-\underset{\underset{NH}{\|}}{C}-NH-CH_2-CH_2-CH_2-CH_2-\underset{H}{\overset{NH_2}{\underset{|}{C}}}-COOH$$
Homoarginine

$$H_2N-CH_2-CH_2-CH_2-CH_2-\underset{H}{\overset{NH_2}{\underset{|}{C}}}-COOH$$
Lysine

Figure 2 *Analogues and antagonists of arginine*

The aromatic non-protein amino acids (Figure 4) include mimosine and 3,4-dihydroxyphenylalanine (DOPA). Mimosine is widely regarded as an analogue of the nutritionally important amino acid, tyrosine and its neurotransmitter derivatives, dopamine and norepinephrine found in the brain. It will be readily apparent from Figure 4 that DOPA, which occurs naturally in the seeds of some leguminous plants,[3] also acts as a precursor of these neurotransmitters.

The structural homology of the lathyrogenic amino acids, BCNA, ODAP, DABA, and BAPN is clearly seen in Figure 1. While these amino acids have not been associated, in structural terms, with the indispensable or dispensable amino acids it is worth noting that at least two lathyrogenic amino acids exert profound effects on the brain metabolism of glutamine. Consequently, some metabolic association with the amino acid neurotransmitter system embracing glutamic acid, γ-aminobutyric acid (GABA), and aspartic acid cannot be ruled out.

Arguably, the most striking structural analogues are those of the indispensble

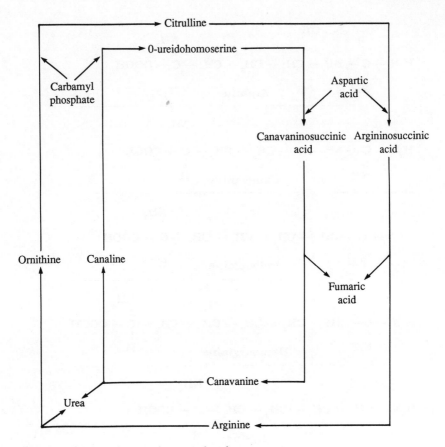

Figure 3 *Some analogues of urea cycle substrates*

amino acid, methionine and its derivatives (Figure 5). When selenium replaces sulphur, a wide spectrum of toxic analogues including selenomethionine, Se-methylselenocysteine, and selenocystathionine arise in certain leguminous plants.[3] Other analogues of the sulphur containing amino acids include SMCO and djenkolic acid.

4 Toxic Properties of Non-protein Amino Acids

Analogues of Urea Cycle Substrates

Complex interactions and antagonisms exist among these analogues in their relationship with arginine. These antagonisms are exemplified by the interaction between the two indispensable amino acids, lysine and arginine. This interaction occurs naturally when diets containing oilseeds from *Brassica napus* or *B. campestris* (rapeseed) are fed to poultry.[1] However, the lysine–arginine antagonism is most clearly demonstrated with the use of crystalline forms of

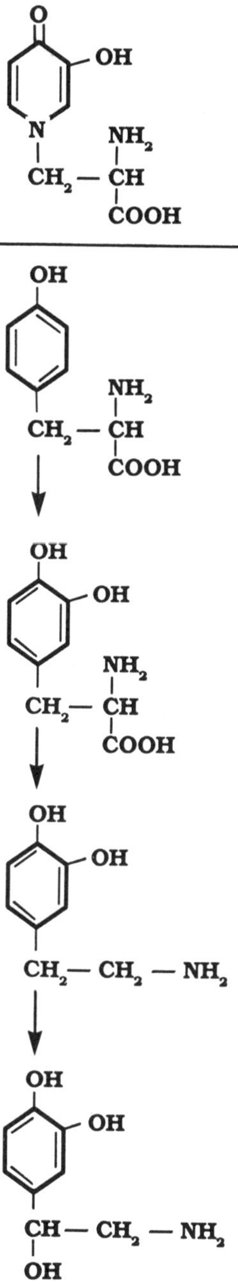

Mimosine

Tyrosine

3,4-Dihydroxyphenylalanine (DOPA)

Dopamine

Norepinephrine

Figure 4 *Structural features of some aromatic amino acids*

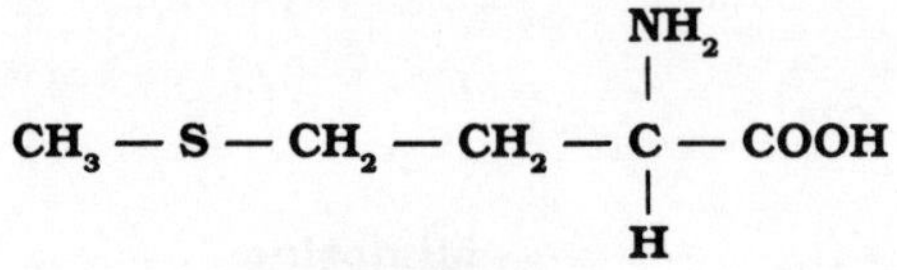

$$CH_3-S-CH_2-CH_2-\underset{H}{\overset{NH_2}{C}}-COOH$$
Methionine

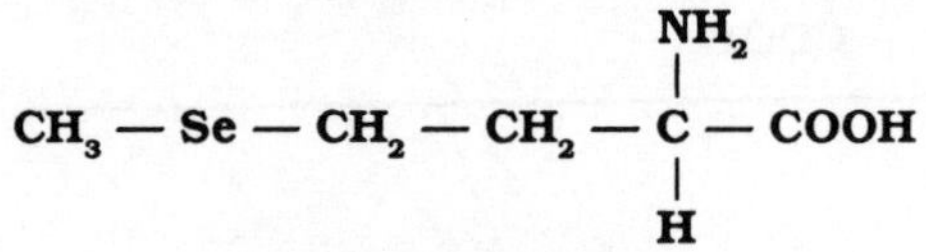

$$CH_3-Se-CH_2-CH_2-\underset{H}{\overset{NH_2}{C}}-COOH$$
Selenomethionine

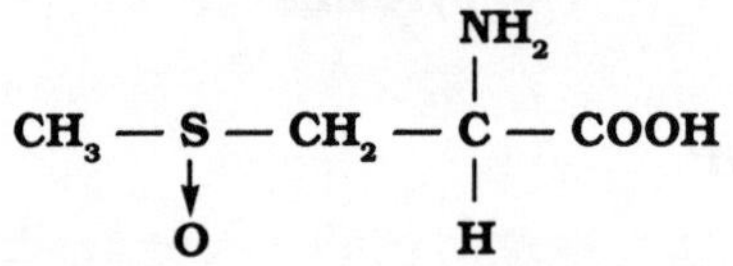

$$CH_3-\underset{\downarrow O}{S}-CH_2-\underset{H}{\overset{NH_2}{C}}-COOH$$
S-Methyl-L-cysteine sulphoxide (SMCO)

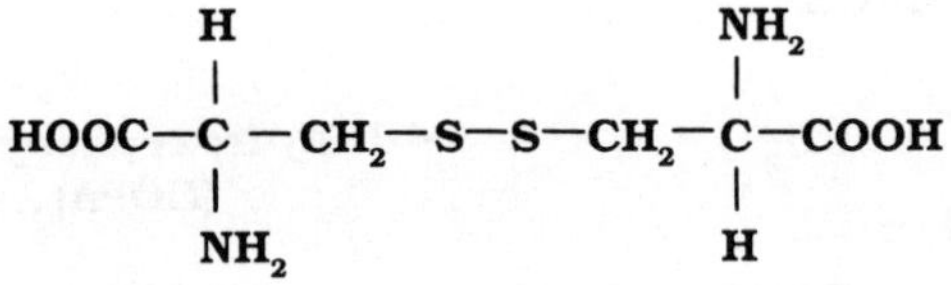

$$HOOC-\underset{NH_2}{\overset{H}{C}}-CH_2-S-S-CH_2-\underset{H}{\overset{NH_2}{C}}-COOH$$
Cystine

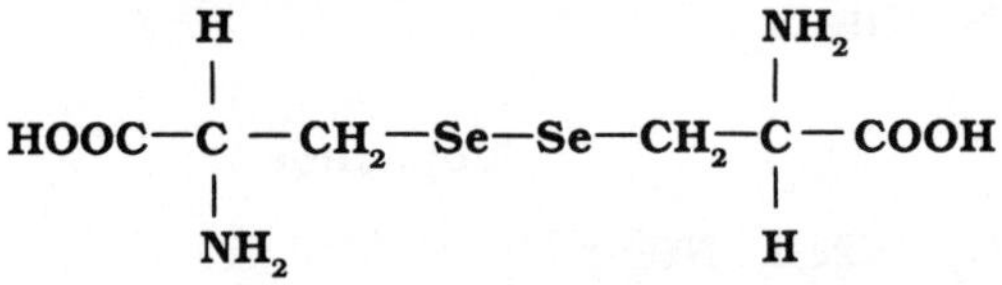

$$HOOC-\underset{NH_2}{\overset{H}{C}}-CH_2-Se-Se-CH_2-\underset{H}{\overset{NH_2}{C}}-COOH$$
Selenocystine

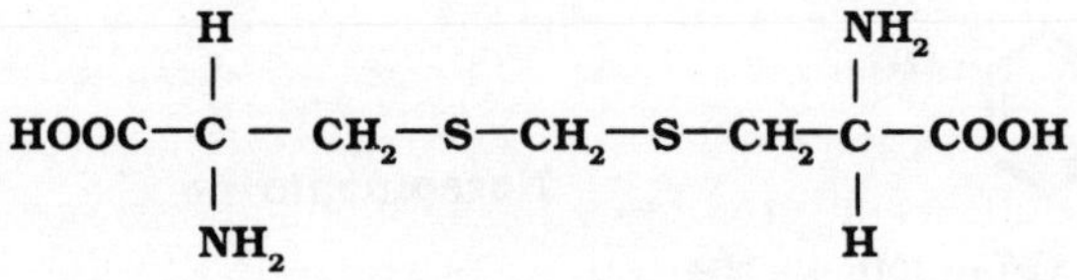

$$HOOC-\underset{NH_2}{\overset{H}{C}}-CH_2-S-CH_2-S-CH_2-\underset{H}{\overset{NH_2}{C}}-COOH$$
Djenkolic acid

Figure 5 *Structural analogues of the sulphur-containing amino acids*

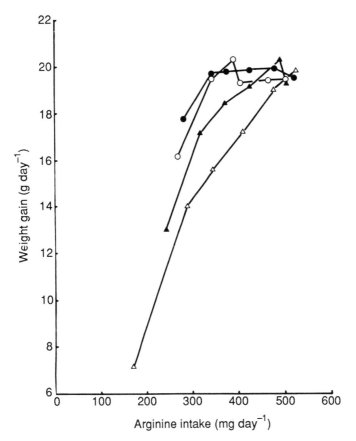

Figure 6 *Growth responses* (g day^{-1}) *and arginine intake* (mg day^{-1}) *of chicks fed* 11.0 (●), 13.5 (○), 16.0 (▲), *and* 18.5 (△) g *lysine* kg^{-1} *diet* (Plotted from data in *Br. Poult. Sci.*, 1970, **11**, 367.)[28]

these amino acids.[28] Excesses of lysine in the diet progressively retard chick growth but the adverse effects are reversed completely by additional supplements of arginine (Figure 6).

While the toxicity of canavanine to micro-organisms and insects is unequivocal,[3] its adverse effects in vertebrates depend upon animal species, age, and dose level. Thus young chicks[5] fed on a diet containing autoclaved seeds of *Canavalia ensiformis* show rapid and severe reductions in growth and efficiency of nitrogen utilization with a dietary canavanine concentration of only 3.7 g kg^{-1}. These effects are largely reproduced on feeding appropriate quantities of pure canavanine.[15] On the other hand, mammalian species are less sensitive except to high doses of canavanine. For example, Hegarty and Pound[14] observed that canavanine was free of any hepatotoxic properties when

[28] J. P. F. D'Mello and D. Lewis, *Br. Poult. Sci.*, 1970, **11**, 367.

40 mg were subcutaneously injected into mice whereas Prete[29] reported dysfunction of the immune system, immunoglobulin deposition, and glomerular lesions in the kidney and decreased survival of auto-immune mice fed canavanine at a dietary concentration of 7.3 g kg^{-1}. More recently, Thomas and Rosenthal[30] concluded that canavanine was only slightly toxic to adult and neonatal rats when administered in a single subcutaneous injection; LD$_{50}$ values were 5.9 and 5.0 g kg^{-1} body weight respectively. However, multiple subcutaneous doses of canavanine evoked severe depressions in growth and food intake particularly in neonates which, in addition, exhibited a marked form of alopecia. Biochemical manifestations in the adult rat included markedly elevated concentrations of ornithine in serum and urine when canavanine was administered at 2 g kg^{-1} body weight; urinary excretion of lysine + histidine and arginine were also enhanced.[31]

The extent to which vertebrate animals are able to metabolize canavanine depends upon the mode of nitrogen excretion. The ornithine–urea cycle (Figure 3) is the principal means whereby mammals dispose of excess amino acid nitrogen, and arginine synthesis also occurs by this pathway. It follows that canavanine metabolism in this species should involve synthesis of canaline, O-ureidohomoserine and canavaninosuccinic acid. The experimental evidence confirms that the major route of canavanine degradation involves arginase-mediated hydrolysis to canaline and urea.[31] In addition, an alternative, albeit minor system has been proposed for the metabolism of canavaninosuccinic acid and canavanine (Figure 7). The former may be reductively cleaved to yield guanidinosuccinic acid and homoserine, while canavanine may undergo transamidination with glycine to form canaline and guanidinoacetic acid.[32]

The total absence of carbamoyl phosphate synthase in tissues of avian species means that the urea cycle is non-functional in these animals. Consequently, birds have no means of synthesizing arginine and they readily succumb to the effects of arginine antagonists such as lysine[28] and canavanine.[5] The presence of residual arginase in the kidney and liver of birds, however, ensures that canavanine is catabolized to urea[5] and canaline. In addition canavanine may also undergo transamidination in avian species as in mammals (Figure 7); but the synthesis of O-ureidohomoserine from canaline is unlikely owing to the absence of carbamoyl phosphate synthase. Consequently, the synthesis of canavaninosuccinic acid, guanidinosuccinic acid, and homoserine from canavanine is unlikely to occur in birds.

Mammalian–avian differences are also seen in the toxicity of *Indigofera spicata* seeds which contain both indospicine, another structural analogue of arginine, as well as canavanine (Table 2). Indospicine is a powerful teratogen, inducing cleavage of the secondary palate and general somatic dwarfism in rats given a single oral dose of an extract of *I. spicata* seeds.[33] In addition, subcu-

---

[29] P. E. Prete, *Can. J. Physiol. Pharmacol.*, 1985, **63**, 843.
[30] D. A. Thomas and G. A. Rosenthal, *Toxicol. Appl. Pharmacol.*, 1987, **91**, 395.
[31] D. A. Thomas and G. A. Rosenthal, *Toxicol. Appl. Pharmacol.*, 1987, **91**, 406.
[32] S. Natelson, A. Koller, H. Y. Tseng, and R. F. Dods, *Clin. Chem.*, 1977, **23**, 960.
[33] J. H. Pearn, *Br. J. Exp. Path.*, 1967, **48**, 620.

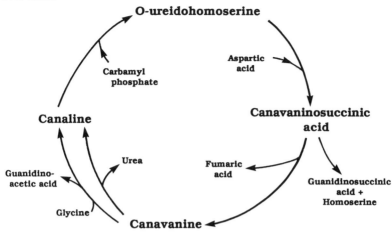

Figure 7 *Mammalian metabolism of canavanine*

taneous injections of indospicine (10 mg kg^{-1} body weight) precipitate fat deposition and cytological changes in the liver of mice within 36 hours.[14] The fat accretion is prevented by simultaneous injections of arginine but not of canavanine. Hegarty and Pound[14] concluded that a substantial part of the hepatoxicity of *I. spicata* seeds may be attributed to the presence of indospicine. Liver damage is also seen in other mammalian species consuming *Indigofera* forage, but not in poultry fed diets containing the seeds of this legume.[34]

The toxicity of homoarginine was established by O'Neal *et al.*[35] who compared its potency with that of DABA. Although less toxic than DABA, homoarginine induced hypersensitivity and mortality in rats adminstered with only 10 mmol kg^{-1} body weight. More recent studies[36,37] indicated that homoarginine not only reduced growth and food intake of rats but also depressed brain concentrations of lysine, ornithine, and arginine. Plasma concentration of lysine, however, increased with homoarginine supplementation, an observation which led to the proposal that homoarginine may be degraded to lysine and urea. In other studies, homoarginine was shown to aggravate the effects of canavanine-induced toxicity in chicks fed diets containing *Canavalia ensiformis*.[15] The complex and ramifying interactions which exist among the analogues of urea cycle substrates are thus well illustrated by the metabolic and nutritional effects of homoarginine.

Mimosine

The deleterious properties of mimosine (Table 3) are extensive and include disruption of reproductive processes, teratogenic effects, loss of hair and wool,

[34] E. H. Britten, A. L. Palafox, M. M. Frodyma, and F. T. Lynd, *Crop Sci.*, 1963, **3**, 415.
[35] R. M. O'Neal, C. H. Chen, C. S. Reynolds, S. K. Meghal, and R. E. Koeppe, *Biochem. J.*, 1968, **106**, 699.
[36] J. K. Tews and A. E. Harper, *J. Nutr.*, 1986, **116**, 1464.
[37] J. K. Tews and A. E. Harper, *J. Nutr.*, 1986, **116**, 1910.

Table 3 Toxic properties of mimosine

Animal type	Method of administration (dose)	Adverse effects
Mice[a]	Dietary (5 g kg^{-1} food)	Normal hair growth
	(10 g kg^{-1} food)	Hair growth inhibited; loss of facial hair
Rat[b]	Dietary (5 g kg^{-1} food)	Irregular oestrus cycling
	(10 g kg^{-1} food)	Termination of oestrus cycling
Rat[c]	Dietary (5–7 g kg^{-1} food)	Teratogenic effects; reduced collagen synthesis; uterine perforations; deformities of cranium, thorax, and pelvis in foetus
Sheep[d]	Single intravenous injections (20–350 g kg^{-1} BW*); or intravenous infusions (10–20 g kg^{-1} BW)	No ill-effects
	Intravenous infusions (8 g over 2 days; 77–96 mg kg^{-1} BW)	Reduced wool strength; defleecing
	(24 g over 4 days; 147 mg kg^{-1} BW)	Reduced wool growth and strength; defleecing; loss of appetite; excessive salivation; organ damage; death
Sheep[e]	Single oral dose (450 or 600 mg kg^{-1} BW)	Shedding of fleece

* Body weight.

[a] R. G. Crounse, J. D. Maxwell, and H. Blank, *Nature (London)*, 1962, **194**, 694.
[b] J. W. Hylin and I. J. Lichton, *Biochem. Pharmacol.*, 1965, **14**, 1167.
[c] S. Dewreede and O. Wayman, *Teratology*, 1970, **3**, 21.
[d] P. J. Reis, D. A. Tunks, and R. E. Chapman, *Aust. J. Biol. Sci.*, 1975, **28**, 69.
[e] P. J. Reis, D. A. Tunks, and M. P. Hegarty, *Aust. J. Biol. Sci.*, 1975, **28**, 495.

(Reproduced by permission from J. P. F. D'Mello, in 'Anti-Nutritional Factors, Potentially Toxic Substances in Plants', ed. J. P. F. D'Mello, C. M. Duffus, and J. H. Duffus, Association of Applied Biologists, Warwick, 1989, p. 29)[4]

and even death.[38–42] Similar effects may be induced on feeding the seeds or foliage of *Leucaena leucocephala* to laboratory animals[38] and ruminants.[43] Thus the defleecing effects in sheep may be precipitated by the administration of crystalline mimosine or by the feeding of *Leucaena* forage. In both cases, defleecing efficacy depends upon the plasma mimosine concentration being maintained above 0.1 mmol l^{-1} for at least 30 hours.[42]

The toxicity of *Leucaena* in ruminants is dependent upon the rate and extent of mimosine degradation by the rumen microflora (Figure 8). During this breakdown, 3,4-DHP is synthesized. However, this product may also arise in

[38] R. G. Crounse, J. D. Maxwell, and H. Blank, *Nature (London)*, 1962, **194**, 694.
[39] J. W. Hylin and I. J. Lichton, *Biochem. Pharmacol.*, 1965, **14**, 1167.
[40] S. Dewreede and O. Wayman, *Teratology*, 1970, **3**, 21.
[41] P. J. Reis, D. A. Tunks, and R. E. Chapman, *Aust. J. Biol. Sci.*, 1975, **28**, 69.
[42] P. J. Reis, D. A. Tunks, and R. E. Chapman, *Aust. J. Biol. Sci.*, 1975, **28**, 495.
[43] M. P. Hegarty, P. G. Schinckel, and R. D. Court, *Aust. J. Agric. Res.*, 1964, **15**, 153.

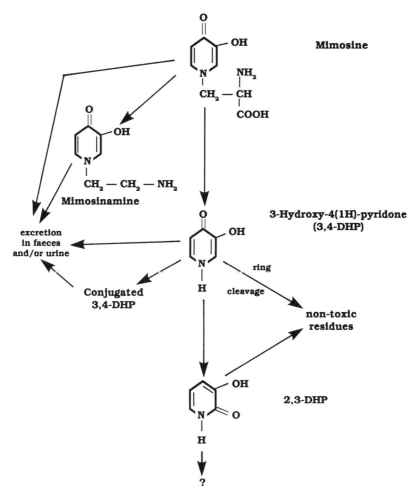

Figure 8 *Mimosine metabolism in the ruminant*

the leaf of the legume itself as a result of post-harvest enzymatic action.[44] It is known that 3,4-DHP is a potent goitrogen,[45] a property not shared with mimosine. Ruminants consuming *Leucaena* macerate the leaves and add alkaline saliva, thus providing optimum conditions favouring autolysis of mimosine to the extent that a significant proportion (0.3) of the amino acid may undergo degradation to 3,4-DHP before even reaching the rumen.[45] Additional quantities of 3,4-DHP arise during rumen fermentation of mimosine and, moreover, another goitrogen and isomer, 2,3-DHP may also be synthesized. Some rumen microbes are also capable of degrading the two forms of DHP to, as yet, unidentified non-toxic products.[46] Despite these transformations, considerable

[44] K. L. Wee and S. S. Wang, *J. Sci. Food Agric.*, 1987, **39**, 195.
[45] R. J. Jones, in 'Plant Toxicology', ed. A. A. Seawright, M. P. Hegarty, L. F. James, and R. F. Keeler, Queensland Poisonous Plants Committee, Yeerongpilly, 1985, p. 111.
[46] R. J. Jones and J. B. Lowry, *Experentia*, 1984, **40**, 1435.

quantities of mimosine and 3,4-DHP may escape degradation to appear in the faeces.[47] Furthermore, conjugated forms of DHP may also appear in the faeces and urine, while the parent amino acid may itself undergo decarboxylation within the tissues of the ruminant to yield mimosinamine which is then excreted in the urine.[42] It will be apparent that the type and severity of manifestations of *Leucaena* toxicity are likely to depend upon the extent to which mimosine is degraded in the rumen. Ruminant animals in Australia lack the requisite bacteria for degradation of DHP and are, consequently, susceptible to conditions such as loss of appetite, goitre, and reductions in blood thyroxine concentrations.[45] However, in Hawaii and in other parts of the tropics,[46,48] ruminants are believed to possess the full complement of bacteria that are required for DHP breakdown which accounts for the absence of *Leucaena* toxicity in these countries.

The toxicity of mimosine for poultry has not been established beyond reasonable doubt. Although the adult cockerel is capable of metabolizing a single oral dose without ill-effects,[49] other studies[15] show that the young chick is sensitive to mimosine to the extent that growth and food intake are impaired with a dietary mimosine concentration of 3.3 g kg^{-1}. The toxicity of mimosine is frequently invoked as the mechanism whereby *Leucaena* leaf meal (LLM) reduces performance of poultry. However, the arguments for such a mode of action are based on indirect evidence and have been questioned by D'Mello and Acamovic.[17] Since LLM is poorly digested,[50] it might be expected that mimosine absorption from LLM diets would be extremely low. This is, indeed, the case,[51] with up to 0.92 of ingested mimosine being excreted by chicks fed on LLM diets (Figure 9). Moreover, mimosine does not appear in the blood of chicks given LLM.[15] However, mimosine might play a dominant role in the toxicity of *Leucaena* seeds. There is some evidence that mimosine absorption from the seed does occur with subsequent deposition in the organs and tissues of poultry.[52]

Lathyrogenic Amino Acids

The deleterious properties of the lathyrogenic amino acids and dipeptides have been reviewed exhaustively[2,3,53] and, consequently only a brief account follows. These components of *Lathyrus* and *Vicia* species are believed to be responsible for the conditions of neurolathyrism in humans and of osteolathyrism, a pathological disorder induced experimentally in laboratory animals by feeding seeds of *Lathyrus odoratus*. Neurolathyrism in humans is characterized by muscular rigidity, weakness, paralysis of leg muscles, and, in extreme cases,

[47] R. Elliot, B. W. Norton, J. T. B. Milton, and C. W. Ford, *Aust. J. Agric. Res.*, 1985, **36**, 867.
[48] M. G. Dominguez-Bello and C. S. Stewart, *Proc. Nutr. Soc.*, 1989, **48**, 168A.
[49] J. A. Springhall, *Nature (London)*, 1965, **207**, 552.
[50] J. P. F. D'Mello and D. Thomas, *Trop. Agric. (Trinidad)*, 1978, **55**, 45.
[51] J. P. F. D'Mello and T. Acamovic, *Anim. Feed Sci. Technol.*, 1982, **7**, 247.
[52] U. Ter Meulen, F. Pucher, M. Szyszka, and E. A. El-Harith, *Arch. Geflugelk.*, 1984, **48**, 41.
[53] D. N. Roy, *Nutr. Abstr. Rev.*, 1981, **51**, 691.

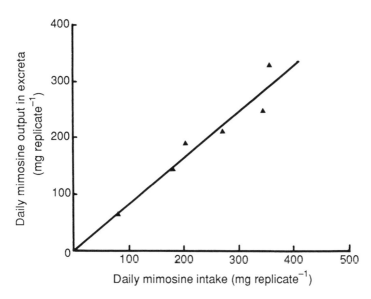

Figure 9 *Mimosine output in excreta of young chicks consuming different quantities of mimosine derived from Leucaena leaf meal*
(Reproduced by permission from *Anim. Feed. Sci. Technol.*, 1982, **7**, 247.)[51]

death.[54] This neurological disorder is endemic in parts of India where the incidence is exacerbated by episodes of famine.[55] The neurotoxicity of BCNA is well established (Table 4) following the sterling efforts of Ressler and co-workers.[56-58] Single doses of BCNA are sufficient to elicit severe metabolic and neurological manifestations of toxicity. Pyridoxal hydrochloride delays the onset of the neurological lesions and increases the LD_{50} dose. Avian species are significantly more susceptible than mammals to BCNA administration.[58] The glutamyl dipeptide of BCNA is considered to be as effective a neurotoxin as the parent amino acid, although method of administration may determine the relative potency.[24] Although the occurrence of BCNA and its dipeptide accounts for the neurotoxic properties of *V. sativa*, the role of these factors in human neurolathyrism has been discounted since this legume generally only occurs as a contaminant of other legume grains such as those of *Lathyrus* species.[53] *V. sativa*, however, may form a component of poultry diets.

There is little doubt that ODAP is the most potent neurotoxic amino

---

[54] G. Padmanaban, in 'Toxic Constituents of Plant Foodstuffs', ed. I. E. Liener, Academic Press, New York, 1980, p. 239.
[55] P. S. Sarma and G. Padmanaban, in 'Toxic Constituents of Plant Foodstuffs', ed. I. E. Liener, Academic Press, New York, 1969, p. 267.
[56] C. Ressler, *J. Biol. Chem.*, 1962, **237**, 733.
[57] C. Ressler, J. Nelson, and M. Pfeffer, *Biochem. Pharmacol.*, 1967, **16**, 2309.
[58] C. Ressler, *Recent Adv. Phytochem.*, 1975, **9**, 151.

Table 4 Toxicity of β-cyanoalanine (BCNA)

Animal species	Method of administration (dose)	Toxicity symptoms	LD_{50} (mg kg^{-1} BW*)
Rat[a,b]	Stomach tube (150 mg kg^{-1} BW), or subcutaneous injection (100–250 mg kg^{-1} BW), or food (10 g kg^{-1} diet)	Cystathioninuria; irreversible hyperactivity; tremors; convulsions; rigidity; prostration; death	135
	Subcutaneous injection (110–250 mg kg^{-1} BW) after pyridoxal. HCl administration		189
	Subcutaneous injection (110–250 mg kg^{-1} BW) with pyridoxal. HCl before and after BCNA administration		225
Chick[c]	Food (0.75 g kg^{-1} diet) or Subcutaneous injection	Convulsions; tetanic spasms of back muscles; death	70

* Body weight.
[a] C. Ressler, *J. Biol. Chem.*, 1962, **237**, 733.
[b] C. Ressler, J. Nelson, and M. Pfeffer, *Biochem. Pharmacol.*, 1967, **16**, 2309.
[c] C. Ressler, *Recent Adv. Phytochem.*, 1975, **9**, 151.

(Reproduced by permission from J. P. F. D'Mello, in 'Anti-Nutritional Factors, Potentially Toxic Substances in Plants', ed. J. P. F. D'Mello, C. M. Duffus, and J. H. Duffus, Association of Applied Biologists, Warwick, 1989, p. 29)[4]

acid[59-61] found in higher plants, being positively associated with neurolathyrism in humans consuming *Lathyrus sativus*. The neurological and biochemical lesions caused by ODAP are summarized in Table 5. The precipitation of these adverse effects is influenced by age of animal and by physiological factors such as the induction of an acidotic state,[55] but not by animal species. The age effect has been attributed to the intervention of a functional blood–brain barrier in adults[54] and to their ability to increase urinary excretion of ODAP (Table 5). The role of ODAP in the aetiology of neurolathyrism in adult humans thus requires elucidation, particularly in view of more recent claims that ODAP does, indeed, enter the central nervous system of mature animals.[53]

Toxicological investigations[35,62,63] with DABA, another neurolathyrogen, reveal increased susceptibility of mammals relative to avian species (Table 6). However, the oxalyl derivative of DABA is neurotoxic to the one-day-old chick.

The osteolathyritic properties of BAPN and its glutamyl dipeptide are well recognized. The nitrile inhibits lysyl oxidase activity thereby preventing ade-

[59] P. S. Cheema, K. Malathi, G. Padamanaban, and P. S. Sarma, *Biochem. J.*, 1969, **112**, 29.
[60] P. R. Adiga, S. L. N. Rao, and P. S. Sarma, *Curr. Sci.*, 1963, **32**, 153.
[61] S. L. N. Rao and P. S. Sarma, *Biochem. Pharmacol.*, 1967, **16**, 218.
[62] C. Ressler, P. A. Redstone, and R. H. Erenberg, *Science*, 1961, **134**, 188.
[63] J. C. Watkins, D. R. Curtis, and T. J. Biscoe, *Nature (London)*, 1966, **211**, 637.

Table 5 Toxic features of β-N-oxalyl-α, β-diaminopropionic acid (ODAP)

Age and species	Intraperitoneal injections (dose)	Manifestations of toxicity
Young rat[a]	(1.4 mmol kg^{-1} BW*)	Tremors; convulsive seizures in 10 min; increased brain [NH$_3$] and [glutamine]; death
Adult rat[a]	(5 mmol kg^{-1} BW)	Absence of ill-effects; no changes in brain biochemistry; 50% of dose excreted in urine
Adult rat[a]	(2.2 mmol kg^{-1} BW); rat rendered 'acidotic'	Convulsive seizures; increased brain [NH$_3$], [urea], [glutamine], and [ODAP]
Chick[b]	(20 mg chick^{-1})	Wry neck; head retraction; convulsions
	(30 mg chick^{-1})	Lethal
Adult fowl[c]	(100 mg bird^{-1})	Absence of ill-effects
Adult fowl[c]	(100 mg bird^{-1}) bird rendered acidotic prior to ODAP administration	Toxic manifestations

* Body weight.
[a] P. S. Cheema, K. Malathi, G. Padmanaban, and P. S. Sarma, *Biochem. J.*, 1969, **112**, 29.
[b] P. R. Adiga, S. L. N. Rao, and P. S. Sarma, *Curr. Sci.*, 1963, **32**, 153.
[c] S. L. N. Rao and P. S. Sarma, *Biochem. Pharmacol.*, 1967, **16**, 218.

(Reproduced by permission from J. P. F. D'Mello, in 'Anti-Nutritional Factors, Potentially Toxic Substances in Plants', ed. J. P. F. D'Mello, C. M. Duffus, and J. H. Duffus, Association of Applied Biologists, Warwick, 1989, p. 29)[4]

quate formation of collagen–elastin cross-linkages and consequently yields connective tissue containing excessively soluble collagen.[58]

Analogues of Sulphur Containing Amino Acids

The debilitating disorders associated with the selenoamino acids are manifestations of acute selenium poisoning.[64] *Se*-Methylselenocysteine and selenocystathionine are implicated in the condition known as 'blind staggers' which develops in range herbivores grazing *Astragalus* and other *Se*-accumulating plants. Affected animals walk about aimlessly, froth at the mouth, and appear to be in considerable physical pain. On the other hand, selenomethionine and selenocysteine induce several deformities and sloughing of the hooves in livestock.[2] The metabolism of selenomethionine (*Se*-Met) has been the subject of recent investigations.[65,66] It is now apparent that *Se*-Met is readily incorporated

[64] T. C. Stadtman, *Science*, 1974, **183**, 915.
[65] I. H. Waschulewski and R. A. Sunde, *J. Nutr.*, 1988, **118**, 367.
[66] J. A. Butler, M. A. Beilstein, and P. D. Whanger, *J. Nutr.*, 1989, **119**, 1001.

Table 6 Toxic effects of a,γ-diaminobutyric acid (DABA)

Animal species	Method of administration (dose)	Lesions
Rat[a]	Stomach tube	Weakness in hind legs; upper extremity tremors; convulsions; death
	Intraperitoneal injection (4.4 mmol kg^{-1} BW*)	Hyperirritability; tremors; convulsions; death; increased [NH$_3$] in blood and brain and [glutamine] in brain; inhibition of ornithine carbamoyltransferase; reduced urea production
Chick[b]	Intraperitoneal injection (3.2 or 6.5 mmol kg^{-1} BW)	Absence of toxic symptoms
	(12.9 mmol kg^{-1} BW)	Increased [glutamine] in brain

* Body weight.

[a] C. Ressler, P. A. Redstone, and R. H. Erenberg, *Science*, 1961, **134**, 188.
[b] R. M. O'Neal, C. H. Chen, C. S. Reynolds, S. K. Meghal, and R. E. Koeppe, *Biochem. J.*, 1968, **106**, 699.

(Reproduced by permission from J. P. F. D'Mello, in 'Anti-Nutritional Factors, Potentially Toxic Substances in Plants', ed. J. P. F. D'Mello, C. M. Duffus, and J. H. Duffus, Association of Applied Biologists, Warwick, 1989, p. 29)[4]

into tissue proteins of animals and that the balance between synthetic and degradative pathways is markedly influenced by methionine status. At adequate intakes of methionine, *Se*-Met is degraded to selenocysteine which may be further catabolized to yield selenide. Alternatively, *Se*-Met may undergo transamination and decarboxylation to release more selenide. The latter compound then becomes available for incorporation into selenoproteins such as glutathione peroxidase. However, when methionine intake is sub-optimal, *Se*-Met is directed towards tissue protein incorporation, thus reducing the availability of selenium for synthesis of glutathione peroxidase.[65]

The toxicity of SMCO in cattle and sheep[26,27] arises after its fermentation by rumen bacteria to dimethyl disulphide (Figure 10). The resulting adverse effects are characterized by a severe haemolytic anaemia which appears within one to three weeks in animals fed mainly or solely on *Brassica* forage. Early signs of this disorder are loss of appetite, reduced milk production, the appearance of refractile, stainable granules (Heinz–Ehrlich bodies) within the erythrocytes, and a decline in blood haemoglobin concentrations. Extensive organ damage generally accompanies this condition, the liver becoming swollen, pale, and necrotic. Critical daily intakes of SMCO range between 15–19 g kg^{-1} body weight irrespective of whether the amino acid is supplied in crystalline form or as a component of *Brassica* forage. Spontaneous but incomplete recovery is observed in survivors continuing to graze the crop, but further fluctuations in blood haemoglobin concentrations may follow. If *Brassica* forage is withdrawn, normal blood chemistry is restored within three to four weeks.[26]

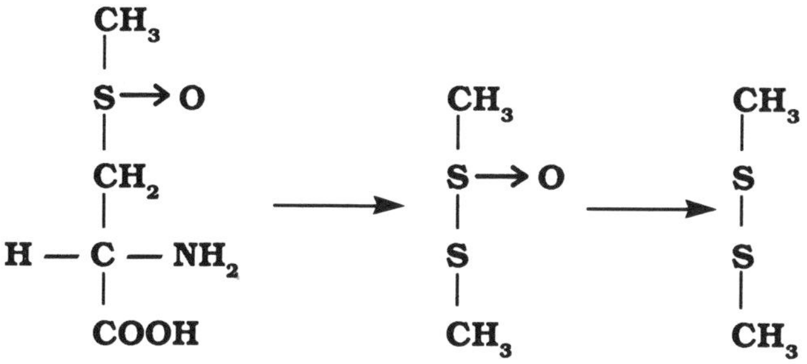

SMCO

Dimethyl disulphide oxide

Dimethyl disulphide

Figure 10 *Fermentation of* S-*methylcysteine sulphoxide (SMCO) in the rumen*

The toxicity of djenkolic acid in humans arises from its poor solubility under acidic conditions after consumption of the djenkol bean. The amino acid readily precipitates in body fluids leading to acute kidney malfunction.[67]

Potential Applications

Despite the negative attributes of the non-protein amino acids, there is increasing evidence that their toxic properties may be harnessed for therapeutic and other economic purposes. Thus, mimosine is endowed with anti-tumour activity[68] and its use as a chemical defleecing agent has also been considered.[41,42] Administration of DOPA as a palliative in the management of Parkinson's disease is widely recognized. Currently, there is some interest in the use of BAPN in reversing the abnormalities of collagen metabolism which occur in diabetic subjects.[69] Collagen solubility is generally decreased in both human and animal diabetics. It has been shown that BAPN is effective in restoring collagen solubility of diabetic rats to normal levels. There is also much interest

[67] A. G. van Veen, in 'Toxicants Occurring Naturally in Foods', National Academy of Sciences, Washington, 1966, p. 174.
[68] W. D. Dewys and T. C. Hall, *Eur. J. Cancer*, 1973, **9**, 281.
[69] S. M. Janakat, P. C. Bates, D. J. Millward, Z. A. Yahya, and J. P. W. Rivers, *Proc. Nutr. Soc.*, 1989, **48**, 137A.

in the use of *Se*-Met in the prevention of selenium deficiency in humans.[70] Both Keshan disease and Kashin–Beck disease respond to selenium administration, and there is evidence that *Se*-Met is more effective and less toxic than inorganic forms of the trace element. However, as with other therapeutic applications, all potential uses of amino acids should undergo critical safety evaluation before commercial exploitation is contemplated.

5 Mechanisms of Toxicity

It may be stated at the outset that the original hypothesis that an individual amino acid exerts its toxic effects via a single mechanism is no longer tenable. The multimodal action of amino acids is best illustrated by the diverse mechanisms proposed for the lysine–arginine antagonism in avian species.[28] Since these animals cannot synthesize arginine, they are particularly sensitive to this interaction. Mammals are less susceptible since they are able to regenerate arginine via a fully functional urea cycle.[71] A significant contributory factor to the antagonism is the increased activity of kidney arginase in chicks fed excess lysine which results in enhanced catabolism of arginine.[72] If arginase activity in birds is suppressed by the use of a specific inhibitor (*α*-aminoisobutyric acid), then they too become less sensitive to the lysine–arginine antagonism. However, other factors contribute to this antagonism. Thus, excess lysine inhibits transamidinase activity in the liver, thereby reducing endogenous synthesis of creatine.[73] Another factor is the increased excretion of arginine in the urine of birds fed excess lysine.[72] Both amino acids share a membrane-bound carrier for transport across cell membranes. Consequently, excess lysine inhibits the binding of arginine to the carrier, thus reducing arginine reabsorption in the kidney and enhancing arginine excretion in the urine. A fourth factor implicated in the lysine–arginine antagonism is the depression in food intake which invariably also occurs in animals fed amino acid imbalanced diets.[72] While the mechanisms responsible for this depression in food intake have yet to be elucidated, changes in brain concentrations of neurotransmitters derived from the indispensable amino acids have been linked to this phenomenon. The neurotransmitters, norepinephrine and dopamine are synthesized from phenylalanine, while tryptophan serves as the precursor of 5-hydroxytryptamine.[74] Competition exists between these amino acid precursors and certain neutral amino acids for uptake across the blood–brain barrier leading to reduced cerebral synthesis of neurotransmitters. Excess lysine also reduces brain concentrations of norepinephrine and dopamine in chicks.[75]

There is now compelling evidence to indicate that the toxic action of canavanine is also mediated via diverse mechanisms (Figure 11). The major route of

---

[70] A. D. Salbe and O. A. Levander, *J. Nutr.*, 1990, **120**, 207.
[71] M. S. Edmonds and D. H. Baker, *J. Nutr.*, 1987, **117**, 1396.
[72] R. E. Austic, in 'Nutrient Requirements of Poultry and Nutritional Research', ed. C. Fisher and K. N. Boorman, Butterworths, London, 1986, p. 59.
[73] R. E. Austic and M. C. Nesheim, *Poult. Sci.*, 1972, **51**, 1098.
[74] H. F. Bradford, 'Chemical Neurobiology', Freeman, New York, 1986, Chapter 4, p. 155.
[75] L. M. Harrison and J. P. F. D'Mello, unpublished observations.

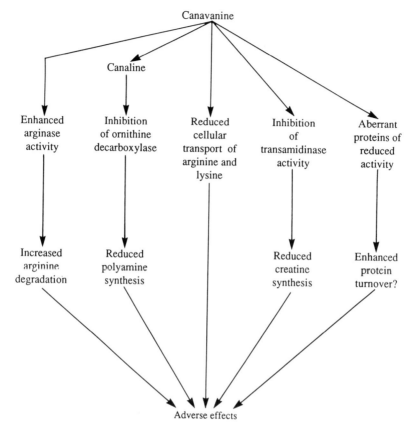

Figure 11 *Diverse mechanisms underlying the toxicity of canavanine*

canavanine metabolism in mammals is via the urea cycle where canavanine undergoes arginase-induced hydrolysis to canaline and urea.[31] This pathway should offer a mechanism for at least partial detoxification of canavanine in ureotelic animals. However, increased urea production has also been observed in avian species following the feeding of heat-treated seeds of *Canavalia ensiformis* which contain canavanine.[5] The increased arginase activity that would inevitably accompany the administration of such diets would also lead to inadvertent losses of arginine in a manner analogous to that seen in the lysine–arginine antagonism. Deleterious consequences would follow owing to the indispensability of arginine for avian species. Canavanine administration elicits striking elevations in the serum and urinary concentrations of ornithine.[31] These effects have been attributed to the synthesis of canaline from canavanine. Canaline forms a covalent complex with pyridoxal phosphate thereby inhibiting the activities of enzymes dependent upon this cofactor. One such enzyme is ornithine decarboxylase, and its inhibition by canaline would account for the substantial accumulation of ornithine in body fluids. Ornithine decarboxylase is a key enzyme in the synthesis of the polyamines putrescine, spermidine, and spermine which are important components involved in the regulation of cell growth and differentiation.[76] Inhibition of polyamine synthesis by canaline thus provides another focal point for the toxic action of canavanine. Recent studies show that canavanine may also act by competing with arginine and lysine for transport across cell membranes. Larbier reported decreases in the true digestibility of arginine and lysine in chicks fed *Canavalia ensiformis*.[77] Other work suggests that canavanine may act, like lysine, by inhibiting transamidinase activity, thus reducing creatine synthesis.[78] It is widely recognized that canavanine can replace arginine during protein synthesis leading to the production of aberrant macromolecules with modified functional properties.[2,3] Studies with rats have demonstrated the incorporation of canavanine into liver enzymes[79] and into brain components yielding proteins of reduced activity.[80] However, other studies with rats show negligible rates of incorporation of canavanine into tissue proteins.[30,31] It is possible that canavanyl proteins are degraded as rapidly as they are synthesized and there is limited but compelling evidence to support this proposition.[81] Any cycle of synthesis and degradation of canavanyl proteins would inevitably increase protein turnover rates, a feature which might, at least in part, explain the poor nitrogen retention efficiencies of animals fed on *Canavalia ensiformis*.[5] Recent work highlights another focal point for the canavanine–arginine antagonism.[82] Mammalian cells are capable of converting arginine to nitric oxide, an intermediate in the biosynthesis of nitrite and nitrate. Although the biological significance of these reactions awaits elucidation, it has been established that canavanine inhibits nitrite and nitrate synthesis in certain cell types, whereas homoarginine is a highly effective substrate. These studies attest to the complex and ramifying interactions which exist among the analogues of arginine.

The canavanine–arginine interaction appears to be analogous in several respects with the well-established antagonism between lysine and arginine. Thus, in both interactions, arginine requirements and urea excretion are enhanced,[5,28] although it should be recognized that a proportion of this additional urea may arise from the degradation of canavanine as well as of arginine by the action of arginase. Furthermore, the failure of supplementary arginine to reduce substantially the blood concentrations of canavanine or lysine is a feature common to both antagonisms. However, differences between these antagonisms are also apparent in that although efficiencies of utilization of food and nitrogen are improved on supplementing excess lysine diets with arginine,[28] corresponding improvements do not occur on adding arginine to diets containing canavanine.[5]

The toxicity of indospicine is mediated via at least three mechanisms. It is established that indospicine acts as a competitive inhibitor of arginase.[83] This

[76] O. Heby and L. Persson, *Trends Biochem. Sci.*, 1990, **15**, 153.
[77] M. Larbier, Proceedings of 6th European Symposium on Poultry Nutrition, Konigslutter, 1987.
[78] J. P. F. D'Mello, A. G. Walker, and E. Noble, *Br. Poult. Sci.*, 1991, **31**, 759.
[79] C. C. Allende and J. E. Allende, *J. Biol. Chem.*, 1964, **239**, 1102.
[80] P. Crine and E. Lemieux, *J. Biol. Chem.*, 1982, **257**, 832.
[81] S. E. Knowles, J. M. Gunn, R. W. Hanson, and F. J. Ballard, *Biochem. J.*, 1975, **146**, 595.
[82] M. A. Marletta, *Trends Biochem. Sci.*, 1989, **14**, 488.
[83] N. P. Madsen and M. P. Hegarty, *Biochem. Pharmacol.*, 1970, **19**, 2391.

property may account for the absence of hepatotoxicity in birds fed seeds of *Indigofera spicata*[34] due to the minor role of arginase in overall nitrogen elimination by uricotelic animals. Mammals, however, succumb to indospicine-induced hepatotoxicity owing to the critical role of arginase in the nitrogen metabolism of this species.[14] In addition, indospicine strongly inhibits amino-acylation of arginine, whereas the charging of leucine remains unaffected.[84] Incorporation of arginine in proteins is therefore impaired. *In vitro* studies[85] show that DNA synthesis is also inhibited by indospicine and that this effect may be reversed to a substantial degree by increasing arginine concentration.

Some general rules may now be formulated. The striking mammalian–avian difference in sensitivity to certain amino acids clearly resides in the disparate mode of elimination of excess nitrogen. Arginase appears to be the common focal point for the toxic action of at least three amino acids. When its activity in tissues is enhanced as, for example, by the administration of lysine or canavanine, toxicity is greater in avian species than in mammals; however, if arginase activity is depressed by the administration of indospicine, mammals succumb more readily to its toxicity.

Competitive inhibition of another key enzyme of the urea cycle may account for an additional mammalian–avian difference with respect to the toxicity of DABA. This amino acid inhibits ornithine carbamoyltransferase[63] causing primary liver dysfunction and secondary brain lesions in rats. The absence of organ damage in birds is consistent with the minor role of urea cycle enzymes in nitrogen excretion by this species.

Despite their diverse nature, a number of non-protein amino acids elicit similar metabolic effects. Thus canaline shares its ability to form complexes with pyridoxal phosphate, with mimosine, and BCNA resulting in the inhibition of certain aminotransferases and decarboxylases.[3] The three amino acids also inhibit the hepatic pyridoxal phosphate-dependent enzyme, cystathionase, a feature which accounts for the cystathioninuria observed in rats on administration of BCNA (Table 4) or of mimosine.[86]

The mode of action of ODAP continues to be an aspect of much debate and uncertainty. However, this issue is likely to receive renewed impetus in view of recent confirmation[87] that another amino acid, β-N-methyl-α, β-diaminopropionic acid (or β-N-methylaminoalanine), structurally analogous to ODAP, also evokes neurological disorders in humans consuming seeds of *Cycas circinalis* (false sago palm) in the Pacific island of Guam. There is now increasing awareness that the aetiology of a number of neurological diseases may be associated with dietary factors.

The toxicity of the selenoamino acids is attributed to their ability to replace the corresponding sulphur amino acids during protein synthesis.[3] In addition, there is now evidence for competition between these groups of amino acids at

---

[84] N. P. Madsen, G. S. Christie, and M. P. Hegarty, *Biochem. Pharmacol.*, 1970, **19**, 853.
[85] G. S. Christie, F. G. De Munk, N. P. Madsen, and M. P. Hegarty, *Pathology*, 1971, **3**, 139.
[86] J. A. Grove, P. D. Ballata, V. Eastmo, and L. R. Hwang, *Nutr. Rep. Int.*, 1978, **17**, 629.
[87] P. S. Spencer, P. B. Nunn, J. Hugon, A. C. Ludolph, S. M. Ross, D. N. Roy, and R. C. Robertson, *Science*, 1987, **237**, 517.

the level of intestinal transport.[88] Thus, *Se*-Met inhibits the transport of methionine across the jejunal brush border membrane of the pig since both amino acids share a common Na^+-dependent transport mechanism.

The toxic action of SMCO is mediated via its ruminal metabolite dimethyl disulphide (Figure 10) which is believed to inactivate key proteins through blockage of sulphydryl groups.[89] Ruminants appear to compensate for this protein inactivation by increasing synthesis of growth hormone and thyroxine which then stimulate the production of replacement proteins.

6 Detoxification

The multimodal action of individual amino acids ensures limited scope for detoxification by animals. Complete mitigation of adverse effects is only possible if the myriad of mechanisms is matched by an equally versatile metabolic strategy in animals. The tropical bruchid beetle (*Caryedes brasiliensis*) embodies this principle in its detoxification of canavanine.[3] In addition to developing enzymes capable of distinguishing between canavanine and arginine during protein synthesis, this insect has also abandoned the normal uricotelic mode of nitrogen elimination used by other invertebrates, in favour of ammoniotelism and ureotelism. Uricotelism inevitably confers dependence upon dietary arginine and, therefore, increased sensitivity to its structural analogues.

Mammals possess a significant capacity to detoxify canavanine[30,31] by virtue of their ureotelism whereas birds succumb readily to relatively low intakes of canavanine.[5] However, considerable amelioration of adverse effects in chicks may be achieved by dietary additions of arginine[5] or creatine,[78] whilst lysine exacerbates the toxicity. Dietary supplementation with arginine is even more effective in eliminating the adverse effects of excess lysine[28] (Figure 6).

The ability of ruminants to detoxify mimosine is subject to geographical differences in rumen ecology.[46] In many parts of the tropics cattle and goats are capable of detoxifying mimosine by the pathways outlined in Figure 8. On the other hand, ruminants in Australia are only able to degrade mimosine as far as DHP. However, recent biotechnological advances have allowed the transfer of DHP degrading bacteria from animals in Hawaii to Australian cattle with complete success. Inoculated cattle rapidly reduce DHP excretion in the urine despite a doubling of *Leucaena* intake.[45] Geographical variations in *Indigofera spicata* toxicity have also been reported, suggesting that any differences in rumen ecology may also be exploited in the detoxification of indospicine.[45]

In the case of SMCO, rumen microbial activity precipitates toxicity through the synthesis of a highly reactive metabolite.[26,27] However, manipulation of rumen function by dietary means may be used to influence the metabolism of the parent amino acid. Thus administration of oral supplements of methionine twice-weekly to lambs grazing kale elicited increases in the rumen pool and plasma concentrations of SMCO, indicating reduced degradation of SMCO in the rumen. Accompanying these changes was the development of a slightly less

[88] S. Wolffram, B. Berger, B. Grenacher, and E. Scharrer, *J. Nutr.*, 1989, **119**, 706.
[89] T. N. Barry, T. R. Manley, C. Redekopp, and T. F. Allsop, *Br. J. Nutr.*, 1985, **54**, 165.

severe manifestation of haemolytic anaemia in methionine supplemented lambs.[27]

Detoxification of crop plants containing non-protein amino acids is currently receiving much attention. The occurrence of legume-derived deleterious amino acid residues and their metabolites in animal products destined for human consumption[52] has concentrated attention on the use of plant breeding techniques to reduce or eliminate toxicity of legumes. There is little doubt that genetic methods have a role to play in this respect. However, it should be recognized that the toxic amino acids may confer insect and fungal resistance to the plant and removal of these compounds might result in agronomic disadvantages. In this respect, it is salutary to note that infestation by psyllid insects has all but decimated the *Leucaena* crop in some countries[90]; attempts to develop low mimosine lines are likely to compound this problem. In addition, the exact identity of the principal toxic agent of plants must be known with some certainty if plant breeding techniques are to succeed. Thus *Leucaena* and *Lathyrus* species contain at least two toxic amino acids (Table 2) and, furthermore, *Leucaena* contains other antinutritional factors which contribute to its toxicity.[17]

Since the toxic non-protein amino acids occur in unconjugated forms, their removal by extraction of plant products with solvents or aqueous solutions is generally an effective method of detoxification. For example, extraction of *Lathyrus* seeds with hot water results in the removal of ODAP.[54] Other studies[91] have shown that if whole seeds of *Canavalia ensiformis* are extracted with dilute $KHCO_3$ solution at 80 °C, concentrations of canavanine decline to negligible levels within 48 hours. After heat-treatment, the extracted seeds may be incorporated in broiler diets at concentrations of up to 280 g kg^{-1} without deleterious effects. The relative ease with which canavanine is extracted from intact beans is noteworthy and may provide the basis for the detoxification of other crop plants containing non-protein amino acids.

Considerable mitigation of adverse effects may be achieved by the controlled feeding of legume or *Brassica* forages containing toxic non-protein amino acids. For example, when sheep are conditioned to *Leucaena* by gradually increasing intake of the legume, mimosine toxicity may be averted.[45] Similarly, current recommendations for the prevention of SMCO toxicity in sheep involve some control of *Brassica* intake, particularly in late winter when SMCO concentrations in forage are maximal.[26] In addition, further control of SMCO intake may be accomplished by feeding *Brassica* forage derived from low-sulphur soils.[92] Such crops have markedly lower SMCO concentrations than those grown on soils with normal sulphur status.

7 Concluding Comments

A wide range of toxic, non-protein amino acids occur naturally in crop plants. Legumes generally contain higher concentrations and a more diverse array of these compounds than other plant species. This diversity is well illustrated by

[90] J. W. Beardsley, *Leucaena Res. Rep.*, 1986, **7**, 2.
[91] J. P. F. D'Mello and A. G. Walker, *Anim. Feed Sci. Technol.*, 1991, **33**, 117.
[92] T. N. Barry, T. R. Manley, K. R. Millar, and R. H. Smith, *J. Agric. Sci.*, 1984, **102**, 635.

the occurrence of canavanine with indospicine in *Indigofera spicata*[8]; mimosine with djenkolic acid in *Leucaena* species[7] and homoarginine with ODAP in *Lathyrus* species.[3]

There are marked variations in the susceptibility of animals to the toxic action of non-protein amino acids. Mammals are more sensitive than avian species to the deleterious effects of DABA and indospicine but birds are more susceptible to BCNA and canavanine toxicity. Age differences are also apparent with young animals exhibiting increased sensitivity to canavanine and ODAP. In ruminants, the presence of symbiotic microflora sometimes confers protection to the host animal by eliciting degradation of a toxic amino acid (*e.g.* mimosine) to harmless compounds. However, in the case of SMCO, the rumen micro-organisms precipitate toxicity through the synthesis of a highly reactive metabolite.

Amino acid toxicity may be exacerbated by a variety of dietary factors. Thus, the growth performance and efficiency of food utilization of chicks fed diets containing canavanine decline further with supplementation of either lysine[5] or homoarginine.[15] The practice of feeding ruminants with forage mixtures composed of *Gliricidia* and *Leucaena*[93] should be monitored in long-term studies in view of a possible deleterious interaction between their constituent amino acids (canavanine and mimosine, respectively). The toxicity of BAPN may be enhanced by administration of nitrofurazone, a drug used in the medication of poultry flocks.[94] Synergism may also occur between amino acids and other toxic components of plants. Such a relationship might exist, for example, between SMCO and the glucosinolates found in forage *Brassica*.[26]

There is substantial evidence to indicate that the toxic action of individual non-protein amino acids is mediated via a diverse array of mechanisms. Thus, canavanine may induce its toxicity by increasing renal and hepatic arginase activities[5,78]; by inhibiting key enzymes dependent upon pyridoxal phosphate through synthesis of canaline[3]; and by competing with arginine for intestinal transport[77] and during protein synthesis.[79,80] In addition, there is evidence to suggest that the effect on arginase activity is likely to be a more important determinant of toxicity in avian species than in mammals.

Recent biotechnological innovations[45,46] have yielded the most significant advance in detoxification of a non-protein amino acid. Complete degradation of mimosine and DHP in Australian cattle fed on *Leucaena* forage has been achieved by manipulation of rumen function. The application of such an approach in the detoxification of other forage plants may now be pursued with renewed impetus.

---

[93] O. B. Smith and M. F. J. Van Houtert, *World Anim. Rev.*, 1987, **62**, 57.
[94] D. N. Roy, S. H. Lipton, H. R. Bird, and F. M. Strong, *Poult. Sci.*, 1961, **40**, 55.

CHAPTER 3

Lectins

GEORGE GRANT

1 Introduction

It has long been known that the growth of young animals fed upon raw legume seeds is poor and that consumption of certain seeds, such as raw kidney bean (*Phaseolus vulgaris*) or raw castor bean (*Ricinus communis*), causes severe gastrointestinal disturbances, rapid weight loss, and with certain species, a high incidence of mortality. However, the link between many of these deleterious effects and the presence of lectins in the diet has only comparatively recently been established. Now, it is clear that many dietary lectins can seriously interfere with the normal body metabolism of animals and man.[1]

Lectins are carbohydrate-binding proteins or glycoproteins. They are of non-immunoglobulin nature and are capable of specific recognition of and reversible binding to carbohydrate moieties of complex glycoconjugates without altering the covalent structure of any of the recognized glycosyl ligands.[2] They were first detected in extracts of castor beans (*Ricinus communis*) by Stillmark in 1888 and since then have been found in a wide range of seeds, vegetables, fruits, and plant tissues.[3-6] They are not exclusive to the plant kingdom and have also been identified in bacteria, mammalian tissues, and cellular slime moulds.[7-9]

---

[1] A. Pusztai, in 'Recent Advances of Research in Anti-nutritional Factors in Legume Seeds', ed. J. Huisman, T. F. B. van der Poel, and I. E. Liener, Pudoc, Wageningen, Netherlands, 1989, p. 17.
[2] J. Kocourek and V. Horesji, *Nature (London)*, 1981, **290**, 188.
[3] M. S. Nachbar and J. D. Oppenheim, *Am. J. Clin. Nutr.*, 1980, **33**, 2338.
[4] J. Kocourek, in 'The Lectins', ed. I. E. Liener, N. Sharon, and I. J. Goldstein, Academic Press, New York, 1986, p. 3.
[5] M. E. Etzler, in 'The Lectins', ed. I. E. Liener, N. Sharon, and I. J. Goldstein, Academic Press, New York, 1986, p. 371.
[6] I. E. Liener, in 'The Lectins', ed. I. E. Liener, N. Sharon, and I. J. Goldstein, Academic Press, New York, 1986, p. 527.
[7] S. H. Barondes, in 'The Lectins', ed. I. E. Liener, N. Sharon, and I. J. Goldstein, Academic Press, New York, 1986, p. 438.
[8] S. H. Barondes, in 'The Lectins', ed. I. E. Liener, N. Sharon, and I. J. Goldstein, Academic Press, New York, 1986, p. 468.
[9] N. Sharon, in 'The Lectins', ed. I. E. Liener, N. Sharon, and I. J. Goldstein, Academic Press, New York, 1986, p. 494.

No clear role for lectins has been established. However, several functions for plant lectins are possible.[5] Lectins may act in a defensive role in that they protect the plant from attack by specific bacteria, fungi, or insects.[10,11] They may also be involved in the packaging and processing of storage materials during seed maturation or germination, modulate cell wall elongation, stimulate mitosis of embryonic cells, or act as storage proteins.[5,12] Furthermore, root lectins appear to be necessary to facilitate rhizobial–legume symbiosis.[13]

2 Detection of Lectins

Lectins (agglutinins) in plant protein extracts are detected by their ability to agglutinate human or animal red blood cells or to precipitate polysaccharides, glycoproteins, or protein–carbohydrate glycoconjugates.[14,15] Various enzyme-linked immunosorbent assay (ELISA) and radioimmunoassay (RIA) procedures have also been developed to quantify the content of specific lectins in plant extracts or food products.[16–18] The method of detection routinely used is that of agglutination of human or animal red blood cells.[14,19] Seed meal or protein is extracted with 0.9% (w/v) saline [sample: buffer 1:20 (w/v)]. Serial twofold dilutions are then made of the extract and to each dilution is added an equal volume of a 5% (v/v) suspension of erythrocytes. The samples are mixed and left at room temperature for up to 2 hours. Subsequently, the extent of erythrocyte clumping is evaluated by eye.[19,20] Alternatively, the optical density of 620 nm of the sample/erythrocyte suspension is measured.[21,22] Since clumped cells rapidly settle out of the suspension, the resultant reduction in the optical density of the erythrocyte suspension can be used as an index of the extent of haemagglutination.

The concentration of the serially diluted seed extract (μg seed meal or μg protein ml^{-1}) at which 50% of all the erythrocytes are agglutinated or at which the absorbance of the erythrocyte suspension is reduced by 50% is defined as that which contains 1 haemagglutinin unit (HU) of activity. From

[10] A. M. R. Gatehouse, S. J. Shackley, K. A. Fenton, and J. Bryden, *J. Sci. Food Agric.*, 1989, **47**, 269.
[11] R. C. Pratt, N. K. Singh, R. E. Shade, L. L. Murdock, and R. A. Bressan, *Plant. Physiol.*, 1990, **93**, 1453.
[12] M. Nsumba-Lubaki and W. J. Peumans, *Plant. Physiol.*, 1986, **80**, 747.
[13] H. Sui-Cheong and S. Malek-Hedayat, in 'Lectins', ed. T. C. Bog-Hansen and D. L. J. Freed, Sigma Chemical Co., St. Louis, 1988, Vol. 6, p. 95.
[14] W. G. Jaffe, in 'Toxic Constituents of Plant Foodstuffs', ed. I. E. Liener, Academic Press, New York, 1980, p. 73.
[15] I. J. Goldstein and R. D. Poretz, in 'The Lectins', ed. I. E. Liener, N. Sharon, and I. J. Goldstein, Academic Press, New York, 1986, p. 35.
[16] C. F. Talbot and M. E. Etzler, *Plant. Physiol.*, 1978, **61**, 847.
[17] M. Hosselet, E. van Dreissche, M. van Poucke, and L. Kanarek, in 'Lectins', ed. T. C. Bog-Hansen and J. Breborowicz, Walter de Gruyter, New York, 1985, Vol. 4, p. 583.
[18] R. I. Prince, B. G. Miller, M. Bailey, E. Telemo, D. Patel, and F. J. Bourne, Proceedings of the British Society for Animal Production, 1988, paper 88.
[19] G. Grant, L. J. More, N. H. McKenzie, J. C. Stewart, and A. Pusztai, *Br. J. Nutr.*, 1983, **50**, 207.
[20] N. Gilboa-Garber, in 'Methods of Enzymology', ed. V. Ginsburg, Academic Press, New York, 1982, Vol. 83, p. 378.
[21] I. E. Liener, *Arch. Biochem. Biophys.*, 1955, **54**, 23.
[22] H. Lis and N. Sharon, in 'Methods of Enzymology', ed. V. Ginsburg, Academic Press, New York, 1972, Vol. 28, p. 360.

this value, the specific haemagglutination activity (HU mg^{-1} meal or protein) is calculated and by comparison with the specific activity of a control sample of purified lectin, an estimate of the total lectin content of the seed extract can be made. The precision of these assays is limited ($\pm$ 1 tube) because of the serial dilution procedure. However, they are rapid and easy to carry out and are thus suitable for routine scanning of samples.

Extracts of certain seeds do not agglutinate normal red blood cells and as a result some of these varieties have been reported to be 'lectin-free'. This is not necessarily the case. Seeds, such as pinto bean (*Phaseolus vulgaris*) and lupin (*Lupinus angustifolius*), contain lectins that only agglutinate erythrocytes which have been pre-treated with pronase or trypsin.[23,24] Thus, to ensure that lectins of this type are detected, seed extracts which exhibit little or no activity against normal erythrocytes should be tested against red blood cells which have been pre-treated with proteolytic enzymes to increase their sensitivity to agglutinins.[14,20,25] A number of lectins require Ca and Mn, bovine serum albumin, or dextran to be added to the dilution buffer to allow maximal expression of their haemagglutinating activity.[26]

3 Assessment *In Vitro* of Potential Oral Toxicity

Seeds such as kidney bean (*Phaseolus vulgaris*), runner bean (*Phaseolus coccineus*), and tepary bean (*Phaseolus acutifolius*) which contain lectins that strongly agglutinate a wide range of different red blood cells including trypsin-treated cattle erythrocytes and normal rabbit, rat, pig, sheep, and human O and AB erythrocytes have a high oral toxicity for rats.[14,19] In contrast, seeds such as peas (*Pisum sativum*) and lentils (*Lens culinaris*) that contain lectins which agglutinate only rabbit and pronase-treated rat erythrocytes are essentially non-toxic.[14,19] Thus, it has been suggested that the potential oral toxicity of a lectin can be assessed on the basis of its reactivity towards animal and human erythrocytes and towards trypsin-treated cattle erythrocytes in particular.[14,19] However, some lectins which react with only a limited number of erythrocyte types have now been shown to severely impair growth when included in diets for rats.[27-29] Thus, although this screening technique can highlight lectins which are potentially highly toxic, it cannot be used to predict which dietary lectins will have little or no deleterious effects.

No existing *in vitro* method can thus reliably be used to predict the potential deleterious effects of a dietary lectin. However, since the extent to which a

---

[23] A. Pusztai, G. Grant, and J. C. Stewart, *Biochim. Biophys. Acta.*, 1981, **671**, 146.
[24] C. S. Kim and K. T. Madhusudhan, *J. Food Sci.*, 1988, **53**, 1234.
[25] H. Lis and N. Sharon, in 'The Lectins', ed. I. E. Liener, N. Sharon, and I. J. Goldstein, Academic Press, New York, 1986, p. 265.
[26] O. Makela, *Ann. Med. Exp. Biol. Fenn.*, 1957, **35**, 3.
[27] L. Manage and K. Sohonie, *Toxicon*, 1972, **10**, 89.
[28] M. Higuchi, I. Tsuchiya, and K. Iwai, *Agric. Biol. Chem.*, 1984, **48**, 695.
[29] G. Grant, S. W. B. Ewen, S. Bardocz, D. S. Brown, P. M. Dorward, W. B. Watt, J. C. Stewart, and A. Pusztai, in 'Recent Advances of Research in Anti-nutritional Factors in Legume Seeds', ed. J. Huisman, T. F. B. van der Poel, and I. E. Liener, Pudoc, Wageningen, Netherlands, 1989, p. 34.

Table 1 Nutritional toxicity of purified seed lectins to rats or mice

	Source	Scientific name	Nominal carbohydrate[o] specificity
Toxic	Kidney bean[a,b]	*Phaseolus vulgaris*	Galβ1,4GlcNAcβ1,2Man
	Black bean[c]	*Phaseolus vulgaris*	Galβ1,4GlcNAcβ1,2Man
	Kintoki bean[d]	*Phaseolus vulgaris*	Galβ1,4GlcNAcβ1,2Man
	Tora-name[e]	*Phaseolus vulgaris*	Galβ1,4GlcNAcβ1,2Man
Growth inhibitory	Soyabean[f,g]	*Glycine max*	a and βGalNAc > a and βGal
	Amaranth[h]	*Amaranthus cruentus*	GalNAc[b]
	Horse gram[i]	*Dolichos biflorus*	GalNAca1,3GalNAc > aGalNAc
	Hyacinth bean[j]	*Dolichos lablab*	Man > GlcNAc > Glc > Gal[q,r]
	Winged bean[k,l]	*Psophocarpus tetragonolobus*	aGalNAc > aGal
	Lima bean[j]	*Phaseolus lunatus*	GalNAca1,3[L-Fuc$a$1,2]Galβ > GalNAc
	Mucuna[m]	*Dioclea grandiflora*	aMan > aGlc > GlcNAc
Essentially non-toxic	Pea[n]	*Pisum sativum*	aMan > aGlc = GlcNAc
	Lentil[n]	*Lens culinaris*	aMan > aGlc > GlcNAc
	Faba bean[o]	*Vicia faba*	aMan > aGlc = GlcNAc
	Mucuna de batata[m]	*Dioclea sclerocarpa*	?

Based on data from:

[a] A. Pusztai and R. Palmer. *J. Sci. Food Agric.*, 1977, **28**, 620.
[b] A. Pusztai, E. M. W. Clarke, G. Grant, and T. P. King. *J. Sci. Food Agric.*, 1981, **32**, 1037.
[c] P. M. Hanovar, C-V. Shih, and I. E. Liener. *J. Nutr.*, 1962, **77**, 109.
[d] T. Hara. I. Tsukamoto, and M. Miyoshi. *J. Nutr. Sci. Vitaminol.*, 1983, **29**, 589.
[e] Y. Furruichi, M. Savada, and T. Takahashi. *Nutr. Rep. Int.*, 1988, **37**, 713.
[f] I. E. Liener. *J. Nutr.*, 1953, **49**, 527.
[g] G. Grant, W. B. Watt, J. C. Stewart, and A. Pusztai. *Med. Sci. Res.*, 1987, **15**, 1197.
[h] A. M. Calderon de la Barca, L. Vazquez-Moreno, M. E. Valencia, M. Cordero, and G. Lopez-Cervantes. 'Lectins', ed. T. C. Bog-Hansen and D. L. G. Freed. Sigma Chemical Co., St. Louis, 1988, Vol. 6, p. 125.
[i] L. Manage, A. Joshi, and K. Sohonie. *Toxicon*, 1972, **10**, 89.
[j] S. Salgarkar and K. Sohonie. *Ind. J. Biochem.*, 1965, **3**, 197.
[k] M. Higuchi, M. Sugo, and K. Iwai. *Agric. Biol. Chem.*, 1983, **47**, 1879.
[l] M. Higuchi, I. Tsuchiya, and K. Iwai. *Agric. Biol. Chem.*, 1984, **48**, 695.
[m] G. Grant, N. H. McKenzie, R. A. Moreira, and A. Pusztai, *Qual. Plant: Plant Foods Hum. Nutr.*, 1986, **36**, 47.
[n] S. Jindal, G. L. Soni, and R. Singh. *J. Plant Foods*, 1982, **4**, 95.
[o] L. A. Rubio, G. Grant, S. Bardocz, P. Dewey, and A. Pusztai. *Br. J. Nutr.*, 1991, (in press).
[p] I. J. Goldstein and R. D. Poretz. 'The Lectins', ed. I. E. Liener, N. Sharon, and I. J. Goldstein, Academic Press, New York, 1986, p. 35.
[q] R.S. Sandhu, S.S. Kamboj, R.S. Reen, R.J. Tatake, D. Subrahmanyam, R. Somasundaram, and S.G. Gangal, 'Lectins', ed. T.C. Bog-Hansen and J. Breborowicz, Walter de Gruyter, Berlin, 1985, Vol. 4, p. 559.
[r] J. Favero, F. Miquel, J. Dornand, M. Janicot, and J-C. Mani, 'Lectins', ed. T. C. Bog-Hansen and E. van Driesche, Walter de Gruyter, Berlin, 1986, Vol. 5, p. 391.

Table 2 *Effects of dietary lectins upon growth and body composition of pair-fed rats*

	Toxic kidney bean lectin[a,d]	Growth inhibitory soyabean lectin[e,g]	Lactalbumin control
Lectin intake (mg rat^{-1} d^{-1})	60	53	0
Weight change (g rat^{-1} d^{-1})	-0.6 ± 0.1	$+0.6 \pm 0.1$	$+1.3 \pm 0.2$
Small intestine Dry weight (mg 100 g^{-1} DBW*)	5510 ± 120	3000 ± 122	2120 ± 127
Pancreas Dry weight (mg 100 g^{-1} DBW)	585 ± 72	629 ± 45	421 ± 40
Hind-limb muscles Soleus (mg 100 g^{-1} DBW)	100 ± 10	94 ± 9	94 ± 8
Plantaris (mg 100 g^{-1} DBW)	142 ± 20	179 ± 2	181 ± 4
Gastrocnemius (mg 100 g^{-1} DBW)	688 ± 60	803 ± 21	820 ± 19
Carcass lipid (g 100 g^{-1} DBW)	12.7 ± 1.4	21.1 ± 1.1	21.1 ± 1.1

* Dry body weight.

Based on data from:
[a] F. Greer, A. C. Brewer, and A. Pusztai, *Br. J. Nutr.*, 1985, **54**, 95.
[b] J. T. A. de Oliviera, A. Pusztai, and G. Grant, *Nutr. Res.*, 1988, **8**, 943.
[c] S. Bardocz, G. Grant, D. S. Brown, H. M. Wallace, S. W. B. Ewen, and A. Pusztai, *Med. Sci. Res.*, 1989, **17**, 143.
[d] S. Bardocz, G. Grant, D. S. Brown, S. W. B. Ewen, and A. Pusztai, *Med. Sci. Res.*, 1989, **17**, 309.
[e] G. Grant, W. B. Watt, J. C. Stewart, and A. Pusztai, *Med. Sci. Res.*, 1987, **15**, 1197.
[f] G. Grant, W. B. Watt, J. C. Stewart, and A. Pusztai, *Med. Sci. Res.*, 1987, **15**, 1355.
[g] G. Grant, J. T. A. de Oliviera, P. M. Dorward, M. G. Annand, M. Annand, and A. Pusztai, *Nutr. Rep. Int.*, 1987, **36**, 373.

the dietary lectin interferes with the body metabolism of animals depends, at least in part, upon the degree to which it binds to the small intestine epithelium *in vivo* and the subsequent changes it causes upon epithelial cell metabolism, studies on the effects of various lectins upon isolated intestinal epithelial cell preparations, or intestinal epithelial cell cultures may possibly provide a basis for a screening method.[1,30-32]

4 Nutritional Studies with Purified Seed Lectins

Lectins have been detected in and isolated from a wide range of seeds but very few pure lectins have been tested for oral toxicity because of the large quantities of each lectin necessary for this type of study. As a result, most studies have been done with seed meals or seed meal extracts. However, since seeds can

[30] H. G. C. J. M. Hendriks, J. F. J. G. Koninkx, M. Draaijer, J. E. van Dijk, J. A. M. Raaijmakers, and J. M. V. M. Mouwen, *Biochim. Biophys. Acta.*, 1987, **905**, 371.
[31] H. G. C. J. M. Hendriks, J. F. J. G. Koninkx, M. Draaijer, J. E. van Dijk, J. A. M. Raaijmakers, and J. M. V. M. Mouwen, *Anim. Prod.*, 1989, **49**, 229.
[32] R. J. Hamer, J. Koninkx, M. G. van Oort, J. Mouwen, and J. Huisman, in 'Recent Advances of Research in Anti-nutritional Factors in Legume Seeds', ed. J. Huisman, T. F. B. van der Poel, and I. E. Liener, Pudoc, Wageningen, Netherlands, 1989, p. 30.

contain a number of potentially deleterious factors, the actions of which may be additive or even synergistic, this can make interpretation of the results from such studies somewhat problematic.[33,34]

Fifteen purified lectins have so far been assessed for oral toxicity towards rats or mice and they can be separated into 3 basic categories (Table 1 and Table 2):
(1) The toxic lectins which inhibit growth when included at a concentration of 3 g lectin kg^{-1} in diets, prevent growth when incorporated at a level of 5–7 g lectin kg^{-1} diet and cause weight loss and a high level of mortality when fed at a level of > 10 g lectin kg^{-1} diet.
(2) The growth inhibitory lectins which considerably reduce but do not prevent growth when they are added to diets at a level of 5–10 g lectin kg^{-1} diet.
(3) The essentially non-toxic lectins which may slightly impair growth if they are incorporated into diets at concentrations of between 7–20 g lectin g^{-1}.

5 Effects of Toxic or Growth Inhibitory Lectins in Rats and Mice

Lectins, unlike most dietary proteins, are at least partially resistant to proteolytic degradation *in vivo*. As a result 40–90% of the dietary intake of kidney bean, kintoki bean (*Phaseolus vulgaris*), or jack bean (*Canavalia ensiformis*) lectins can survive passage through the gastrointestinal tract and can be recovered in the faeces of experimental animals.[35–37] Lectins from winged bean (*Psophocarpus tetragonolobus*), mucuna (*Dioclea grandiflora*), soyabean (*Glycine max*), faba bean (*Vicia faba*), wheat germ (*Triticum vulgare*), pea, and tomato (*Lycopersicon esculentum*) also survive gut passage.[28,38–44] Thus, it seems likely that nutritionally significant amounts of most dietary lectins will survive in the small intestine of animals or man in a intact and fully reactive form.

If the appropriate carbohydrate structures are present in the intestinal lumen, lectins will bind to them (Table 3). Thus, kidney bean, kintoki bean, winged bean, soyabean, and tomato lectins bind to epithelial cells on the mid and upper surface of villi particularly in duodenal and jejeunal regions of the small

---

[33] I. E. Liener, 'Toxic Constituents of Plant Foodstuffs', ed. I. E. Liener, Academic Press, New York, 1980.
[34] P. R. Cheeke and L. R. Schull, 'Natural Toxicants in Feeds and Poisonous Plants', ed. P. R. Cheeke and L. R. Schull, Avi Publishing Co., Connecticut, 1985.
[35] V. C. Sgarbieri, E. M. W. Clarke, and A. Pusztai, *J. Sci. Food Agric.*, 1982, **33**, 881.
[36] T. Hara, Y. Mukunoki, I. Tsukamoto, M. Miyoshi, and K. Hasegawa, *J. Nutr. Sci. Vitaminol.*, 1984, **30**, 381.
[37] S. Nakata and T. Kimura, *J. Nutr.*, 1985, **115**, 1621.
[38] P. G. Brady, A. M. Vannier, and J. G. Banwell, *Gastroenterol.*, 1978, **75**, 236.
[39] D. C. Kilpatrick, A. Pusztai, G. Grant, C. Graham, and S. W. B. Ewen, *FEBS Lett.*, 1985, **185**, 299.
[40] G. Grant, N. H. McKenzie, R. A. Moreira, and A. Pusztai, *Qual. Plant. Plant Foods Hum. Nutr.*, 1986, **36**, 47.
[41] G. Bertrand, B. Seve, D. J. Gallant, and R. Tome, *Sci. Aliments.*, 1988, **8(2)**, 187.
[42] M. S. Shet, S. M. Madiaiah, and R. Nazeer-Ahamed, *Ind. J. Exp. Biol.*, 1989, **27(1)**, 58.
[43] A. Pusztai, S. W. B. Ewen, G. Grant, W. J. Peumans, E. J. M. van Damme, L. Rubio, and S. Bardocz, *Digestion*, 1990, **46**, 308.
[44] L. A. Rubio, G. Grant, S. Bardocz, P. Dewey, and A. Pusztai, *Br. J. Nutr.*, 1991, (in press).

Table 3 *Survival of ^{125}I-labelled kidney bean lectins in the small intestine 3 hours after being given as a single dose to rats by gastric intubation and the extent of uptake into systemic circulation (all values expressed as percentage of dose reaching the small intestine)*

	Kidney bean[a] lectin	Tomato[b] lectin
Small intestine		
Lumen contents	16.0	15.8
Tissue-bound	8.7	8.0
Caecum and colon	0.6	15.0
Blood (6 ml)	4.6	0.2
Liver	0.6	0.7
Kidneys	0.2	0.1

Based on data from:
[a] D. C. Kilpatrick, A. Pusztai, G. Grant, C. Graham, and S. W. B. Ewen, *FEBS Lett.*, 1985, **185**, 299.
[b] A. Pusztai, F. Greer, and G. Grant, *Biochem. Soc. Trans.*, 1989, **17**, 481.

intestine.[36,39,41-43,45,46] In contrast, pea and faba bean lectins do not adhere to the intestinal epithelium although they do survive in the gut lumen.[41,43,44,47]

Lectin, bound to the surface of enterocytes *in vivo*, is taken into the cells by endocytosis and can subsequently be found within the cytoplasm in multivesicular bodies, endosomes, and lysosomes.[48] Small amounts of any protein can be swept into the enterocytes in this manner but in animals with a fully mature gut these internalized proteins are degraded in the lysosomal system and therefore no systemic transfer of intact protein occurs.[49] However, this is not the case with lectins. Some of the lectin survives passage through the cytoplasm of the enterocyte and is subsequently released systematically perhaps by exocytosis into the intercellular space.[39,50]

The interaction of toxic or growth inhibitory lectins with the small intestine epithelium *in vivo* results in severe disruption to the metabolism of the small intestine (Table 2). Kidney bean or soyabean lectins both induce extensive, rapid, and progressive increases in the length and weight of the small intestine.[51-53] Indeed, with rats given kidney bean lectin, the absolute weight of their small intestine can double within 5-7 days despite the fact that the rats are

[45] T. P. King, A. Pusztai, and E. M. W. Clarke, *Histochem. J.*, 1980, **12**, 201.
[46] J. G. Banwell, C. R. Abromovsky, F. L. Weber, R. Howard, D. H. Boldt, *Dig. Dis. Sci.*, 1984, **29**, 921.
[47] R. Begbie and T. P. King, in 'Lectins', ed. T. C. Bog-Hansen and J. Breborowicz, Walter de Gruyter, New York, 1985, Vol. 4, p. 15.
[48] T. P. King, A. Pusztai, G. Grant, and D. Slater, *Histochem. J.*, 1986, **18**, 413.
[49] W. A. Walker, in 'Physiology of the Gastrointestinal Tract', ed. L. R. Johnson, Raven Press, New York, 1981, p. 1271.
[50] A. Pusztai, F. Greer, and G. Grant, *Biochem. Soc. Trans.*, 1989, **17**, 481.
[51] F. Greer, A. C. Brewer, and A. Pusztai, *Br. J. Nutr.*, 1985, **54**, 95.
[52] J. T. A. de Oliveira, A. Pusztai, and G. Grant, *Nutr. Res.*, 1988, **8**, 943.
[53] G. Grant, W. B. Watt, J. C. Stewart, and A. Pusztai, *Med. Sci. Res.*, 1987, **15**, 1355.

rapidly losing body weight.[54] This intestinal growth is due to both hypertrophy and hyperplasia.[51,53,55] Thus, a two-fold increase in the number of cells in the proliferative crypts of Lieberkuhn and a 25 % increase in overall size of these cells is evident.[54] Considerable thickening of the smooth muscle layer is also apparent.[54]

Protein synthesis rates in the small intestine mucosa increase within 2 hours of exposure of rats to kidney bean lectin and significant increases in total RNA and protein contents are evident within 24 hours.[56,57] Furthermore, these changes appear to coincide and perhaps are preceded by an accumulation of polyamines (putrescine, spermidine, spermine) in the tissue.[54,56,58] Similar metabolic changes occur during proliferation induced by hormones or growth factors.[59] Therefore it is likely that the lectin induces crypt cell proliferation by a comparable mechanism. The lectin does not however bind to the luminal surface of the crypt cells.[56] In addition, hormones and growth factors stimulate crypt cell proliferation by binding to receptors on the basolateral membrane of the cells. Thus, the lectin may by binding to gut endocrine cells on the mid or upper villus cause the release of trophic hormones which subsequently stimulate cell division in the crypts of Lieberkuhn. Alternatively, systematically absorbed lectin may mimic the effects endogenous growth factors by binding to appropriate receptors on the basolateral membrane of crypt cells and thereby stimulate cell proliferation.

Toxic and growth inhibitory lectins also stimulate rapid pancreatic growth (Table 2). Initially, the enlargement is primarily due to hypertrophy.[29,58] Subsequently, after 10–16 days, cellular hyperplasia also occurs.[29,60] The increase in pancreatic size and RNA and protein contents also coincides and may to some extent be preceded by accumulation of polyamines in the tissue.[61] The mechanism by which this enlargement occurs is unknown. However, it may be the result of release of trophic hormones, such as cholecystokinin, due to the interaction of lectins with endocrine cells in the small intestine.[62] Alternatively, it may be a direct effect of systematically absorbed lectin.

Stimulation of intestinal and pancreatic growth by lectins greatly increases the requirement of these tissues for nutrients. Within 5–7 days, approximately 40–60% of the dietary intake of nitrogen of rats given kidney bean lectin is required to support the additional intestinal growth.[63] As the intestine further enlarges, its requirement for nitrogen increases and eventually exceeds that of

[54] A. Pusztai, G. Grant, D. S. Brown, S. W. B. Ewen, and S. Bardocz, *Med. Sci. Res.*, 1988, **16**, 1283.
[55] H. Tajri, R. M. Klein, E. Lebenthal, and P. C. Lee, *Dig. Dis. Sci.*, 1989, **33**, 1364.
[56] A. Pusztai, G. Grant, L. M. Williams, D. S. Brown, S. W. B. Ewen, and S. Bardocz, *Med. Sci. Res.*, 1989, **17**, 215.
[57] R. M. Palmer A. Pusztai, P. Bain, and G. Grant, *Comp. Biochem. Physiol.*, 1987, **88C**, 179.
[58] S. Bardocz, G. Grant, D. S. Brown, H. M. Wallace, S. W. B. Ewen, and A. Pusztai, *Med. Sci. Res.*, 1989, **17**, 143.
[59] G. D. Luk and P. Yang, *Gut*, 1987, **28(S1)**, 95.
[60] G. Grant, W. B. Watt, J. C. Stewart, and A. Pusztai, *Med. Sci. Res.*, 1987, **15**, 1197.
[61] S. Bardocz, G. Grant, D. S. Brown, S. W. B. Ewen, and A.Pusztai, *Med. Sci. Res.*, 1989, **17**, 309.
[62] T. Fushiki and K. Iwai, *FASEB J.*, 1989, **3**, 121.
[63] A. Pusztai, G. Grant, S. W. B. Ewen, D. S. Brown and S. Bardocz, *Proc. Nutr. Soc.*, 1990, **49**, 144A.

dietary intake.[63] Thus, the wasteful growth of the small intestine and pancreas induced by kidney bean and soyabean lectins greatly restricts the amount of nutrients available to the animal to maintain and facilitate growth of other body tissues.

Disruption to the small intestine absorptive epithelium, abnormal development of microvilli, and elevated rates of cell senescence and loss occur on the mid and upper regions of the villi to which the toxic and growth inhibitory lectins bind.[28,29,42,64–72] These changes in the intestinal epithelium appear rapidly and in some cases are evident after only 1 hour of exposure to the lectin.[71,73] This would suggest that the cellular damage is a direct result of the interaction of lectin with the cells. However, since kidney bean lectin taken up through the absorptive enterocytes and released into the lamina propria apparently directly triggers degranulation of mast cells, the disruption of the epithelial surface may, in part, be a result of this response.[74–76] The gut damage is reversible and 24 hours after withdrawal of the dietary lectin no abnormalities are apparent on the villi.[45,73,77]

The activities of several intestinal and brush border enzymes expressed primarily on the mid and upper villus, such as enterokinase, leucine aminopeptidase, alkaline phosphatase, maltase, and sucrase, are reduced in animals given toxic lectins.[26,69,72,78–82] Lipid absorption is also apparently lowered.[81] These effects are probably due to the disruption of the absorptive epithelium.

A major change in the bacterial population of the small intestine is mediated by the toxic kidney bean lectin. The total number of coliforms in the small intestine, luminal and tissue-bound anaerobes and aerobes, is greatly

[64] T. P. King, A. Pusztai, and E. M. W. Clarke, *J. Comp. Pathol.*, 1982, **92**, 357.
[65] T. Hara, I. Tsukamoto, and M. Miyoshi, *J. Nutr. Sci. Vitaminol.*, 1983, **29**, 589.
[66] S. Jindal, G. L. Soni, and R. Singh, *Nutr. Rep. Int.*, 1984, **29**, 95.
[67] M. A. Rossi, J. Mancini, and F. M. Lajolo, *Br. J. Exp. Path.*, 1984, **65**, 117.
[68] J. G. Banwell, R. Howard, D. Cooper, and J. W. Costerton, *Appl. Environ. Microbiol.*, 1985, **50**, 68.
[69] J. M. Rouanet, J. Lafont, M. Chalet, A. Creppy, and P. Besancon, *Nutr. Rep. Int.*, 1985, **31**, 237.
[70] A. M. Calderon de la Barca, L. Vasquez-Moreno, M. E. Valencia, M. Cordero, and B. G. Lopez-Cervantes, in 'Lectins', eds. T. C. Bog-Hansen and D. L. J. Freed, Sigma Chemical Co., St. Louis, 1988, Vol. 6, p. 125.
[71] A. Sjolander, K. Magnusson, and S. Latkovic, *Cell Struc. Funct.*, 1986, **11**, 285.
[72] J. Lafont, J. M. Rouanet, J. Gabrion, J. L. Zambonino Infante, and P. Besancon, *Digestion*, 1988, **41**, 83.
[73] T. P. King, A. Pusztai, and E. M. W. Clarke, *J. Comp. Pathol.*, 1980, **90**, 585.
[74] A. Pusztai, F. Greer, M. Silva Lima, A. Prouvost-Danon, and T. P. King, in 'Chemical Taxonomy, Molecular Biology and Function of Plant Lectins', Alan R. Liss Inc., New York, 1983, p. 271.
[75] F. Greer and A. Pusztai, *Digestion*, 1985, **32**, 42.
[76] M. H. Perdue, J. F. Forstner, N. W. Roomi, and D. G. Gall, *Am. J. Physiol.*, 1984, **247**, G632.
[77] M. D. Weinman, C. H. Allan, J. S. Trier, and S. J. Hagan, *Gastroenterology*, 1989, **97**, 1193.
[78] J. L. Madara and J. S. Trier, in 'Physiology of the Gastrointestinal Tract', ed. L. R. Johnson, Raven Press, New York, 1987, p. 1209.
[79] S. Nakata and T. Kimura, *Agric. Biol. Chem.*, 1986, **50**, 645.
[80] T. Kimura, S. Nakata, Y. Harada, and A. Yoshida, *J. Nutr. Sci. Vitaminol.*, 1986, **32**, 101.
[81] J. G. Banwell, D. H. Boldt, J. Meyers, F. L. Weber, B. Miller, and R. Howard, *Gastroenterology*, 1983, **84**, 506.
[82] J. M. Rouanet, P. Besancon, and J. Lafont, *Experientia*, 1983, **39**, 1356.

increased.[68,73,81,83–86] In particular, tissue-associated *E. coli* can be detected in considerable numbers even in the upper part of the jejunum. This alteration in the gut flora appears to play a significant role in the overall toxicity of kidney bean lectin since germ-free animals given high levels of this lectin maintain their initial body weight and survive in contrast to the rapid weight loss and death observed with conventional animals.[68,81,83,84,87,88] No other dietary lectins have been directly studied for their effect upon rat small intestine flora. However, jack bean lectin may also affect gut bacterial population since this lectin was found to promote adherence of *Samonella typhimurium* to the mucosa of rat intestine.[89]

Production and secretion of mucus by goblet cells is greatly elevated in animals given kidney bean lectin.[45,51,71,73] A similar effect occurs with jack bean and wheat germ lectins but not with soyabean lectin.[29,53,71] Goblet cell numbers are not however increased.[56] Therefore, this elevated output of mucus is the result of a hypertrophic response. Lectin binds to and is taken into goblet cells and therefore the effect may be due to the direct action of lectin upon the goblet cells. It may also be, in part, a consequence of the lectin-mediated degranulation of mast cells.[74–76]

Lectin-related damage to the absorptive epithelium and the reduction in the activities of some of the enzymes of terminal digestion were thought to result in severe malabsorption of dietary nutrients and thus cause the rapid weight loss and death observed in animals given toxic lectins. However, this is not the case since digestion and absorption of dietary carbohydrate, lipid and proteins, other than the lectin, are only slightly impaired.[35,90] Furthermore, the growth inhibitory lectins also severely disrupt the absorptive epithelium but do not cause rapid weight loss.[29,65,70] Kidney bean lectin also mediates mucus hypersecretion, high rates of epithelial cell turnover and leakage of serum proteins into the gut lumen.[75] This constitutes a considerable net loss to the animal since much of this endogenous material cannot be efficiently reutilized.[91] However, the net absorption of nitrogen by lectin-fed animals is still equivalent to > 70 % of dietary intake and absorption of dry matter > 80 % of dietary intake. Thus, malabsorption of dietary nutrients and elevated outputs of endogenous materials exacerbate the deleterious effects of lectins but are not a primary cause of lectin toxicity.

Toxic and growth inhibitory lectins induce rapid and wasteful growth of both the small intestine and the pancreas and this thereby limits the proportion of nutrients available to support growth of other tissues. Reduced absorption of

[83] D. J. Jayne-Williams and D. Hewitt, *J. Appl. Bact.*, 1972, **35**, 331.
[84] D. J. Jayne-Williams and C. D. Burgess, *J. Appl. Bact.*, 1974, **37**, 149.
[85] G. G. Untawale and J. McGinnis, *Poult. Sci.*, 1979, **58**, 928.
[86] A. B. Wilson, T. P. King, E. M. W. Clarke, and A. Pusztai, *J. Comp. Pathol.*, 1980, **90**, 597.
[87] E. A. S. Rattray, R. Palmer, and A. Pusztai, *J. Sci. Food Agric.*, 1974, **25**, 1035.
[88] D. Hewitt, M. E. Coates, M. L. Kakade, and I. E. Liener, *Br. J. Nutr.*, 1973, **29**, 423.
[89] R. L. Abud, B. L. Lindquist, R. L. Ernst, J. M. Merrick, E. Lebenthal, and P. C. Lee, *Proc. Soc. Exp. Biol. Med.*, 1989, **192**, 81.
[90] A. Pusztai, E. M. W. Clarke, G. Grant, and T. P. King, *J.Sci. Food Agric.*, 1981, **32**, 1037.
[91] A. Allan, in 'Physiology of the Gastrointestinal Tract', ed. L. R. Johnson, Raven Press, New York, 1981, p. 617.

nutrients, increased excretion of endogenous material, particularly nitrogen, into the intestinal lumen and competition for nutrients by bacteria in the gut further exacerbate the problem. However, although these combined effects may explain the growth inhibition caused by some lectins, they cannot account for the very rapid weight loss and high incidence of mortality found with animals given highly toxic lectins since this (the majority of rats die within 3–4 days) occurs well before lectin-induced intestinal and pancreatic growth is sufficiently extensive to seriously impair nutrient availability. Thus, the toxic lectins must interfere with other additional aspects of body metabolism.

6 Systemic Effects of Purified Lectins

Extensive loss of muscle occurs upon feeding of rats with kidney bean lectin (Table 2). The gastrocnemius and plantaris hind-limb muscles of rats drop in the weight (g 100 g^{-1} body weight) by 20–25 % within 10 days.[52,57] The protein and RNA contents of the tissues also decline by a similar amount.[57] Since these muscles are thought to be reflective of skeletal muscle in general, the changes in their weight indicate that an extensive loss of skeletal muscle occurs as a result of consumption of the toxic kidney bean lectin.

The decline in gastrocnemius and plantaris muscle weights is evident within 2 days of exposure to kidney bean lectin.[57] In contrast, the weights (g 100 g^{-1} body weight) of these muscles in animals fed on a protein-free diet are unaffected after 10 days as are muscle weights in soyabean lectin-fed animals.[53,57] Thus, the loss of skeletal muscle is not due to reduced availability of nitrogen. It seems to be the result of a reduction in the rate of muscle protein synthesis mediated indirectly by the lectin.[57] The rate of protein synthesis is low, that of protein degradation remains normal, and as a result there is a net catabolism of muscle protein.[57]

Rapid depletion of adipose lipids occurs in animals given the toxic kidney bean lectin (Table 3). As a result, the body lipid content of rats decreases by 60–70% within 10 days.[92,93] Body reserves of glycogen, primarily of liver glycogen, are also quickly depleted.[92,93]

The body reserves of lipid, glycogen and protein are utilized to generate glucose during periods of low food availability or intake. Therefore, lectin-mediated depletion of these reserves considerably limits the ability of the rat to respond to such conditions. As a result, if rats given kidney bean lectin are deprived of a dietary source of glucose by being fasted overnight, their blood glucose levels fall dramatically and the rats become hypoglycaemic.[93,94] Diets in which raw kidney bean forms the sole source of protein may have very high levels of lectin (up to 20 g lectin kg^{-1} diet) and are therefore liable to cause extremely rapid depletion of the body reserves. Since rats given these diets also generally have a low food intake, they are likely to quickly

[92] J. T. A. de Oliveira, PhD Thesis, University of Aberdeen, Aberdeen, 1986.
[93] G. Grant, J. T. A. de Oliveira, P. M. Dorward, M. G. Annand, M. Waldron, and A. Pusztai, *Nutr. Rep. Int.*, 1987, **36**, 763.
[94] H. F. Hintz, D. E. Hogue, and L. Krook, *J. Nutr.*, 1967, **93**, 77.

Table 4 *Comparison of the ability of dietary lectins to induce production of circulating anti-lectin antibodies (+ or −) and the nominal carbohydrate specificity of the lectins*

Lectin	Orally-induced anti-lectin antibodies	Lectin nominal[f] carbohydrate specificity
Kidney bean[a,b]	+	Galβ1,4GlcNAcβ1,2Man
Pinto bean[b]	+	?
Wheatgerm[c]	+	GlcNAc(β1,4GlcNAc)$_{1,2}$ > βGlcNAc > Neu5Ac
Jack bean[d]	+	αMan > αGlc > αGlcNAc
Lentil[d]	+	αMan > αGlc > αGlcNAc
Pea[b]	−	αMan > αGlc = αGlcNAc
Peanut[d]	+	Galβ1,3GalNAc > α and βGal
Soyabean[d]	+	α and βGalNAc > α and βGal
Tomato[e]	?*	GlcNAc(β1,4GlcNAc)$_{1-3}$

* Tomato lectin is taken up systemically.
? Unknown.

Based on data from:
[a] A. Pusztai, E. M. W. Clarke, G. Grant, and T. P. King, *J. Sci. Food Agric.*, 1981, **32**, 1037.
[b] R. Begbie and T. P. King, 'Lectins', ed. T. C. Bog-Hansen and J. Breborowicz, Walter de Gruyter, Berlin, 1985, Vol. 4, p. 15.
[c] L. Sollid, J. Kolberg, H. Scott, J. Ek, O. Fausa, and P. Brandtzaeg, *Clin. Exp. Immunol.*, 1986, **63**, 95.
[d] H. J. de Aizpurua and G. J. Russell-Jones, *J. Exp. Med.*, 1988, **167**, 440.
[e] D. C. Kilpatrick, A. Pusztai, G. Grant, C. Graham, and S. W. B. Ewen, *FEBS Lett.*, 1985, **185**, 229.
[f] I. J. Goldstein and R. D. Poretz, 'The Lectins', ed. I. E. Liener, N. Sharon, and I. J. Goldstein, Academic Press, New York, 1986, pp. 35.

become hypoglycaemic, and die. This may, in part, explain the rapid demise of experimental animals given diets containing high levels of toxic lectins.

Kidney bean lectins induce a slight enlargement of the liver and a considerable reduction in the size of the thymus.[51,52] The physiological implications of these changes are unknown.

Appreciable amounts of intact and fully reactive lectin rapidly appear in the systemic circulation and in the liver and kidneys after oral exposure of rats to kidney bean lectin.[50,90,95] Subsequently, the animals develop a strong humoral antibody (IgG type) response exclusively against the lectin.[47,90,96] Furthermore, the systemic uptake of the lectin appears to be continuous during the period of exposure to the lectin since the antibody titre increases progressively with time and upon repeated exposure. This suggests a failure of the local (gut) secretory immune system to prevent the uptake of the lectin.[97,98]

Systemic uptake of lectin and development of high titres of circulating

[95] A. Pusztai, *Report of the Rowett Research Institute*, 1980, **36**, 110.
[96] P. E. V. Williams, A. Pusztai, A. MacDaermid, and G. M. Innes, *Anim. Feed Sci. Technol.*, 1984, **12**, 1.
[97] F. M. Greer, PhD Thesis, University of Aberdeen, Aberdeen, 1983.
[98] A. Pusztai, in 'Lectins', ed. T. C. Bog-Hansen and E. van Dreissche, Walter de Gruyter, Berlin, 1986, Vol. 5, p. 317.

anti-lectin antibodies occurs with a wide range of different orally supplied lectins whether they be toxic, growth inhibitory, or essentially non-toxic (Table 4). Therefore, continuous exposure to any particular dietary lectin is liable to lead to development of anti-lectin antibodies, formation of circulating immune complexes, and subsequently to the deposition of these complexes in the kidneys.[99] Over a prolonged period, this may lead to severe kidney disorders.[99] Therefore, in the long-term, consumption of even essentially non-toxic lectins may have deleterious consequences.

High circulating levels of lectin are found in animals fed upon kidney bean lectin (Table 3). Small amounts are also found in the liver and kidneys. In contrast, very little circulating lectin is found in animals given tomato lectin orally although the lectin is taken up systemically since significant amounts are present in the liver and kidneys. This suggests that the rate of uptake of the toxic lectin is much higher than that of other lectins and that the liver and kidneys are unable to sequester all of the toxic lectin. The physiological implications of this are unknown. However, it seems likely that the rapid depletion of body reserves found in animals given toxic lectins is linked with the high levels of circulating lectin in these animals.

The major interference with systemic metabolism is possibly due indirectly to the modifications in the gut flora mediated by the toxic lectins.[86] Bacterial proliferation and increased adherence of bacteria to the small intestine epithelium may increase gut permeability and therefore lead to rapid systemic uptake of lectin. This would be consistent with the preliminary findings that the rate of systemic uptake of dietary kidney bean lectin by germ-free rats is much lower than that by conventional rats and may, in part, explain the lower toxicity of kidney bean lectins to germ-free animals.[87,100]

7 Other Lectins

Dietary jack bean lectin (Concanavalin A; Con A; nominal specificity αMan > αGlc > GlcNAc) has effects upon Japanese quail similar to those observed with dietary kidney bean lectin in the same species.[84,101] Thus, it is extremely toxic when fed at high concentrations to conventional Japanese quail but is much less deleterious towards germ-free equivalents.[101] Furthermore, when given as a large single dose to rats, Con A causes reductions in small intestine epithelium enzyme activities and induces damage to the villi and microvilli similar to that found with kidney bean lectin.[37,79,71,102,103] However, the lectin is not poisonous when fed in quantity to dogs or rabbits (as reported by Jayne-Williams).[101] In addition, whilst Con A was originally found to be highly toxic when injected into rabbits, later studies suggest that this was due to

[99] J. R. Anderson and I. A. R. More in 'Muir's Textbook of Pathology', ed. J. R. Anderson, Edward Arnold (Publishers) Ltd., London, 1980, p. 805.
[100] A. Pusztai, G. Grant, T. P. King, and E. M. W. Clarke, in 'Recent Advances in Animal Nutrition', ed. W. Haresign and D. J. A. Cole, Butterworth, London, p. 47.
[101] D. J. Jayne-Williams, *Nature (London), New Biol.*, 1973, **243**, 150.
[102] V. Lorenzsonn and W. A. Olsen, *Gastroenterology*, 1982, **82**, 838.
[103] A. C. de Oliveira, B. Vidal, and V. C. Sgarbieri, *J. Nutr. Sci. Vitaminol.*, 1989, **35**, 315.

a contaminant in the preparation and that pure Con A is atoxic.[104-106] Thus, although raw jack bean meal is highly toxic when included in diets for rats, quail, or cattle, evidence regarding the potential role of Con A in these deleterious effects is contradictory.[101,107,108]

The mannose-specific lectin from mucuna (*Dioclea grandiflora*), a seed species related to *Canavalia*, slightly inhibited but did not prevent the growth of rats when it was included at a level of 0.44 g kg^{-1} in their diets.[40] This would suggest that mucuna lectin and possibly the closely related Con A may cause anti-nutritional effects in rats similar to those observed in rats given soyabean lectin and that these lectins are thus growth inhibitory but not toxic.

The mannose-specific lectin from field bean (*Vicia faba*) is essentially non-toxic to young rapidly growing rats causing only very slight growth inhibition and no significant damage to the small intestine epithelium when included in their diets at a concentration of 0.7 g kg^{-1}.[44] However, there is evidence to indicate that this lectin may be quite deleterious to chicks.[109] This suggests that there could be considerable species variation in the response of animals to orally-supplied mannose-specific lectins and this may explain the contradictory findings made with Con A. It also highlights the need to treat all lectins in foodstuffs as being potentially toxic.

Tepary bean has a high lectin content and seed extracts cause ultrastructural changes in the small intestine epithelium of rats similar to those observed in rats given the toxic kidney bean lectin.[110] The purified lectins show extensive immunological cross-reactivity with kidney bean lectins and as with kidney bean lectins, are toxic for bruchid beetles.[11,111] Runner bean also contains high levels of lectins which exhibit extensive immunological cross-reactivity with kidney bean lectins.[19] Thus, it seems likely that the lectins are, at least in part, responsible for the high oral toxicity of these beans for rats.[19,112] It has also been suggested that the lectin in guar seeds (*Cyamopsis tetragonolobus*) is in part responsible for the toxicity of these seeds for rats.[113]

Wheat germ, wheat gluten, tomatoes, and peanuts (*Arachis hypogaea*), which are all consumed in considerable quantities in a raw or roasted form as part of the human diet, contain significant levels of lectin.[3,38,114] Considerable quantities of both wheatgerm and tomato lectins survive passage through the gastrointestinal tract in a fully reactive form, interact with the intestinal epithelium, are apparently internalized into enterocytes and are subsequently

[104] J. B. Sumner and S. F. Howell, *Bacteriol.*, 1936, **32**, 227.
[105] C. Dennison, R. H. Stead, and G. V. Quicke, *Agroplante*, 1971, **3**, 27.
[106] C. R. Carlini and J. A. Guimares, *Toxicon*, 1981, **19**, 667.
[107] A. Orru and V. C. Demel, *Quad. Nutr.*, 1941, **7**, 273.
[108] D. K. Shone, *Rhod. Agric. J.*, 1961, **58**, 18.
[109] L. A. Rubio, A. Brenes, and M. Castano, *Br. Poult. Sci.*, 1989, **30**, 101.
[110] A. Sotelo, A. Gonzalez-Licea, M. T. Gonzalez-Garza, E. Velasco, and A. Feria-Velasco, *Nutr. Rep. Int.*, 1983, **27**, 329.
[111] A. Pusztai, W. B. Watt, and J. C. Stewart, *Phytochemistry*, 1987, **26**, 1009.
[112] A. Sotelo, M. E. Arteaga, M. I. Frias, and M. T. Gonzalez-Garza, *Qual. Plant. Plant Foods Hum. Nutr.*, 1980, **30**, 79.
[113] L. P. Rajput, K. Narashimha Murthy, and S. Ramamani, *J. Food Sci.*, 1987, **52**, 1755.
[114] J. M. Concon, D. S. Newbury, and S. N. Eades, *J. Agric. Food Chem.*, 1983, **31**, 939.

released into systemic circulation.[38,39,102] Peanut lectin also appears to survive and be taken up systemically.[115] In addition, wheatgerm and wheat gluten lectins have been implicated in the pathogenesis of coeliac disease.[116,118] However, no nutritional testing of any of these lectins in a purified form has been reported and therefore the implications of their prolonged consumption remains unknown.

8 Castor Bean (*Ricinus communis*) Lectins

Raw or defatted castor bean meal is highly toxic when fed to rats, chickens, cattle, pigs, or humans.[119,126] The bean contains two lectins: ricinus haemagglutinin (nominal specificity βGal > αGal ≫ GalNAc) which is a potent agglutinin but a weak cytotoxin and ricin (nominal specificity α and βGal > GalNAc) which is a weak haemagglutinin but an extremely potent cytotoxin.[45,127,128] Although no nutritional studies on the purified lectins have been carried out, it is generally assumed that the potent toxicity of dietary castor bean is due to the presence of ricin.[129,130]

Similar protein synthesis-inhibiting lectins have been found in jequirity bean (*Abrus precatorius*), pear melon seeds (*Momordica charantia*), sunhemp seeds (*Crotolaria juncea*), mistletoe (*Viscum album*), vegetable sponge (*Luffa cylindrica*), and modecca flower roots (*Adenia digitata*).[45,131,132]

9 Inactivation of Lectins

The oral toxicity of kidney bean lectin can be abolished by denaturation of the protein and under practical conditions this is usually achieved by heat-treatment of seeds or seed meal.[52] However, kidney bean lectins are extremely

[115] H. J. de Aizpurua and G. J. Russell-Jones, *J. Exp. Med.*, 1988, **167**, 440.
[116] M. M. Weiser and A. P. Douglas, *Lancet*, 1976, **2**, 567.
[117] E. Kottgen, B. Volk, F. Kluge, and W. Gerok, *Biochim. Biophys. Res. Commun.*, 1982, **109**, 168.
[118] J. Kolberg and L. Sollid, *Biochim. Biophys. Res. Commun.*, 1985, **130**, 867.
[119] R. Borchers, *Poult. Sci.*, 1949, **28**, 568.
[120] T. Greary, *Vet. Rec.*, 1950, **62**, 472.
[121] P. F. Polit and V. C. Sgarbieri, *J. Agric. Food Chem.*, 1976, **24**, 795.
[122] A. U. Okorie, F. O. I. Anugwa, C. G. Anamelechi, and J. Nwaiwu, *Nutr. Rep. Int.*, 1985, **32**, 659.
[123] A. Rauber and J. Heard, *Vet. Hum. Toxicol.*, 1985, **27(6)**, 498.
[124] G. P. Wedin, J. S. Neal, G. W. Everson, and E. P. Krenzelok, *Am. J. Emerg. Med.*, 1986, **4(3)**, 259.
[125] A. U. Okorie and F. O. I. Anugwa, *Plant Foods Hum. Nutr.*, 1987, **37**, 97.
[126] O. Belzunegui, A. B. Charles, R. Hernandez, and E. Maravi Petri, *Med. Clin (Barc.)*, 1988, **90(17)**, 716.
[127] S. Olnes, in 'Methods in Enzymology', ed. V. Ginsburg, Academic Press, New York, Vol. 50C, p. 330.
[128] Y. Endo, K. Mitsui, M. Motizuki, and K. Tsurugi, *J. Biol. Chem.*, 1987, **262**, 5908.
[129] M. Ishiguro, M. Mitarai, H. Haradai, I. Sekine, I. Nishimori, and M. Kikutani, *Chem. Pharm. Bull.*, 1983, **31(9)**, 3222.
[130] M. Ishiguro, H. Haradai, O. Ichiki, I. Sekine, I. Nishimori, and M. Kikutani, *Chem. Pharm. Bull.*, 1984, **32(8)**, 3141.
[131] S. Olnes and A. Pihl, *Eur. J. Biochem.*, 1973, **35**, 170.
[132] H. Franz, P. Ziska, and A. Kindt, *Biochem. J.*, 1981, **195**, 481.

Table 5 *Effects of aqueous heat-treatment or autoclaving of legume seeds upon haemagglutinating activity and growth and net protein utilization by rats*

	Pre-treatment	Lectin activity (HU mg^{-1} protein)	Weight* change (g rat^{-1} d^{-1})	Net protein* utilization
Kidney bean[a]	none	1000	rats died	rats died
	70 °C, 6 h	1000	rats died	rats died
	80 °C, 3 h	200	− 2.0	negative
	100 °C, 10 m	0	+ 0.8	45
Winged bean[b,c]	none	200	− 1.7	negative
	100 °C, 30 m	0	+ 1.4	59
	autoclaved, 15 m	10	+ 0.4	38
Soyabean[d,e]	none	60	− 0.2	38
	100 °C, 10 m	0	+ 2.0	55
Tepary bean[f]	none	500	ND	ND
	80 °C, 3 h	50	ND	ND
	95 °C, 1 h	1	ND	ND
Hyacinth bean[g,h]	none	ND	− 1.0	ND
	autoclaved, 5 m	ND	− 0.1	ND

* The diets were not supplemented with any essential amino acids.
ND, not determined.

Based on data from:
[a] G. Grant, L. J. More, N. H. McKenzie, and A. Pusztai, *J. Sci. Food Agric.*, 1982, **33**, 1324.
[b] S. S. Kadam, R. R. Smithard, M. D. Eyre, and D. G. Armstrong, *J. Sci. Food Agric.*, 1987, **39**, 267.
[c] S. S. Kadam and R. R. Smithard, *Qual. Plant. Plant Foods Hum. Nutr.*, 1987, **37**, 151.
[d] G. Grant, L. J. More, N. H. McKenzie, J. C. Stewart, and A. Pusztai, *J. Sci. Food Agric.*, 1983, **50**, 207.
[e] G. Grant, N. H. McKenzie, W. B. Watt, J. C. Stewart, P. M. Dorward, and A. Pusztai, *J. Sci. Food Agric.*, 1986, **37**, 1001.
[f] S. A. R. Kabbara, I. R. Abbas, J. C. Scheerens, A. M. Tinsley, and J. W. Berry, *Qual. Plant. Plant Foods Hum. Nutr.*, 1986, **36**, 295.
[g] W. G. Jaffe, *Proc. Soc. Exp. Biol. Med.*, 1949, **71**, 398.
[h] K. Phadke and K. Sohonie, *J. Sci. Ind. Res.*, 1961, **21C**, 178.

resistant to dry or even moist heating.[133–137] Haemagglutinin levels in the beans are unaffected by aqueous heat-treatment at 70 °C for 6 hours and even at 80 °C significant amounts of lectin survive after 3 or 6 hours (Table 5). Indeed, fully hydrated seeds need to be cooked in water at 100 °C for 10–20 minutes to ensure complete elimination of the lectin activity (Table 5). Treatment of unhydrated seeds or meals either by this procedure, by oven-, microwave-, or infrared-heating, or by autoclaving is far less effective

[133] H. J. H. de Muelenaere, *Nature (London)*, 1964, **201**, 1029.
[134] G. Grant, L. J. More, N. H. McKenzie, and A. Pusztai, *J. Sci. Food Agric.*, 1982, **33**, 1324.
[135] D. G. Coffey, M. A. Uebersax, G. L. Hosfield, and J. R. Brunner, *J. Food Sci.*, 1985, **50**, 78.
[136] M. Lowgren and I. E. Liener, *Qual. Plant. Plant Foods Hum. Nutr.*, 1986, **36**, 147.
[137] N. V. Dhurandhar and K. C. Chang, *J. Food Sci.*, 1990, **55**, 470.

for reducing the lectin content than is aqueous heat-treatment of fully hydrated seeds.[134,138,139]

Once the lectin activity has been eliminated, kidney beans included in diets as the sole source of protein support rat growth (Table 5). Furthermore, when they are supplemented with some of the essential amino acids to bring the dietary levels up to the requirements for rats the growth and net protein utilization values obtained are close to that achieved with a good quality protein source such as lactalbumin.[90]

Lectin activities in kintoki bean, winged bean, tepary bean, soyabean, field bean, lupinseed, and other seeds can also be substantially reduced or eliminated by rigorous heat-treatment procedures (Table 5).[24,140–143] This results in a considerable increase in growth and net protein utilization values obtained with animals fed these seeds as their sole source of protein (Table 5). However, this improvement is in many cases due only in part to the elimination of the lectin activity since the seeds contain a number of additional anti-nutritional factors.[33,34]

The potent toxin ricin is not readily inactivated by dry heat-treatment but the haemagglutinating- and protein synthesis inhibiting-activities are apparently eliminated by heating under moist conditions.[144,145] Various aqueous heating and combined extraction and heat-treatment procedures have therefore been developed to abolish the toxicity of castor bean.[119,121,144–148] A dry roasting method has also recently been used.[122,125] However, although these methods abolish the overt toxicity of the seed meal, the growth and food conversion rates obtained with chicks or rats given pre-treated meals are still much poorer than that obtained with animals given a good dietary protein.[119,121,125,146] This may suggest that small amounts of active ricin may still remain in the pre-treated products or that other heat-stable anti-nutritional factors are present in the seed.

10 Implications of Dietary Lectins for Humans

Beans form a significant part of the human diet throughout the world. In the developing countries, they are generally purchased in the raw dry form and are

---

[138] S. S. Kadam, R. R. Smithard, M. D. Eyre, and D. G. Armstrong, *J. Sci. Food Agric.*, 1987, **39**, 267.
[139] T. F. B. van der Poel, J. Blonk, D. J. van Zuilichem, and M. G. van Oort, *J. Sci. Food Agric.*, 1990, **53**, 215.
[140] M. Miyoshi, M. Nakabayashi, T. Hara, T. Yawata, I. Tsukamoto, and Y. Hamaguchi, *J. Nutr. Sci. Vitaminol.*, 1982, **28**, 255.
[141] J. M. Wallace and M. Friedman, *Nutr. Rep. Int.*, 1985, **32**, 743.
[142] I. E. Liener, *J. Am. Oil Chem. Soc.*, 1981, **58(3)**, 406.
[143] S. S. Kantha and J. W. Erdman, *Nutr. Rep. Int.*, 1988, **38(2)**, 423.
[144] A. C. Mottola, G. O. Kohler, and R. T. Prescott, *Feedstuffs*, 1967, **39**, 20.
[145] G. Fuller, H. G. Walker, A. C. Mottola, D. D. Kuzmicky, G. O. Kohler, and P. Vohra, *J. Am. Oil Chem. Soc.*, 1971, **48**, 616.
[146] L. Vilhjalmsdottir and A. Fisher, *J. Nutr.*, 1971, **101**, 1185.
[147] A. C. Mottola, B. Mackey, and V. Herring, *J. Am. Oil Chem. Soc.*, 1971, **48**, 510.
[148] A. C. Mottola, B. Mackey, H. G. Walker, and G. O. Kohler, *J. Am. Oil Chem. Soc.*, 1972, **49**, 662.

subsequently detoxified by traditional soaking and cooking procedures.[149,150] In contrast, for several decades most of the beans eaten by the urban population in developed countries, such as the UK and USA, have prior to purchase been commercially processed to eliminate anti-nutrients. As a result, there is a general lack of public awareness in these countries of the deleterious effects of these seeds when eaten uncooked or of the pre-treatment procedures necessary to render them safe for consumption.

A recent expansion in vegetarianism and in consumption of 'natural healthy foods' has led to increased availability and use of raw beans in the UK and to a considerable number of cases of food poisoning caused by consumption of kidney beans being reported.[151-153] The symptoms generally are nausea and vomiting one to three hours after ingestion of the seeds, abdominal distension, and diarrhoea.[152] These cases occur due to consumption of beans which have not been cooked at all or which have been heated in low temperature 'slow cookers' in which the cooking temperatures ($< 75\,°C$) are insufficient to inactivate the toxic lectins and emphasize the need to ensure that the population are made aware that beans need to be fully hydrated and boiled for at least 10–20 minutes before they are eaten or added to any other dish.[134,151-153]

A major outbreak of food poisoning due to consumption of partly cooked beans occurred in Berlin during the blockade of the city in 1948.[154] Fuel for cooking was in short supply and as a result flaked beans air-lifted into the city were only lightly heated before being eaten.[154] This highlights a potential problem. In developing countries, in particular areas suffering from drought or flood, only limited amounts of fuel for cooking are available. Thus, if beans form part of the diet, it is possible that people are being exposed to significant levels of toxic dietary lectins over a prolonged period of time.[150] The physiological consequences of this are unknown.

Cases of poisoning due to consumption of the highly toxic raw castor bean have been reported.[123,124] However, the incidence is small due to the limited availability of raw castor bean.

Soyabean, tomato, and wheat germ lectins appear to survive passage through the human gut, to interact with the intestinal epithelium, to be taken up into circulation, and to induce antibody production.[40,155,156] Thus, many of the dietary lectins, growth inhibitory and essentially non-toxic, to which we are regularly exposed are liable to induce similar responses.[3] Although wheatgerm agglutinin has been implicated in the aetiology of coeliac disease, the implications of long-term dietary exposure to this or other growth inhibitory or essentially non-toxic lectins or indeed to low levels of toxic lectins for intestinal

[149] W. R. Aykroyd and J. Doughty, in 'Legumes in Human Nutrition', FAO Nutritional Studies (19), FAO, Rome, 1964.
[150] R. Korte, *Ecol. Food Nutr.*, 1972, **1**, 303.
[151] Public Health Laboratory Service, *Br. Med. J.*, 1976, **2**, 1268.
[152] N. N. Noah, A. E. Bender, G. B. Reaidi, and R. J. Gilbert, *Br. Med. J.*, 1980, **281**, 236.
[153] A. E. Bender and G. B. Reaidi, *J. Plant Foods*, 1982, **4**, 15.
[154] C. Greibel, *Z. Lebensm.-Unters.-Forsch.*, 1950, **90**, 191.
[155] S. Freier, E. Lebenthal, M. Freier, P. C. Shah, B. H. Park, and P. C. Lee, *Immunology*, 1983, **4**, 69.
[156] L. Sollid, J. Kolberg, H. Scott, J. Ek, O. Fausa, and P. Brandtzaeg, *Clin. Exp. Immunol.*, 1986, **63**, 95.

and systemic metabolism in humans remains unknown. However, since the accumulation of immune complexes in the kidneys can lead to tissue damage, it is possible that prolonged exposure to even essentially non-toxic dietary lectins could be deleterious to humans. Thus, in general all dietary lectins should be treated as potentially harmful and should where possible be denatured before they are consumed.

11 Summary

Lectins are ubiquitous in the plant kingdom being present in nutritionally significant amounts in seeds, fruits, and plant tissues. Most are highly resistant to proteolytic degradation *in vivo* and survive gut passage in an intact and fully reactive form. If appropriate carbohydrate receptors are available, a proportion of the lectin in the lumen binds to the epithelial surface of the small intestine, is taken into the enterocytes by endocytosis, and is subsequently released systemically. As a result, some dietary lectins seriously interfere with the body metabolism of man and animals.

Toxic and growth inhibitory dietary lectins induce rapid and wasteful growth of the small intestine and pancreas in rats and mice. Furthermore, they cause disruption to the absorptive epithelium and stimulate secretion of endogenous nitrogen. These combined effects greatly reduce the amounts of nutrients available to support the growth in other tissues. In addition, the toxic lectins cause rapid loss of body muscle, lipid, and glycogen. Indeed, this depletion of body reserves may be the primary reason for the rapid demise of rats and mice given toxic lectins since they are unable to maintain glucose homeostasis and become hypoglycaemic.

The essentially non-toxic lectins have no overtly deleterious effects when fed to rats or mice. However, there is evidence that some animal species are susceptible to the action of these lectins. Thus, all dietary lectins should be considered as potentially deleterious and should where possible be fully inactivated prior to consumption.

12 Addendum

It has recently been reported that purified peanut lectin induces cellular hyperplasia in the proximal, mid, and distal small intestine of rats when it is included at a level of 2 g lectin kg^{-1} in their diet.[157] However, despite its potent effect upon the gut, the lectin does not reduce the food intake or growth rate when given at this dietary concentration.[157] It has also been found that taro tuber (*Colocasia antiquorum*) lectin, which appears to be exceptionally resistant to heat denaturation, significantly reduces growth rate when it is given at high dietary levels to mice.[158,159] These findings are of importance since both lectin sources are regularly consumed as part of the human diet.

[157] L. Henney, E. M. Ahmed, D. E. George, K. J. Kao, and H. S. Sitren, *J. Nutr. Sci. Vitaminol.*, 1990, **36**, 599.
[158] Y-J. Seo, S. Une, I. Tsukamoto, and M. Miyoshi, *J. Nutr. Sci. Vitaminol.*, 1990, **36**, 277.
[159] Y-J. Seo, S. Kamitani, I. Tsukamoto, and M. Miyoshi, *Nippon Kasei Gakkaishi*, 1989, **40**, 805.

CHAPTER 4

Proteinase Inhibitors

GRENVILLE NORTON

1 Introduction

From time immemorial plants have been important sources of food and feed for man and his livestock. The demands for such material have increased dramatically in recent years particularly with respect to protein. If the predicted increase in the world's population occurs, an increased proportion of the protein requirements of man will have to be met by proteins of plant origin. Plant proteins already constitute a major source of dietary protein for the people in many of the less well-developed countries.

Shifts in the dietary habits of people in the developed countries have also resulted in the consumption of more plant foods. On the one hand, the swing to health foods has led to a large increase in the direct consumption of plant components. On the other extreme, the consumption of convenience foods often incorporating processed plant components has also increased.

Legume seeds and cereals have always supplied protein and energy in animal rations. This use is extending as the availability of quality protein sources such as fish and animal meals declines or is precluded due to environmental or hygiene pressures. Plant protein sources, usually seed meals, *e.g.* soyabean are being utilized increasingly in feed for young livestock at critical growth stages. Other outlets include pet foods, foods for fish rearing and farming, and rations for fur-producing animals.

Thus plant materials in various forms are being consumed in foods and feeds in greater quantities and for prolonged periods often throughout life. Unfortunately most plants contain a number of anti-nutritional factors (ANF) including protein proteinase inhibitors (PI). PI, therefore, are present in varying amounts in animal feed and human food. Since their discovery, PI have been extensively studied particularly with respect to their nutritional significance in animals and physico-chemical properties. This overview will consider some of the more important inhibitors of serine- and metallo-proteinases with respect to their distribution, classification and characteristics, assay methods, biosynthesis, nutritional and physiological significance, and functions in the plant.

2 Definitions

The proteinases and peptidases or more accurately the endo- and exo-peptidases respectively comprise the two distinct groups of proteinases and peptide hydrolases widely distributed in nature. These enzymes have been classified into four types on the basis of the chemical nature of the groups responsible for their catalytic activity namely, serine proteinases (EC 3.4.21), sulphydryl (thiol) proteinases (EC 3.4.22), acidic (carboxyl) proteinases (EC 3.4.28), and metallo-proteinases (EC 3.4.34). Free active-site protein inhibitors of these proteinases also recognize these groupings and are likewise divided into four corresponding groups.

All PI irrespective of class possess certain common features. The inhibition is strictly competitive and the complex formed by the inhibitor and proteinase lacks enzymatic activity towards all substrates. Without exception, individual reactive sites on the inhibitor can inhibit only proteinases belonging to one of the four mechanistic classes.

Nature of the Reactive Site and Mechanism of Action

The mechanism by which most PI interact with the proteinases they inhibit is agreed and is designated the standard mechanism. Much of the supporting evidence for this has been obtained by Laskowski and co-workers with the soyabean trypsin inhibitors (Kunitz) (STI) and trypsin and other serine proteinases.[1,2] Interaction between the inhibitor reactive site and the enzyme (catalytic) site results in the formation of a proteinase-inhibitor complex devoid of enzymatic activity. At or near the reactive site of the inhibitor is an amino acid residue that is specifically recognized by the primary substrate binding site of the cognate enzyme. This amino acid, designated P_1, corresponds to either lysine or arginine in trypsin inhibitors, tyrosine, phenylalanine, tryptophan, leucine, or methionine in chymotrypsin inhibitors, and alanine and serine in elastase inhibitors, although such allocations are not exclusive. P_1' is an amino acid which is adjacent and joined to P_1 by the reactive site peptide bond, which hydrolysed during the formation of the inhibitor–proteinase complex. Other amino acid residues around the reactive site are numbered P_2, P_3, *etc.* and P_2', P_3', *etc.* accordingly.[3] Since the reactive site peptide bond of the virgin inhibitor is normally encompassed in at least one disulphide loop, the inhibitor fragments formed by the hydrolysis of the bond cannot dissociate and remain in close proximity to each other in the modified inhibitor since conformational changes are minimal. A schematic representation of the reactive site of the virgin and modified proteinaceous proteinase inhibitor is presented below (modified from Laskowski).[2]

---

[1] M. Laskowski, in 'Nutritional and Toxicological Significance of Enzyme Inhibitors in Foods', ed. M. Friedman, Plenum Press, New York, 1983, p. 1.
[2] M. Laskowski and I. Kato, *Ann. Rev. Biochem.*, 1980, **49**, 593.
[3] I. Schechter and A. Berger, *Biochem. Biophys. Res. Commun.*, 1967, **27**, 157.

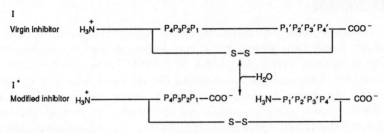

The above discussion refers to single-headed inhibitors (one reactive site). Double-headed inhibitors have two reactive sites on the molecule. Multi-headed inhibitors have several reactive sites on the same molecule and this is generally achieved by the non-covalent association of several peptide chains each with a single reactive site. For example, the potato PI inhibitor I is a pentamer which inhibits five molecules of chymotrypsin without dissociating. Double- and multi-headedness will be considered in more detail with the individual inhibitors.

The Standard Mechanism

The standard mechanism was proposed by Laskowski and co-workers on the basis of detailed studies involving STI and trypsin in catalytic amounts.[2] Subsequently this mechanism has been verified for many other proteinaceous PI. Inhibitors obeying this mechanism are highly specific limited proteolysis substrates for target enzymes. As indicated above the reactive site peptide bond on the inhibitor (Arg-63-Ile-64 STI) specifically interacts with the active (catalytic) site of the enzyme resulting in the virgin inhibitor (reactive site peptide bond intact) being modified (reactive site peptide bond hydrolysed). The equilibrium constant of this reaction is almost unity at neutral pH. Dissociation of the complex by a sudden pH change yields predominantly the virgin inhibitor indicating that the reactive site peptide bond is resynthesized. Because the same stable complex is formed between the enzyme and either the virgin or modified inhibitors, both are equally strong inhibitors of the proteinase. Generally the rate of formation of the complex is much slower with the modified inhibitor than with the virgin inhibitor. That the reactive site peptide bond is hydrolysed has been verified by the fact that removal of either the newly formed carboxyl-terminal residue (P_1) or the amino terminal residue (P_1') eliminates inhibitory activity. Controlled dissociation of the proteinase–inhibitor complex prepared with the modified inhibitor resulted in the release of the enzyme and predominantly the virgin inhibitor indicating that the reactive site peptide bond is reformed. Acylation of the newly formed amino terminal residue (P_1') prevents the formation of the virgin inhibitor and renders the inhibitor inactive.

The following mechanism was proposed for the PI interaction[2]:

$$E + I \leftrightarrow L \leftrightarrow C \leftrightarrow X \leftrightarrow L^* \leftrightarrow E + I^*$$

where E is the proteinase, I and I* are virgin and modified inhibitors respectively, C is the stable inhibitor complex, X is a relatively long-lived intermediate in the E + I* reactions, and L and L* are loose, non-covalent, rapidly dissociable complexes of E with I and I* respectively. The existence of these intermediates is well proven.

The k_{cat}/K_m for the hydrolysis of the reactive site peptide bond by the specific proteinase at neutral pH is very high, 10^{-4}–10^{-6} M^{-1} s^{-1} compared with 10^{-3} M^{-1} s^{-1} for normal substrates.[2] Individual values for k_{cat} and K_m however, are many orders of magnitude lower than those for normal substrates. Thus, their hydrolysis is extremely slow at the normal concentrations used and neutral pH, and the system behaves as if it were a simple equilibrium between the enzyme and inhibitor and the complex. The equilibrium constant for the enzyme-inhibitor association (K_a) is extremely high ranging from 10^7–10^{13} M^{-1}.

X-Ray crystallographic studies of the enzyme-inhibitor complexes (C) including STI and trypsin, have confirmed that the reactive site of the inhibitor reacts with the active site of the enzyme in a substrate like manner.[2,4] Contact is restricted to small but highly complementary parts of the inhibitor and enzyme enabling the formation of van der Waals interactions, hydrogen bonds, and salt bridges, which confer stability on the complex. In the case of the STI, only 12 amino acid residues including those of the reactive peptide bond Arg-63-Ile-64, of the 181 comprising the inhibitor, are involved in the complex formation which occurs with minimal conformation change in either molecule. High resolution X-ray work has revealed that the reactive site peptide bond remains intact in the complex. The carbonyl carbon of the reactive site peptide bond (P_1 residue) is partially tetrahedral as opposed to completely trigonal in the uncomplexed state. This partial tetrahedral distortion is a consequence of attraction between the carbonyl oxygen of the inhibitor to the oxyanion binding pocket of the enzyme which contributes to the stability of the enzyme-inhibitor complex.

In general, complex formation occurs with very little conformational adaptation of the inhibitor. Also the conformation of amino acid residues in the PI interacting with the proteinase is similar in various inhibitors notwithstanding the fact that the inhibitors themselves are not conformationally similar.[2,5]

Inhibitor Families

PI of the serine proteinases have been classified into 10 families.[2] The criteria on which this classification was based were: extensive sequence homology amongst its members; topological relationships between disulphide bridges and the location of the reactive site; interaction between the inhibitor reactive site and the catalytic site occurs according to the standard mechanism. Subsequently the list was extended to 13 inhibitor families including the 6 plant families.[1] These families are listed below:

[4] R. M. Sweet, H. T. Wright, J. Janin, C. H. Chothia, and D. M. Blow, *Biochemistry*, 1974, **13**, 4212.
[5] M. Richardson, *Food Chem.*, 1981, **6**, 235.

Soyabean trypsin inhibitor (Kunitz) family
Soyabean proteinase inhibitor (Bowman–Birk) family
Potato I inhibitor family
Potato II inhibitor family
Barley trypsin inhibitor family
Squash inhibitor family

3 Distribution, Occurrence, and Subcellular Location of PI

PI enjoy a wide distribution in the plant kingdom but because of their nutritional significance in plant comestibles attention has been concentrated on the Leguminosae, Graminae, and Solanaceae which are major food sources.[6,7] Although the seed is most frequently used as a source of PI, these ANF are located in other plant organs such as leaves, tubers, corms, bulbs, and fruits.

In most legume seeds, PI appears to be concentrated in the outer layers of the cotyledon although in the field bean (*Vicia faba*) the concentration in the testa is twice that in the cotyledon.[6]

The exact subcellular location of the PI remains uncertain. Biochemical investigations have indicated that the inhibitors appear to be mainly located in the cytosol of legume cotyledonary cells.[8] More sensitive immunocytochemical techniques using fluorescent antibodies have verified this.[9] Recently the precise ultrastructural location of STI and BBI (AA) in thin sections of soyabean has been determined using the immunogold method.[10,11] The STI appeared to be located predominantly in the cell walls with lesser amounts in the protein bodies, the cytosol between the oleosomes and the nuclei of cotyledonary and embryonic cells whereas the BBI (AA) was found in all protein bodies, nuclei, and to a lesser extent cytosol. No BBI (AA) was present in the cell wall but some was located in the intercellular space.

In the cereal seed, the trypsin-, chymotrypsin-, and endogenous proteinase-inhibitors are located in the embryo and the two endosperm tissues (aleurone layer and starchy endosperm) but primarily in the latter.[7]

In the potato tuber, PI (PI-I, PI-II, PCI, and CPI) were concentrated in the apical cortex with intermediate levels in the lateral cortex and equally low levels in the stem-end cortex and pith.[12] The location of these inhibitors within the potato tuber cell is not known, but the wound induced inhibitors in tomato

---

[6] I. E. Liener and M. L. Kakade, in 'Toxic Constituents of Plant Foodstuffs', ed. I. E. Liener, Academic Press, New York, 1980, p. 7.
[7] S. Boisen, *Acta Agric. Scand.*, 1983, **33**, 369.
[8] B. Baumgartner and M. J. Chrispeels, *Plant Physiol.*, 1976, **58**, 1.
[9] M. J. Chrispeels and B. Baumgartner, *Plant Physiol.*, 1978, **61**, 617.
[10] M. Horisberger and M. Tacchini-Vonlanthen, *Histochemistry*, 1983, **77**, 37.
[11] M. Horisberger and M. Tacchini-Vonlanthen, *Histochemistry*, 1983, **77**, 313.
[12] C. A. Ryan and G. M. Hass, in 'Antinutrients and Natural Toxicants in Foods', ed. R. L. Ory, Food and Nutrition Press Inc., Westport, 1981, p. 169.

and potato leaves have been shown by means of ferritin-labelled antibodies to accumulate in the central vacuole.[13,14]

4 Classification, Characteristics, and Physico-chemical Properties of PI

Legume Seed PI

Soyabean Trypsin Inhibitor (Kunitz) Family

Soyabean Trypsin Inhibitor (Kunitz) (STI). The STI was the first plant PI to be isolated in a partially crystalline form and characterized by Kunitz in the 1940s.[15] Not surprisingly STI have been investigated extensively and information on the isolation, purification, assay, and characteristics such as specificity, chemical and physical properties, stability, and kinetics have been assembled in reviews.[5,6,16–20] Within a given soyabean variety the trypsin inhibitors (TI) represent a heterogeneous group consisting of a number of isoinhibitors with a range of M_r (18–24 kd), isoelectric points (pH 3.5–4.5), amino acid compositions, and stability. The STI represents the major component of these inhibitors with a M_r approximately 20 kd.

Three genetically different variants of the STI have been isolated and designated Ti^a, Ti^b, and Ti^c. These variants or isoinhibitors have different electrophoretic mobilities, association equilibrium constants for trypsin, and hydrolysis equilibrium constants for the hydrolysis of the reactive peptide bond. The most commonly occurring form, Ti^a, has been equated with the original STI which is commercially available.[21] The three variants of STI are inherited at co-dominant multiple alleles while the gene for the lack of the inhibitor, seed designated Ti, is inherited as a recessive allele to the other three. All three variants of STI are single polypeptides, (M_r 22 kd) consisting of 181 residues and two disulphide bonds with a single reactive site located at Arg-63-Ile-64 (Table 1).[22–25] The complete amino acid sequences of the three variants of STI have been determined and the three dimensional structure of STI (Ti^a)-trypsin complex described (Figures 1 and 2).[4,22,24]

STI are primarily inhibitors of trypsin from a wide range of species including

---

[13] M. Walker-Simmons and C. A. Ryan, *Plant Physiol.*, 1977, **60**, 61.
[14] H. Hollander-Czytko, J. K. Andersen, and C. A. Ryan, *Plant Physiol.*, 1985, **78**, 76.
[15] M. Kunitz, *J. Gen. Physiol.*, 1947, **30**, 291.
[16] Y. Birk, *Methods Enzymol.*, 1976, **46**, 695.
[17] Y. Birk, *Int. J. Peptide Protein Res.*, 1985, **25**, 113.
[18] Y. Birk, in 'Hydrolytic Enzymes', ed. A. Neuberger and K. Brocklehurst, Elsevier, Amsterdam, New York and Oxford, 1987, pp. 257.
[19] B. Kassell, *Methods Enzymol.*, 1970, **19**, 839.
[20] M. Richardson, *Phytochemistry*, 1977, **16**, 159.
[21] R. C. Freed and D. S. Ryan, *Biochim. Biophys. Acta*, 1980, **624**, 562.
[22] S. H. Kim, S. Hara, S. Hase, T. Ikenaka, H. Toda, K. Kitamura, and N. Kaizuma, *J. Biochem.*, 1985, **98**, 435.
[23] T. Koide and T. Ikenaka, *Eur. J. Biochem.*, 1973, **32**, 401.
[24] T. Koide and T. Ikenaka, *Eur. J. Biochem.*, 1973, **32**, 417.
[25] T. Koide, S. Tsunasawa, and T. Ikenaka, *Eur. J. Biochem.*, 1973, **32**, 408.

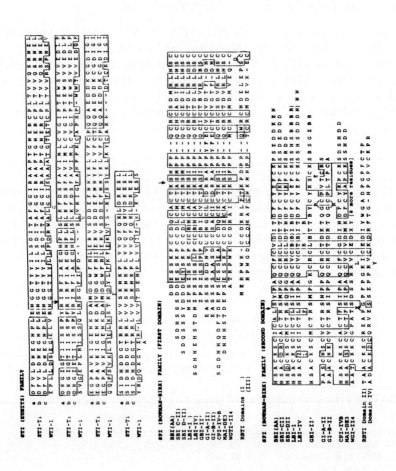

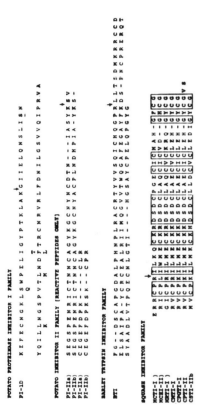

Figure 1 *Comparisons of Primary Sequences of Plant PI*
(STI–Ti—ref. 22, *J. Biochem.*, 1985, **98**, 435: WTI—ref. 34, *J. Biochem.*, 1983, **94**, 849: BBI(AA)—ref. 39, *J. Biochem.*, 1972, **71**, 839: SBI-CII & D-II, ref. 50, *J. Biochem.*, 1978, **83**, 737: LBI—ref. 33, Bayer Symposium V Proteinase Inhibitors, 1974, pp. 344: GBI—ref. 55, *J. Biol. Chem.*, 1975, **250**, 4261: GI—ref. 57, *J. Biochem.*, 1983, **94**, 589: CPI—ref. 65, *Plant. Mol. Biol.*, 1989, **13**, 701: MAI—ref. 66, *Eur. J. Biochem.*, 1979, **97**, 85: WGI—ref. 68, *J. Biochem.*, 1986, **100**, 975: RBTI—ref. 69, *J. Biochem.*, 1987, **102**, 297: PI-I—ref. 71, *FEBS Lett.*, 1974, **45**, 11: PI-II—ref. 76, *J. Biochem.*, 1976, **79**, 381: BTI—ref. 84, *J. Biol. Chem.*, 1983, **258**, 7998: MCTT & MCEI—ref. 89, *J. Biochem.*, 1989, **105**, 88: MRTI—ref. 90, *Phytochemistry*, 1984, **23**, 1401: CMTI, CPGTI, CPTI, CSTI—ref. 88, *Biochim. Biophys. Res. Commun.*, 1985, **126**, 646)
(Homologous sequences are enclosed in boxes. ↓ Reactive site)

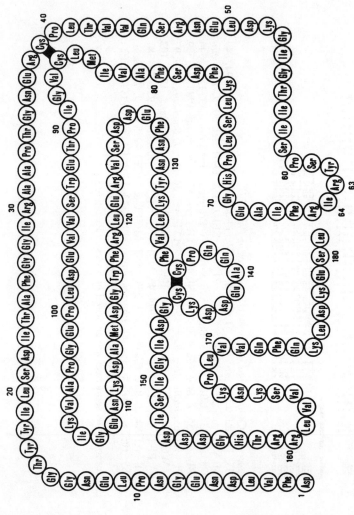

Figure 2 *Structure of Soyabean Trypsin Inhibitor (Kunitz)*
(Reproduced from T. Koide and T. Ikenaka, *Eur. J. Biochem.*, 1973, **32**, 417)

cow, human, salmon, stingray, barracuda, and turkey.[19] Bovine α- and β-chymotrypsin, human plasmin, cocoonase, and plasma kallikrein are also inhibited to various degrees. The thiol proteinases, pepsin, thrombin (human and porcine), and bovine and porcine organ kallikreins are not inhibited. At acid pH (1.5–2.0), pepsin hydrolyses the inhibitor. Since STI inhibits trypsin stoichiometrically to form a stable complex, it is designated a single headed inhibitor. This inhibition is competitive with synthetic substrates (benzoyl L-arginine ethyl ester, BAEE and α-N-benzoyl-DL-arginine-p-nitroanilide, BAPNA) but with protein substrates the type of inhibition is unclear. Although STI (Ti^a) inhibits chymotrypsin more weakly than trypsin, the highly purified inhibitor has been found to bind two molecules of chymotrypsin with comparable affinity to form a ternary complex.[26] Similarly a ternary complex is formed between chymotrypsin and the trypsin–STI complex. It is claimed therefore, that STI is double-headed.

The specificity of the PI is largely determined by amino acids (P_1–P_1') comprising the reactive site. With the STI (Ti^a) it has been shown that the reactive site amino acids, Arg-63-Ile-64 can be modified by chemical and enzymatic means without loss of activity.[27,28] Replacement of Arg-63 (P_1) by lysine resulted in the retention of the antitrypsin specificity albeit that the complex dissociated more slowly, whereas replacement of Arg-63 by phenylalanine had little effect on the specificity towards trypsin or chymotrypsin. Replacement of Arg-63 by tryptophan converted STI into a strong chymotrypsin inhibitor. Chemical removal of Ile-64 (P_1') inactivated STI but replacement of this amino acid with alanine, leucine, or glycine did not affect either specificity or the activity of the inhibitor. Insertion of an additional amino acid between Arg-63-Ile-64 such as alanine, leucine, or glycine abolished the inhibitory activity and converted STI into a substrate.

Other Members of the STI Family. Inhibitors homologous with the STI (Ti^a) have been isolated and characterized from other legume seeds. These include the Silk tree (*Albizzia julibrissin*),[29] Acacia (*Acacia elata*),[30] Winged bean (*Psophocarpus tetragonolobus*),[31–34] and *Erythrina* species.[35–37] Details of the amino acid sequences around the reactive site and specificity are given in Table 1. A number of isoinhibitors were identified in each species but of particular significance were the specific chymotrypsin inhibitors isolated from the silk

[26] B. Bosterling and U. Quast. *Biochim. Biophys. Acta*, 1981, **657**, 58.
[27] D. Kowalski and M. Laskowski, *Biochemistry*, 1976, **15**, 1309.
[28] D. Kowalski, T. R. Leary, R. E. McKee, R. W. Sealock, D. Wang, and M. Laskowski, in 'Bayer Symposium V. Proteinase Inhibitors', ed. H. Fritz, H. Tschesche, L. J. Greene, and E. Truscheit, Springer-Verlag, Berlin, 1974, p. 311.
[29] S. Odani, S. Odani, T. Ono, and T. Ikenaka, *J. Biochem.*, 1979, **86**, 1795.
[30] A. A. Kortt and M. A. Jermyn, *Eur. J. Biochem.*, 1981, **115**, 551.
[31] A. A. Kortt, *Biochim. Biophys. Acta*, 1979, **577**, 371.
[32] A. A. Kortt, *Biochim. Biophys. Acta*, 1980, **624**, 237.
[33] A. A. Kortt, *Biochim. Biophys. Acta*, 1981, **657**, 212.
[34] M. Yamamoto, S. Hara, and T. Ikenaka, *J. Biochem.*, 1983, **94**, 849.
[35] F. J. Joubert, *Int. J. Biochem.*, 1982, **14**, 187.
[36] F. J. Joubert, *Phytochemistry*, 1988, **27**, 1297.
[37] F. J. Joubert, E. H. Merrifield, and E. B. D. Dowdle, *Int. J. Biochem.*, 1987, **19**, 601.

Table 1 Amino Acid Sequences Around the Reactive Site of Various Plant Proteinase Inhibitors

Inhibitor	P_3	P_2	P_1	P_1'	P_2'	P_3' [a]	Enzymes
STI (Kunitz) Family							
Glycine max STI[b]	Ser	Tyr	Arg	Ile	Arg	Phe	T
Psophocarpus tetragonolobus WBI[c]	Ser	Tyr	Arg	Ile	Arg	Phe	T
Erythrina caffra [d]	—	—	Arg	Ser	Ala	Phe	T/C
Erythrina latissima	—	—	Arg	Ser	Thr	Phe	t-PA
SPI (Bowman–Birk) Family							
Glycine max BBI (AA)[e]	(Cys	Thr	Lys	Ser	Asn	Pro	T
	(Cys	Ala	Leu	Ser	Tyr	Pro	C
Phaseolus lunatus LBI[f]	(Cys	Thr	Lys	Ser	Ile	Pro	T
	(Cys	Thr	Leu	Ser	Ile	Pro	C
Phaseolus vulgaris GBI[g]	(Cys	Thr	Ala	Ser	Ile	Pro	E
	(Cys	Thr	Arg	Ser	Met	Pro	T
Arachis hypogaea GI[h]	(Cys	Asp	Arg	Arg	Ala	Pro	T/C
	(Cys	Thr	Arg	Ser	Asn	Pro	T
Macrotyloma axillare MAI[i]	(Cys	Thr	Lys	Ser	Ile	pro	T
	(Cys	Thr	Phe	Ser	Ile	Pro	C
Vigna unguiculata CPI[j]	(Cys	Thr	Lys	Ser	Ile	Pro	T
	(Cys	Thr	Phe	Ser	Ile	Pro	C
Triticum aestivum WGTI[k]	Cys	Thr	Arg	Ser	Ile	Pro	T
Rice Bran and RBTI[l]	Leu	Thr	Lys	Pro	Asp	Pro[I]	T
Domains I and III respectively	Cys	Asp	Lys	Met	Asp	Pro[III]	T
Potato Inhibitor I Family							
Solanum tuberosum PI-I[m]	Val	(Thr	Leu	Asp	Try	Arg	C
		(Met					

Potato Inhibitor II Family							
Solanum tuberosum	PI-II[a]	Ser	Tyr	Lys	Ser	Val	Cys	T
Barley Trypsin Inhibitor Family							
Hordeum vulgare	BTI[a]	Gly	Pro	Arg	Leu	Leu	Thr	T
Squash Inhibitor Family							
Cucurbita maxima	CMTI[p]	Cys	Pro	Arg	Ile	Leu	Met	T
Cucurbita pepo	CPTI[p]	Cys	Pro	Lys	Ile	Leu	Met	T
Cucumus sativus	CSTI-IV[p]	Cys	Pro	Arg	Ile	Leu	Met	T/HF

T—trypsin; C—chymotrypsin; E—elastase; t-PA—tissue-plasminogen activator

[a] I. Schechter and A. Berger, *Biochem. Biophys. Res. Commun.*, 1967, **27**, 157.
[b] T. Koide and T. Ikenaka, *Eur. J. Biochem.*, 1973, **32**, 417.
[c] C. Yamamoto, S. Hara, and T. Ikenaka, *J. Biochem.*, 1983, **94**, 849.
[d] F. J. Joubert, E. H. Merrifield, and E. B. D. Dowdle, *Int. J. Biochem.*, 1987, **19**, 601.
[e] S. Odani and T. Ikenaka, *J. Biochem.*, 1978, **83**, 737.
[f] F. C. Stevens, S. Wuerz, and J. Krahn, in 'Bayer Symposium V. Proteinase Inhibitors', ed. H. Fritz, H. Tschesche, L. J. Greene, and E. Truscheit, Springer-Verlag, Berlin, 1974, p. 344.
[g] K. A. Wilson and M. Laskowski, *J. Biol. Chem.*, 1973, **248**, 756.
[h] S. Norioka and T. Ikenaka, *J. Biochem.*, 1983, **94**, 589.
[i] F. J. Joubert, H. Kruger, G. S. Townshend, and D. P. Botes, *Eur. J. Biochem.*, 1979, **97**, 85.
[j] V. A. Hilder, R. F. Barker, R. A. Samour, A. M. R. Gatehouse, and J. A. Gatehouse, *Plant Mol. Biol.*, 1989, **13**, 701.
[k] S. Odani, T. Koide, and T. Ono, *J. Biochem.*, 1986, **100**, 975.
[l] M. Tashiro, K. Hashino, M. Shiozaki, F. Ibuki, and Z. Maki, *J. Biochem.*, 1987, **102**, 297.
[m] M. Richardson and L. Cossins, *FEBS Lett.*, 1974, **45**, 11.
[n] T. Iwasaki, T. Koyohara, and M. Yoshikawa, *J. Biochem.*, 1976, **79**, 381.
[o] S. Odani, T. Koide, and T. Ono, *J. Biol. Chem.*, 1983, **258**, 7998.
[p] M. Wieczorek, J. Otlewski, J. Cook, K. Parks, J. Leluk, A. Wilimowski-Pelc, A. Polanowski, T. Wilusz, and M. Laskowski, *Biochem. Biophys. Res. Commun.*, 1985, **126**, 646.

tree, winged beans, and *Erythrina* species some of which formed complexes of enzyme-inhibitor in the ratio 2:1. In addition to the specific trypsin and chymotrypsin inhibitors, dual inhibitors with activity towards trypsin and chymotrypsin and occasionally tissue plasmogens were obtained from each species. Only the two trypsin isoinhibitors from winged bean (WTI-IA and IB) have been completely sequenced. About 50% homology distributed uniformly along the polypeptide was found between the sequences of these inhibitors and STI (Ti^a) through 11 deletions and 2 insertions were observed in the former compared with the latter (Figure 1).[34] The amino acid sequence of WTI-IA differed from that of WTI-IB with respect to residues 73 and 152. Partial sequences of the amino-terminus of other members of the STI family (Acacia inhibitors I and II: Silk tree inhibitors A-IC and B-II; *Erythrina* species DE-3) revealed about 50% homology with that of STI (Ti^a).

A highly purified PI, (M_r 20 kd) and resembling the STI but specific for subtilisin, which it inhibited in a stoichiometric manner, has been isolated from barley.[38] No sequence data are available for this inhibitor to date.

Soyabean Proteinase Inhibitor (Bowman–Birk) Family

Soyabean PI (Bowman–Birk) BBI. A PI with properties quite distinct from those of STI was described initially by Bowman in 1944 and subsequently purified and characterized by Birk and co-workers.[16-18] Designated the Bowman–Birk inhibitor, (Inhibitor AA) (BBI) this inhibitor provided the name and prototype for a family of inhibitors that are ubiquitous in legumes and occur in other species. Several comprehensive reviews on the classical BBI have appeared recently.[17,18]

The classical BBI (Inhibitor AA) differs from the STI in a number of respects.[6] BBI is a relatively small molecule (M_r 8 kd). Earlier values of M_r around 20 kd have been attributed to the fact that in aqueous solutions, the inhibitor undergoes self association which is concentration dependent. BBI is a single polypeptide comprising 71 amino acids including 14 half-cystine residues which form 7 disulphide bonds cross-linking the molecule. It is devoid of tryptophan and glycine. The complete amino-acid sequence and covalent structure was determined by Odani and Ikenaka (Figures 1 and 3).[39,40] BBI is a double-headed inhibitor with independent reactive sites for trypsin and chymotrypsin (Table 1). BBI is extremely stable towards heat, acids, and alkalis due to the disulphide bonds conferring structural stability on the molecule.

The classical BBI like all inhibitors belonging to this family, consists of two tandem homology regions on the same polypeptide chain each with an independent reactive site (Table 1, Figure 1). Thus BBI will form a 1:1 complex with either trypsin or chymotrypsin and a ternary complex with both. All complexes dissociate into their active enzyme(s) and inhibitors below pH 5.0. The trypsin–inhibitor complex is devoid of trypsin activity and is inactive towards trypsin

[38] M. Yoshikawa, T. Iwasaki, M. Fujii, and M. Oogaki, *J. Biochem.*, 1976, **79**, 765.
[39] S. Odani and T. Ikenaka, *J. Biochem.*, 1972, **71**, 839.
[40] S. Odani and T. Ikenaka, *J. Biochem.*, 1973, **74**, 697.

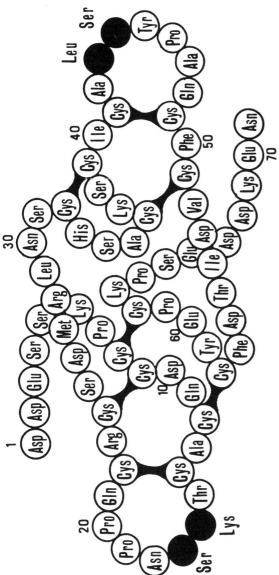

Figure 3 Structure of the Soyabean Proteinase Inhibitor (Bowman–Birk Inhibitor) (Reproduced from S. Odani and T. Ikenaka, J. Biochem., 1973, **74**, 697)

but inhibits chymotrypsin. The reverse is true of the chymotrypsin–inhibitor complex. The trypsin and chymotrypsin reactive sites are located at Lys-16-Ser-17 and Leu-43-Ser-44 respectively and both are located within a nanopeptide loop formed by a single disulphide bond. The amino acids around the reactive sites are similar to each other and to the other inhibitors within the family. BBI undergoes limited proteolysis specific to either trypsin or chymotrypsin and clearly obeys the standard mechanism. Modification of the virgin BBI by trypsin did not affect either the trypsin or chymotrypsin inhibition.[41] Treatment of the trypsin-modified BBI with carboxypeptidase B resulted in the removal of the new carboxyl-terminal lysine with the elimination of antitryptic activity. Chymotryptic inhibitor activity was unaffected. Similar treatment with chymotrypsin abolished the anti-chymotrypsin activity whilst retaining the trypsin inhibitory capacity.[18] Such modifications convert the double-headed BBI into a single-headed inhibitor with activity towards either trypsin or chymotrypsin.

The specific role of active site amino acid residues has been investigated by replacement of Ser-44 (P_1-chymotrypsin reactive site) in the chymotrypsin modified BBI.[42] Removal of the Ser-44 residue by Edman degradation and chemical replacement with serine, alanine, threonine, valine, leucine, or glycine led to the formation of 6 different BBI. Chymotrypsin inhibitory activity was as listed in decreasing order (Gly-44 derivative essentially inactive). An isoinhibitor of BBI, BBI-CII behaved in an identical manner.[43] Since BBI is essentially a symmetrical molecule consisting of two homologous domains each possessing a reactive site for either trypsin or chymotrypsin, cleavage at or around its centre should produce two fragments with trypsin and chymotrypsin activity respectively. Odani and Ikenaka cleaved the inhibitor with CNBr at Met-27 followed by hydrolysis with pepsin to produce two fragments containing 38 and 29 amino acid residues with trypsin and chymotrypsin inhibitory activity respectively.[44,45] The two separated domains were less stable than the native BBI suggesting that the stability of the legume seed double-headed inhibitors was acquired by duplicating an ancestral single-headed structure.

The three-dimensional structure of BBI has not been reported, but that of the peanut inhibitor A-II, a member of the BBI family was reported recently.[46] The inhibitor molecules are a tetramer with a 222 local symmetry in the crystals. Each monomer has an elongated shape 45 × 15 × 15 Å and consists of two domains which have similar three dimensional structures. The two independent proteinase binding sites protrude from the molecule on opposite sides.

Since the initial isolation and characterization of the classical BBI (Inhibitor AA) a number of isoinhibitors have been identified. Odani and Ikenaka isolated 5 Bowman–Birk type isoinhibitors (A, B, C-II, D-II, and E-1) from the

[41] Y. Birk, A. Gertler, and S. Khalef, *Biochim. Biophys. Acta*, 1967, **147**, 402.
[42] S. Odani and T. Ikenaka, *J. Biochem.*, 1978, **84**, 1.
[43] T. Kurokawa, S. Hara, T. Teshima, and T. Ikenaka, *J. Biochem.*, 1987, **102**, 621.
[44] S. Odani and T. Ikenaka, *J. Biochem.*, 1973, **74**, 857.
[45] S. Odani and T. Ikenaka, *J. Biochem.*, 1978, **83**, 747.
[46] A. Suzuki, T. Tsunogae, I. Tanaka, T. Yamane, T. Ashida, S. Norioka, S. Hara, and T. Ikenaka, *J. Biochem.*, 1987, **101**, 267.

Japanese soyabean cultivar, Sode-Furi.[47] The major inhibitor A was identical to the classical BBI (AA) and B only differed from this in chromatographic behaviour. C-II inhibited either trypsin or chymotrypsin at the same reactive site and elastase at a second reactive site. Inhibitors D-II and E-I inhibited trypsin in a non-stoichiometric inhibitor–enzyme ratio (1:1.4). Subsequently the amino acid sequences of C-II, D-II, and E-I (partial) were determined and compared with BBI (AA) and other members of the BBI family (Figure 1 and Table 1).[48–50] C-II was found to be homologous with the classical BBI (AA) but more closely related to the garden bean inhibitor with the reactive sites for elastase and trypsin/chymotrypsin identified as Ala-22-Ser-23 and Arg-49-Ser-50 respectively. D-II, although highly homologous with other members of the BBI family, had Arg at both inhibitor reactive sites. Lack of chymotrypsin inhibitory activity in D-II compared with C-II appeared to be due to the replacement of Met (C-II) with Gln (D-II) at position P_2'. Five BBI-type isoinhibitors (PI-I-V) have been isolated in a purified state from soyabean cultivar Tracey and partially characterized.[51] PI-V was identical to the classical BBI (AA). PI-I-IV were immunologically closely related but distinct from PI-V. Ten Bowman–Birk isoinhibitors have been purified from several soyabean cultivars and designated BBSTI-E', -E, -D, -C', -C, -B', -B, -A", A' in decreasing order of electrophoretic mobility.[52] These isoinhibitors were classified into 4 subgroups. Subgroup I included BBSTI-E', -E, and -D which were similar to or identical with the classical BBI. BBSTI-E was identified as the classical BBI. Subgroup II containing inhibitors C', C, and A" resembled subgroup I in amino acid composition but were weak inhibitors of trypsin and corresponded to D-II and E-I above.[50] The weakest trypsin inhibitors B and B' in subgroup III were related to E immunologically. Subgroup IV inhibitors, A and A', although cross-reacting with anti-BBSTI-E, did not resemble other members of BBI family. These inhibitors containing 200–210 amino acid residues (M_r 20 kd), had high glycine and low half cystine contents and were designated a separate class status called the glycine-rich soyabean trypsin inhibitors.

Since the various isoinhibitors differ in the extent they inhibit trypsin, chymotrypsin, and elastase, part of the nutritional variation of soyabean cultivars may be attributable to the relative proportions of the isoinhibitors present in the seed. Clearly information on these isoinhibitors and their molecular biology will assume increasing importance in future breeding programmes.

Lima Bean PI (LBI). LBI have been studied extensively and reviewed.[6,16,18,19] These inhibitors are very similar to the classical BBI (AA) with respect to M_r (8–10 kd), high half-cystine content, stability and physical properties, complex

---

[47] S. Odani and T. Ikenaka, *J. Biochem.*, 1977, **82**, 1513.
[48] S. Odani and T. Ikenaka, *J. Biochem.*, 1976, **80**, 641.
[49] S. Odani and T. Ikenaka, *J. Biochem.*, 1977, **82**, 1523.
[50] S. Odani and T. Ikenaka, *J. Biochem.*, 1978, **83**, 737.
[51] D. L. R. Hwang, K. T. Davis Lin, W. K. Yang, and D. E. Foard, *Biochim. Biophys. Acta*, 1977, **495**, 369.
[52] A. L. Tan-Wilson, J. C. Chen, M. C. Duggan, C. Chapman, R. S. Obach, and K. A. Wilson, *J. Agric. Food Chem.*, 1987, **35**, 974.

formation, and properties of the enzyme–inhibitor complex. At least 4 isoinhibitors have been isolated and all are double-headed inhibiting trypsin and chymotrypsin in a stoichiometric manner. Two isoinhibitors, LBI-I and IV have been sequenced and the reactive sites identified as Lys-Ser and Leu-Ser for trypsin and chymotrypsin respectively.[53] Like the classical BBI (AA), LBI consist of two homologous peptide regions, one domain containing the trypsin reactive site and the other the chymotrypsin reactive site. A high degree of homology exists between LBI-I and IV and the classical BBI (AA).

Garden Bean PI (GBI). Within the *Phaseolus vulgaris* are a number of different cultivated types including Garden bean, French bean, Navy bean, and Pinto bean. PI have been isolated from all these types and all have the characteristics of BBI (AA).[6] In summary all are low molecular weight (8–10 kd), rich in half-cystine and double-headed inhibitors of trypsin and chymotrypsin. Three isoinhibitors I, II, and IIIb have been isolated from the Garden bean.[54,55] All have M_r between 8–9 kd but differ slightly in amino acid composition and specificity. Isoinhibitor IIIb inhibited trypsin and chymotrypsin simultaneously and independently and resembled BBI (AA) whereas isoinhibitors I and II were essentially inactive against chymotrypsin. The major isoinhibitor II or II' in different cultivars, was a strong inhibitor of elastase and trypsin. Apart from 5 internal residues, the amino acid sequence of GBI-II' has been determined and the reactive sites identified as Ala-17-Ser-18 and Arg-44-Ser-45 for elastase and trypsin respectively (Table 1 and Figure 1). Normally inhibitors in the BBI family have the trypsin reactive site located in the first domain and the chymotrypsin one in the second one.

Groundnut (Peanut) PI (GI). The earlier work on these inhibitors has been reviewed comprehensively.[6,16–18] Tixier in 1968 first purified and characterized an inhibitor from peanut (about 17 kd) which was active against trypsin, chymotrypsin, and plasmin. An inhibitor of trypsin and chymotrypsin isolated from groundnut (M_r 8 kd) was unusual in containing arginine at the trypsin reactive site.[16] Removal of the carboxyl-terminal arginine residue in the trypsin modified inhibitor did not eliminate the trypsin activity and it was concluded that this could not be the trypsin reactive site. It was suggested that the chymotrypsin inhibitory site coincided with or was in the vicinity of the trypsin-inhibitor site and that the molecule was not strictly double-headed.

Five PI (A-I, A-II, B-I, B-II, and B-III) have been isolated from peanut and characterized.[55a] All the inhibitors were similar to the classical BBI (AA) with respect to molecular weight and half-cystine content. Each isoinhibitor inhibited trypsin and chymotrypsin in the ratios 1:2 and 1:1 respectively. The ternary trypsin–inhibitor complex was inactive towards chymotrypsin but the binary chymotrypsin–inhibitor complex inhibited trypsin with the slow release of chymotrypsin. The inhibitory activities of GI towards these two proteinases

[53] F. C. Stevens, S. Wuerz, and J. Krahn, in 'Bayer Symposium V. Proteinase Inhibitors', ed. H. Fritz, H. Tschesche, L. J. Greene, and E. Truscheit, Springer-Verlag, Berlin, 1974, p. 344.
[54] K. A. Wilson and M. Laskowski, *J. Biol. Chem.*, 1973, **248**, 756.
[55] K. A. Wilson and M. Laskowski, *J. Biol. Chem.*, 1975, **250**, 4261.
[55a] S. Norioka, K. Omichi, and T. Ikenaka, *J. Biochem.*, 1982, **91**, 1427.

were not independent of each other. All 5 isoinhibitors have been completely sequenced.[56,57] Isoinhibitors A-I, A-II, B-I, and B-III (A-II group) had the same amino acid sequences except in the N-terminal regions suggesting that all could be derived from A-II or a nascent inhibitor with the longer N-terminal amino acid sequence by proteolysis (Figure 1). Isoinhibitor B-II has 16 amino acid replacements compared with the sequence of A-II and is probably the product of a different gene from that of the A-II group. In all GI the first trypsin reactive site was Arg-13-Arg-14 and the second Arg-42-Ser-43 (GI-B-II) (Table 1). The first trypsin reactive site is also the chymotrypsin reactive site.[58] The three dimensional structure of GI and the participation of disulphide bridges in the maintenance of inhibitory activity has been described.[46,59]

Cowpea PI (CPI). Two types of PI have been isolated from cowpea by means of sequential affinity chromatography on immobilized trypsin (BEPTI) and chymotrypsin (BEPCI) and having anti-tryptic or chymotryptic and tryptic activity respectively.[60] Both inhibitors had M_r near 8 kd (reducing conditions) were rich in half-cystine and were clearly related to the classical BBI (AA).[61] In solution, the inhibitors existed as dimeric or tetrameric forms.[62] The dimeric form of the inhibitor (M_r 16–17 kd) interacted with either trypsin or chymotrypsin and trypsin. The dimer of BEPTI inhibited 2 molecules of trypsin, *i.e.* 1 molecule enzyme per monomer, while the BEPCI dimer inhibited 1 molecule each of trypsin and chymotrypsin presumably each being bound on different monomer components.[63] Half-site reactivity was proposed to explain this enzyme-inhibitor dimer complex formation. Subsequently Gatehouse and coworkers found that the PI purified by affinity chromatography on immobilized trypsin and chymotrypsin had M_r around 17 kd and consisted of two monomers (M_r around 8.5 kd) linked by disulphide bonds.[64] Each monomer possessed a single reactive site for either trypsin or chymotrypsin. Therefore the native inhibitors inhibited either trypsin or trypsin and chymotrypsin. Both inhibitor types (trypsin predominantly) consisted of several isoinhibitors. Recently the amino acid sequence of the cowpea trypsin inhibitor was determined (Table 1 and Figure 1).[65]

Other Legume Seed PI Belonging to the BBI Family. Numerous inhibitors of the BBI type have been isolated from other legume species and collated.[6,17,18] The complete amino acid sequences of two PI (DE-3 and DE-4) from *Macrotyloma axillare* have been determined (Table 1 and Figure 1).[66] Likewise, the amino

[56] S. Norioka and T. Ikenaka, *J. Biochem.*, 1983, **93**, 479.
[57] S. Norioka and T. Ikenaka, *J. Biochem.*, 1983, **94**, 589.
[58] S. Norioka and T. Ikenaka, *J. Biochem.*, 1984, **96**, 1155.
[59] S. Norioka, T. Kurokawa, and T. Ikenaka, *J. Biochem.*, 1987, **101**, 713.
[60] L. S. Gennis and C. R. Cantor, *J. Biol. Chem.*, 1976, **251**, 734.
[61] L. S. Gennis and C. R. Cantor, *J. Biol. Chem.*, 1976, **251**, 741.
[62] L. S. Gennis and C. R. Cantor, *J. Biol. Chem.*, 1976, **251**, 747.
[63] L. S. Gennis and C. R. Cantor, *J. Biol. Chem.*, 1976, **251**, 754.
[64] A. M. R. Gatehouse, J. A. Gatehouse, and D. Boulter, *Phytochemistry*, 1980, **19**, 751.
[65] V. A. Hilder, R. F. Barker, R. A. Samour, A. M. R. Gatehouse, J. A. Gatehouse, and D. Boulter, *Plant Mol. Biol.*, 1989, **13**, 701.
[66] F. J. Joubert, H. Kruger, G. S. Townshend, and D. P. Botes, *Eur. J. Biochem.*, 1979, **97**, 85.

acid sequences of proteinase inhibitors I-A and I-A' from Adzuki beans (*Phaseolus angularis*) have been reported and compared with inhibitor II from the same seed.[67]

Non-legume Seed PI Belonging to the BBI Family—Wheat Germ Trypsin Inhibitors. Two types of trypsin inhibitor, Group I (M_r 14.5 kd) and Group II (M_r 7 kd) have been isolated from wheat germ by affinity chromatography on immobilized trypsin and exclusion, ion-exchange, and reversed phase chromatography.[68] Trypsin was inhibited stoichiometrically by inhibitors of Groups I and II in a 2 and 1 ratio respectively. Both groups contain a number of isoinhibitors. Sequence analysis of inhibitors (I-2b, II-4 and 5) indicated a high degree of homology and showed Inhibitors I to be duplicated forms of inhibitors II (Figure 1 and Table 1). Both inhibitor groups were found to be highly homologous to the classical BBI (AA). Group II inhibitors represent examples of single-headed inhibitors corresponding to one inhibitory domain of the BBI.

Non-legume Seed PI Belonging to the BBI Family—Rice Bran Trypsin Inhibitors. The complete amino acid sequence of a double-headed trypsin inhibitor (RBTI) from rice bran has been determined (Figure 1).[69] This inhibitor consisted of 133 amino acid residues including 18 half-cystine residues (9 disulphide bridges). Two reactive sites were at Lys-17-Pro-18 and Lys-83-Met-84. RBTI was found to be composed of 4 domains, with domains I and III and II and IV being homologous with the first and second domains of the classical BBI (AA) respectively. Thus RBTI could be regarded as a duplicated form of the BBI.

Evolutionary Aspects. The amino acid sequences of a number of BB-type proteinase inhibitors have been compared and in all cases apart from the peanut inhibitors A-II and B-II, the amino acid in the P_1' position of the reactive site in domain 1 was serine.[57] Prior to this discovery it was assumed the P_1' serine residue at this reactive site was invariate since replacement with other amino acids resulted in a significant decrease in inhibitor activity.[50]

Double-headed inhibitors of the BBI type are thought to have evolved from an ancestral single-headed inhibitor by internal gene duplication and subsequent mutation causing differences in the reactive site residues. The soyabean BBI D-II, with the same residues at both reactive sites, represents a primitive form of double-headed inhibitor and points to the prototype inhibitor in legumes and possibly other plants being a single-headed inhibitor of trypsin having arginine as the P_1 residue at the reactive site.[48-50] The effectivity of the inhibitors of proteinases appears to be dependent upon long polypeptide chains.[18] Duplication of a monovalent inhibitor served two purposes in that it furnished a second reactive site and increased the potential of the PI to inhibit the proteinase. A phylogenic tree of legume seed double-headed inhibitors has

---

[67] T. Kiyohara, K. Yokoto, Y. Masaki, O. Matsui, T. Iwasaki, and M. Yoshikawa, *J. Biochem.*, 1981, **90**, 721.

[68] S. Odani, T. Koide, and T. Ono, *J. Biochem.*, 1986, **100**, 975.

[69] M. Tashiro, K. Hashino, M. Shiozaki, F. Ibuki, and Z. Maki, *J. Biochem.*, 1987, **102**, 297.

been constructed based on the matrix of amino acid differences of their amino acid sequences to demonstrate the molecular evolution.[57]

Potato and Similar PI

PI can account for up to 15–25% of the soluble protein of the potato tuber.[6] Many inhibitors have been isolated from potato and related species which have a wide range of specificity against serine endopeptidases, metallo carboxypeptidases, papain, microbial proteinases, and kallikreins. Potato PI that have been purified and characterized fall into 4 main categories: Potato Inhibitor I, Potato Inhibitor II, Carboxypeptidase Inhibitor, and Polypeptide Inhibitors of serine proteinases.

Potato PI I Family

Potato inhibitor I family, also known as potato chymotrypsin inhibitor I, is a potent inhibitor of chymotrypsin but is weakly active against trypsin. Subtilisin, pronase, and some other alkaline microbial proteinases are also inhibited. Inhibitors in this family are noncovalent pentamers (M_r 41 kd) of 4 different subunit types designated A, B, C, and D.[70] Each protomer is a single polypeptide consisting of 70–71 amino acids (M_r about 8 kd) with a single interchain disulphide bridge.[71] Individual subunits exhibit microheterogeneity due to amino acid substitution at various positions within the polypeptide chain. Numerous isoinhibitors have been identified by isoelectric focusing. Each pentamer inhibits 5 molecules of chymotrypsin and it may be concluded that each protomer possesses a reactive site. The amino acid sequences of the protomers A, B, C, and D were found to exhibit considerable homology with each other but not with any other proteinase inhibitors (Figure 1 and Table 1).[71-73] The chymotrypsin reactive site was identified as Met-/Leu-47-Asp-48 peptide bond. PI-1 which accumulates in the leaves of wounded tomato and potato plants has been shown to be similar to the potato inhibitor I with respect to subunit molecular weight, composition, and specificity towards serine proteinases.[70] Only two isoinhibitors have been identified in tomato.

Potato PI II Family

Inhibitor II has a M_r around 21 kd and consists of dimers of 4 distinctly different protomers (M_r 10.5 kd). These protomers are not homologous with those constituting potato inhibitor I.[74] Each protomer is a heterogeneous

---

[70] G. Plunkett, D. F. Senear, G. Zuroske, and C. A. Ryan, *Arch. Biochem. Biophys.*, 1982, **213**, 463.
[71] M. Richardson and L. Cossins, *FEBS Lett.*, 1974, **45**, 11.
[72] M. Richardson, *Biochem. J.*, 1974, **137**, 101.
[73] M. Richardson, R. D. J. Barker, R. T. McMillan, and L. M. Cossins, *Phytochemistry*, 1977, **16**, 837.
[74] J. Bryant, T. Green, T. Gurusaddaish, and C. A. Ryan, *Biochemistry*, 1976, **15**, 3418.

mixture of isoinhibitors which possess a reactive site for chymotrypsin but whose antitryptic activity varies. Despite varietal differences in inhibitor II patterns all such preparations were found to be immunologically related.[6] Two isoinhibitor forms, IIa and IIb, isolated from potato resembled each other and potato inhibitor II in many chemical and physicochemical properties.[75] Inhibitor IIa inhibited trypsin strongly followed by chymotrypsin and subtilisin, Inhibitor IIb, however, inhibited both chymotrypsin and subtilisin equally strongly but was inactive towards trypsin. Active fragments of both isoinhibitors have been sequenced (Figure 1 and Table 1).[76,77] An active fragment from IIa (M_r about 4.8 kd) remained strongly inhibitory towards trypsin but inhibited chymotrypsin weakly. A similar fragment from IIb retained activity against chymotrypsin and subtilisin. The reactive site peptide bond was identified as Lys-Ser. Proteinase inhibitor II isolated from the leaves of wounded tomato and potato plants has been found to be very similar to potato inhibitor II.[70]

Carboxypeptidase Inhibitors A and B

A polypeptide that specifically inhibited the pancreatic carboxypeptidases A and B was isolated and characterized by Ryan and co-workers.[12,78-80] The carboxypeptidase inhibitor was found to consist of three isoinhibitors CPI-I, -II, and -III with CPI-II predominating. CPI-II consisted of equal amounts of variant peptides containing either 38 or 39 amino acid residues, M_r 4.2 and 4.35 kd respectively. The amino acid sequence of CPI-I differs from that of CPI-II with respect to two positions (Ala and Gly replacing Ser-30 and Arg-32 respectively) (Figure 1). CPI-III has an identical sequence to CPI-II apart from lacking the amino-terminal pyrrolidone carboxylic acid. Each isoinhibitor contains 6 half-cystine in the form of 3 disulphide bonds.[81] The amino acid sequence of CPI-II is homologous with that of the tomato fruit carboxypeptidase inhibitor.

Polypeptide Inhibitors of Serine Proteinases

Three polypeptide inhibitors of the pancreatic serine proteinases have been isolated from potato tubers and purified to electrophoretic homogeneity.[82,83] Polypeptide chymotrypsin inhibitors I and II (PCI-I and -II) with M_r 5.4 and

[75] T. Iwasaki, T. Kiyohara, and M. Yoshikawa, *J. Biochem.*, 1972, **72**, 1029.
[76] T. Iwasaki, T. Kiyohara, and M. Yoshikawa, *J. Biochem.*, 1976, **79**, 381.
[77] T. Iwasaki, J. Wada, T. Kiyohara, and M. Yoshikawa, *J. Biochem.*, 1977, **82**, 991.
[78] G. M. Hass and C. A. Ryan, *Methods Enzymol.*, 1981, **80**, 778.
[79] J. M. Rancour and C. A. Ryan, *Arch. Biochem. Biophys.*, 1968, **125**, 380.
[80] C. A. Ryan, G. M. Hass, and R. W. Kuhn, *J. Biol. Chem.*, 1974, **249**, 5495.
[81] T. R. Leary, D. T. Grahn, H. Neurath, and G. M. Hass, *Biochemistry*, 1979, **18**, 2252.
[82] G. M. Hass, R. Venkatakrishnam, and C. A. Ryan, *Proc. Natl. Acad. Sci., (USA)*, 1976, **73**, 1941.
[83] G. Pearce, L. Sy, C. Russell, C. A. Ryan, and G. M. Hass, *Arch. Biochem. Biophys.*, 1982, **213**, 456.

6.4 respectively are devoid of methionine and tryptophan but contain 6 half-cystine residues as 3 disulphide bridges. Both inhibitors inhibit chymotrypsin and elastase strongly. Polypeptide trypsin inhibitor (PTI) M_r 5.3 kd is devoid of methionine, histidine, and tryptophan but contains 8 half-cystine residues as 4 disulphide bridges. PTI inhibits trypsin very strongly, chymotrypsin moderately, but is inactive towards elastase. In immunological double diffusion assays all three polypeptide inhibitors exhibit a high degree of immunological identity with each other, Potato Inhibitor II, and a fragment of this inhibitor prepared according to Iwasaki and co-workers.[77] The amino acid sequence of PCI-I reveals considerable homology with those of CPI and fragments of Potato Inhibitor II.

Barley Trypsin Inhibitor Family

Barley Trypsin Inhibitor (BTI)

This trypsin inhibitor was shown to be a single polypeptide (M_r 13.3 kd) consisting of 121 amino acid residues. The molecule contained 10 half-cystine residues as 5 disulphide bonds.[84] On the basis of amino acid sequence, the inhibitor could not be included in any of the established families of PI and therefore represented a new inhibitor family (Figure 1). The reactive site of the inhibitor was tentatively assigned to Arg-33-Leu-34. The amino acid sequence of the BTI, however, exhibited considerable homology with that of the wheat α-amylase inhibitor and the bifunctional α-amylase-trypsin inhibitor of ragi (Indian Finger Millet).[85] The latter inhibitor was found to be a single polypeptide (M_r 13.3 kd) consisting of 122 amino acid residues with 10 half-cystine residues as 5 disulphide bonds. Two reactive sites for trypsin were identified at Arg-34-Leu-35 which corresponds to the reactive site of the BTI and Arg-91-Ala-92.

Corn Trypsin Inhibitor

An inhibitor of trypsin and activated Hageman Factor (Factor XIIa) has been isolated by means of affinity chromatography and sequenced.[86] This inhibitor proved to be a single polypeptide consisting of 112 amino acid residues including 10 half-cystine- (5 disulphide bonds), proline-, and 3 tryptophan-residues. The reactive site was located at the Arg-36-Leu-37 peptide bond. This inhibitor had extensive sequence homology with the millet bifunctional inhibitor.[18,87]

---

[84] S. Odani, T. Koide, and T. Ono, *J. Biol. Chem.*, 1983, **258**, 7998.
[85] F. A. P. Campos and M. Richardson, *FEBS Lett.*, 1983, **152**, 300.
[86] W. C. Mahoney, M. A. Hermodson, B. Jones, D. D. Powers, R. S. Corfmann, and G. R. Reeck, *J. Biol. Chem.*, 1984, **259**, 8412.
[87] M. Richardson, S. Valdes-Rodriguez, and A. Blanco-Labra, *Nature (London)*, 1987, **327**, 432.

Other Cereal PI

Some inhibitors have been discussed previously (Legume seed PI). For a detailed discussion on the occurrence, properties, physiological role, and nutritional significance of cereal proteinase inhibitors in cereals reference should be made to Boisen.[7]

Squash Family of PI

PI of bovine trypsin and Hageman Factor Fragment (HF$_F$, β-Factor XIIa) have been isolated from the seeds of a number of species of the Cucurbitaceae including squash, summer squash, zucchini, cucumber, bitter gourd, and *Momordica repens*.[88-90] These inhibitors consist of around 29–30 amino acid residues with 3 disulphide bridges and are considered to be the smallest inhibitors of the serine proteinases known (M_r 3.3–4.0 kd). The amino acid sequences of a number of inhibitors have been determined including the squash (pumpkin) (*Cucurbita maxima*), summer squash (*Cucurbita pepo*), zucchini (*Cucurbita pepo var Giromontia*), cucumber (*Cucumis sativus*),[88] *Momordica repens*[90] and bitter gourd (*Momordica charantia*).[89] A system of nomenclature was proposed for this family of inhibitors in which the Latin names of each species were used to avoid confusion. Thus the squash (*Cucurbita maxima*) trypsin inhibitor is abbreviated to CMTI and summer squash (*C. pepo*) trypsin inhibitor CPTI.[88] The isoinhibitors are assigned Roman numerals in the first instance and supplemented with lower case letters as necessary. This system has been applied by other workers and is used here.[89] The amino acid sequences of the inhibitors in this family are highly homologous (Figure 1). The reactive site amino acid residues (P_1 and P_1') are occupied by Lys- or Arg-5-Ile-6 respectively with the exception of the bitter gourd (*Momordica charantia*) elastase inhibitor [MCEI-I] which has a Leu residue at P_1 (Table 1).[89] Since the location of the half-cystine residues in the polypeptide chain of all the squash family of inhibitors is identical, it is highly likely that the disulphide bonds occupy the same positions as in MCTI-II. Studies on proteinase inhibitor variants of CMTI-III, revealed that substitution of the reactive site P_1 arginine residue with different amino acids produced variants with different specificities. Variants with Arg, Lys, Leu, Ala, Phe, or Met at P_1 inhibited trypsin, those with Val, Ile, Gly, Leu, Ala, Phe, or Met inhibited human leukocyte elastase, while those with Leu, Ala, Phe, or Met inhibited cathepsin G and chymotrypsin.[91] Such variants have potential as agents for controlling inflammatory cell

---

[88] M. Wieczorek, J. Otlewski, J. Cook, K. Parks, J. Leluk, A. Wilimowska-Pelc, A. Polanowski, T. Wilusz, and M. Laskowski, *Biochem. Biophys. Res. Commun.*, 1985, **126**, 646.
[89] S. Hara, J. Makino, and T. Ikenaka, *J. Biochem.*, 1989, **105**, 88.
[90] F. J. Joubert, *Phytochemistry*, 1984, **23**, 1401.
[91] C. A. McWherter, W. F. Walkenhorst, E. J. Campbell, and G. I. Glover, *Biochemistry*, 1989, **28**, 5708.

proteolytic injury. The X-ray structure of the binary complex of CMTI-I and bovine trypsin has been determined.[92]

5 Assay Procedures for PI

Routine procedures for the assay of proteinase inhibitors in biological materials, foods, and feeds are mainly based on the decrease of enzymatic hydrolysis of either natural or synthetic substrates by the inhibitors.[16,19] Although natural substrates such as casein or haemoglobin are still used, synthetic substrates such as α-N-benzoyl-DL-arginine p-nitroanilide (BAPNA) and α-N-benzoyl-L-tyrosine ethyl ester (BTEE) are most frequently used in conjunction with bovine trypsin and chymotrypsin respectively.[93,94] Assays based on BAPNA have been developed primarily for soya and soya products.[93,94] Over the years this method has undergone considerable modification and refinement to improve its accuracy and reproducibility, especially with heated soya products.[95-100] The above assay was adapted for the routine estimation of trypsin inhibitors in large numbers of a wide range of foods including those derived from soya.[100] A detailed review of the content of PI in plant foods including methods of assay, appeared recently.[101]

The above assay procedures measure the total anti-tryptic activity in tissue or food extracts and include contributions from polyphenols and fatty acids in addition to proteinaceous PI. Specific methods have to be applied for the assay of proteinaceous PI especially at low concentrations. Affinity chromatography has been employed for such assays in seeds and seed flour preparations.[102]

Immunological techniques provide specific, sensitive, and accurate procedures for the assay of PI. Following investigations on the interaction of monoclonal antibodies with STI, enzyme-linked immunoassays (ELISA) were developed for STI and BBI (AA) using monoclonal antibodies.[103-105] The respective monoclonal antibodies were absolutely specific for the native forms of either STI (isoforms Ti^a and Ti^b) or BBI (AA). Inhibitors denatured by heat and other processes were discriminated against by the antibodies. These assays

[92] W. Bode, H. J. Greyling, R. Huber, J. Otlewski, and T. Wilusz, *FEBS Lett.*, 1989, **242**, 285.
[93] B. F. Erlanger, N. Kowowsky, and W. Cohen, *Arch. Biochem. Biophys.*, 1961, **95**, 271.
[94] K. A. Walsh and P. E. Wilcox, *Methods Enzymol.*, 1970, **19**, 31.
[95] M. L. Kakade, N. Simons, and I. E. Liener, *Cereal Chem.*, 1969, **46**, 518.
[96] M. L. Kakade, J. J. Rackis, J. E. McGhee, and G. Puski, *Cereal Chem.*, 1974, **51**, 376.
[97] G. E. Hamerstrand, L. T. Black, and J. D. Glover, *Cereal Chem.*, 1981, **58**, 42.
[98] W. L. Lehnhardt and H. G. Dills, *J. Am. Oil Chem. Soc.*, 1984, **61**, 691.
[99] J. J. Rackis, J. E. McGhee, I. E. Leiner, M. L. Kakade, and G. Puski, *Cereal Sci. Today*, 1974, **19**, 513.
[100] C. Smith, W. Van Megen, L. Twaalfhoven, and C. Hitchcock, *J. Sci. Food Agric.*, 1980, **31**, 341.
[101] J. J. Rackis, W. J. Wolf, and E. C. Baker, in 'Nutritional and Toxicological Significance of Enzyme Inhibitors in Foods', ed. M. Friedman, Plenum Press, New York, 1986, p. 185.
[102] J. P. Roozen and J. De Groot, in 'Recent Advances of Research in Antinutritional Factors in Legume Seeds', ed. J. Huisman, T. F. B. van der Poel, and I. E. Liener, Pudoc, Wageningen, 1989, p. 114.
[103] D. L. Brandon, S. Haque, and M. Friedman, *J. Agric. Food Chem.*, 1987, **35**, 195.
[104] D. L. Brandon, A. H. Bates, and M. Friedman, *J. Food Sci.*, 1988, **53**, 102.
[105] D. L. Brandon, A. H. Bates, and M. Friedman, *J. Agric. Food Chem.*, 1989, **37**, 1192.

should prove invaluable tools for monitoring residual levels of STI and BBI in soya, soya products, and in plant breeding studies.

Sensitive but non-specific techniques have been developed for the screening of large numbers of samples for trypsin and chymotrypsin inhibitor activity.[106,107] These procedures involve the use of trypsin- or chymotrypsin-conjugates incorporated into a suitable support (agarose gel slabs or agar films) onto which the inhibitor solution or extract can be applied and these allowed to diffuse into or be moved electrophoretically (rocket electrophoresis) into the gel. Enzyme inhibition is revealed as colourless areas on a coloured background by the use of suitable chromogenic substrates (N-acetyl-DL-phenylalanine-β-naphthyl ester). Since as little as 1 ng STI can be detected, such procedures should be useful in breeding programmes and monitoring fractions obtained in the separation of the inhibitors.

6 Biosynthesis and Molecular Biology of PI

Studies on the biosynthesis of PI have been restricted to potato and soyabean. mRNA(poly(A)$^+$-RNA) from the leaves of wounded and unwounded tomato plants were translated *in vitro* using the rabbit reticulocyte lysate system.[108] Although both poly(A)$^+$-RNA directed the incorporation of equivalent amounts of ^{35}S-methionine into TCA precipitable proteins only that from the wounded leaves directed the synthesis of proteins that could be immunoprecipitated with antisera for Inhibitors I and II. These inhibitors were synthesized *in vitro* as precursors with M_r approximately 2 kd larger than those found in the leaves. The additional polypeptide sequences of these preproteins represented signal peptides which were removed prior to or during the sequestration of the inhibitors into the vacuoles.

mRNA(poly(A)$^+$-RNA) for lectin and STI were isolated from immature seeds of soyabean.[109] Polysomes specific for these proteins were obtained by immunoadsorption techniques using monospecific antibodies against either lectin or STI. Poly(A)$^+$-RNA (STI) were translated *in vitro* (rabbit reticulocyte lysate system) and the products immunoprecipitated with antisera for STI. This polypeptide M_r 23.8 kd was approximately 2 kd larger than the native STI and it was concluded that *in vivo* the nascent STI polypeptide undergoes processing subsequent to synthesis. STI complementary DNA (cDNA) hybridized to a 770 nucleotide message in blotting experiments with total poly(A)$^+$-RNA. After subtracting 100 nucleotides for the poly(A) tail, the STI RNA could code for a polypeptide M_r 27 kd. Compared with the nascent inhibitor polypeptide (M_r 23.8 kd) synthesized *in vitro* it may be concluded that the untranslatable regions account for around 12 % of the message.

Poly(A)$^+$-RNA from immature soyabean seeds, consisting of approximately

---

[106] J. A. Gatehouse and A. M. R. Gatehouse, *Anal. Biochem.*, 1979, **98**, 438.
[107] I. Kourteva, R. W. Sleigh, and S. Hjerten, *Anal. Biochem.*, 1987, **162**, 345.
[108] C. E. Nelson and C. A. Ryan, *Proc. Natl. Acad. Sci.*, (*USA*), 1980, **77**, 1975.
[109] L. O. Vodkin, *Plant Physiol.*, 1981, **68**, 766.

500 nucleotides, directed the synthesis of BBI (AA) and PI I–IV in the wheat germ cell-free system.[110] The polypeptides synthesized were immunoprecipitated with antisera for the above inhibitors and examined by SDS–PAGE. The nascent inhibitors, whose identities were confirmed by sequence analysis, co-migrated with dimeric or trimeric forms of the native inhibitors. Since the nascent polypeptides could not be dissociated into the monomeric forms of the inhibitors by reduction and alkylation, it was concluded that BBI and related inhibitors were synthesized as multimeric forms. Recently sequence analysis of cDNA clones containing information for proteinase inhibitors IV and C-II of soyabean revealed that these inhibitors were synthesized as precursors with a short peptide leader sequence.[111]

Copy DNA (cDNA) libraries were constructed in the plasmid pUC9 from poly(A)$^+$ mRNA from wounded and unwounded tomato leaves. Since only the RNA from the wounded tissues directed the synthesis of inhibitor I and II *in vitro*, the difference between the two cDNA populations enabled the cDNA coding for the wound inducible poly(A)$^+$ mRNA to be identified.[112,113] Tomato leaf inhibitor I cDNA was 571 bp in length and contained an open reading frame coding for a polypeptide of 111 amino acids. This was 42 amino acids longer than Inhibitor I isolated from tomato leaves or potato tubers. The Inhibitor I polypeptide deduced from cDNA was considered to be preproprotein containing a 23 amino acid leader sequence followed by a 19 amino acid pro-sequence containing 9 charged amino acids which were lost during or after compartmentation. Inhibitor II cDNA was 693 bp in length and coded for a preprotein of 148 amino acids which included a 25 amino acid leader sequence. The sequence of 123 amino acids revealed that Inhibitor II consisted of 2 domains that arose by gene duplication and elongation. One domain possessed the trypsin reactive site and the other the chymotrypsin. Using the inhibitor I and II cDNA as hybridization probes, the temporal induction of the inhibitor mRNA was followed in response to wounding.[114] A single wound on a lower leaf resulted in the accumulation of inhibitors I and II in the upper adjacent but unwounded leaves within 4–6 hours. The corresponding mRNA appeared approximately 2 hours before the proteins and obtained a maximum after about 8 hours, thereafter it decayed with a half-life of 10–12 hours. Repeated wounding, three times hourly, to simulate insect wounding accentuated the rates of accumulation of inhibitor mRNA and proteins. Thus genes coding for Inhibitors I and II were rapidly switched on by signals (PIIF) released on wounding and reinforced by repeated wounding.

Inhibitor I and II cDNA were used as probes to identify genomic clones from

---

[110] D. E. Foard, P. A. Gutay, B. Ladin, R. N. Beachy, and B. A. Larkins, *Plant Mol. Biol.*, 1982, **1**, 227.
[111] P. E. Joudrier, D. E. Foard, L. A. Floener, and B. A. Larkins, *Plant Mol. Biol.*, 1987, **10**, 35.
[112] J. S. Graham, G. Pearce, J. Merryweather, K. Titani, L. Ericsson, and C. A. Ryan, *J. Biol. Chem.*, 1985, **260**, 6561.
[113] J. S. Graham, G. Pearce, J. Merryweather, K. Titani, L. Ericsson, and C. A. Ryan, *J. Biol. Chem.*, 1985, **260**, 6555.
[114] J. S. Graham, G. Hall, G. Pearce, and C. A. Ryan, *Planta*, 1986, **169**, 399.

both potato and tomato.[115,116] Both the tomato and potato wound-inducible Inhibitor I genes were found to be members of a small (10) gene family. Nucleotide sequences of the tomato and potato genes exhibited in excess of 90 % identity in the coding region, but the potato Inhibitor I prepro-protein lacked 12 nucleotides present in the tomato gene at the junction where the prosequence is cleaved to produce the native protein. A wound-inducible potato II inhibitor gene has been isolated which is part of a small multigene family (8–10 genes).[117] The corresponding tomato Inhibitor II gene, although similar to the potato gene, differed from this in having 4 rather than 2 domains.[118] Tobacco plants have been transformed with the intact potato Inhibitor II gene and the leaves of such plants respond to wounding by producing mRNA coding for this inhibitor.

cDNA clones constructed from size selected mRNA from immature soyabean seeds have been used to identify genes encoding for the classical BBI (AA).[119] The gene was found not to be highly reiterated in the genome. DNA sequence analysis of the gene encoding BBI isolated from a soyabean genomic library, revealed that it was similar but not identical to that of the cDNA clone and contained no intervening sequences. The protein and gene sequences of the cowpea Bowman–Birk type trypsin inhibitor and chymotrypsin–trypsin inhibitors have been compared.[120] Regions of high conservation and high divergence within the 5' leader, mature protein and 3' noncoding regions of the BBI and in the genes encoding them were observed in different members of this family.

7 Inactivation of PI

Most raw plant protein and carbohydrate sources such as legume seeds, cereals, and potatoes contain a number of ANF including PI which reduce the nutritional value of the food in feed. In many instances inactivation or elimination of the PI leads to a marked improvement in nutritional value of these foods or feeds. Various procedures are available for this purpose including heat treatments, reducing agents, protein fractionation, and alcohol extraction. In the main, these techniques have been developed for the soyabean but the principles involved can be applied to the inactivation of ANF in most plant protein sources.[6,101,121–123]

---

[115] J. S. Lee, W. E. Brown, G. Pearce, T. W. Dreher, J. S. Graham, K. G. Ahern, G. D. Pearson, and C. A. Ryan, *Proc. Natl. Acad. Sci., (USA)*, 1986, **83**, 7277.
[116] T. E. Cleveland, R. Thornburg, and C. A. Ryan, *Plant Mol. Biol.*, 1987, **8**, 199.
[117] R. W. Thornburg, T. E. Cleveland, and C. A. Ryan, *Proc. Natl. Acad. Sci., (USA)*, 1987, **84**, 744.
[118] C. A. Ryan and G. An, *Plant Cell Environ.*, 1988, **11**, 345.
[119] R. W. Hammond, D. E. Foard, and B. A. Larkins, *J. Biol. Chem.*, 1984, **259**, 9883.
[120] V. A. Hilder, A. M. R. Gatehouse, S. E. Sheerman, R. F. Barker, and D. Boulter, *Nature (London)*, 1987, **330**, 160.
[121] R. A. Burns, *J. Food Sci. Prot.*, 1987, **50**, 161.
[122] M. Friedman and M. R. Gumbmann, in 'Nutritional and Toxicological Significance of Enzyme Inhibitors in Foods', ed. M. Friedman, Plenum Press, New York, 1986, p. 357.
[123] M. Friedman and M. R. Gumbmann, *J. Food Sci.*, 1986, **51**, 1239.

Heat Treatment

It is widely accepted that most PI in plant sources are inactivated by heat even though experimental proof in specific materials is sometimes lacking, conflicting, or incomplete. Further, not all seed PI mirror the soyabean types with respect to inactivation by heat.

Moist heat treatment with live steams (toasting in the oilseed processing industry) has been widely used commercially to inactivate PI and other ANF in oilseeds and legume seed protein sources.[6,101] This process also improves the nutritional value of the protein *per se*. Dry heat is relatively ineffective. Loss of PI activity is a function of a combination of temperature, duration of heating, particle size, and moisture conditions used in the processing, all of which have to be controlled precisely to optimize the nutritional value of the material. Excessive heating can have deleterious effects on the nutritive value of soyabean. Heating soyabean meals under mildly alkaline conditions accelerates the inactivation of the inhibitors but has detrimental effects on protein quality. Isolated inhibitors [purified STI and BBI ($\Lambda\Lambda$)] behave differently towards heat than those *in situ* (meals).[124]

Information on the inactivation of PI in other species is scarce and variable and even conflicting. Autoclaving for 15–30 minutes essentially eliminated the trypsin inhibitor activity in most pulse legume seed meals.[101] The PI of Cowpea, Red gram, and French bean, however, proved to be relatively insensitive to heat deactivation. Cowpea in particular appears to have especially refractive PI since autoclaving seed meals in water for prolonged periods had no significant effect on activity levels. (Unpublished observations, this laboratory).

Isolated PI from cereals are relatively heat stable whereas those present in seed meals or flours are inactivated rapidly.[7] In processed cereal products, only rye bread has been found to contain appreciable amounts of trypsin inhibitor activity. Of the 3 major PI found in potato, only the CPI is completely heat insensitive.[125] Potato PI-I and -II were completely inactivated by heating at 100 °C for 3 minutes.

In most plant materials PI are present in different molecular forms (isoinhibitors) and these frequently have different heat sensitivities. In soyabean, the transition temperatures for the irreversible denaturation of the isoinhibitors Ti^a, Ti^b, and Ti^c were found to be 64 °C, 62 °C, and 57 °C respectively. Notwithstanding differences such as these together with the variations in the stability of the typical inhibitors of the individual inhibitor families, the general consensus is that most PI will be almost destroyed by normal processing or cooking. Thus all cooked and processed foods will have small residual amounts of PI in them.[100,101,126]

---

[124] C. M. DiPietro and I. R. Liener, *J. Agric. Food Chem.*, 1989, **37**, 39.
[125] D. Y. Huang, B. G. Swanson, and C. A. Ryan, *J. Food Sci.*, 1981, **46**, 287.
[126] B. H. Doell, C. J. Ebden, and C. A. Smith, *Qual. Plant. Plant Foods Hum. Nutr.*, 1982, **31**, 139.

Reducing Agents

Moist heat treatment (45–75 °C for 60 minutes) of raw soyabean flour with L-cysteine or N-acetyl-L-cysteine resulted in the introduction of new half-cystine residues into the sulphur-poor proteins and improved the nutritional value (PER in rats).[122,123,127,128] Modification of the proteins is due to the formation of mixed disulphide bonds between the added thiols and PI and storage proteins resulting in the inactivation of the former and increased digestibility of the latter. Moist heat treatment (75 °C for 60 minutes) with sodium sulphite was found to be more effective than cysteine in inactivating the PI possibly through rearrangement of disulphide bonds and configurational changes of these proteins without changes in the amino acid composition.[122,123] The commercial feasibility of the improvement of legume seed proteins by means of disulphide bond interchange still has to be explored.

Germination

There appears to be little correlation between trypsin inhibitor activity in seeds as affected by germination and nutritional value.[101] For instance, although an improvement in the nutritional value of soyabean has been observed during germination, this was not accompanied by a decrease in trypsin inhibitor activity.[129] The nutritional value of seeds is not always improved during germination and occasionally the nutritional value is improved despite an accompanying increase in trypsin inhibitor activity levels.[6]

Breeding

Breeding, probably in association with genetic engineering, to eliminate PI could eventually afford the most acceptable and economical means of improving the nutritional quality of legume and other seeds. Certain lines of soyabean devoid of STI (A_2) and containing 30–35 % less trypsin inhibitor activity than commercial varieties (Amsoy 71) have been tested in feeding trials with chicks and pigs.[130,131] The nutritional values (weight gain/feed intake) of the defatted raw meals for the lines lacking STI were higher than that for raw Amsoy but inferior to the toasted meal. Soyabean lines lacking STI still contain BBI which are more heat stable and resistant to digestion in the stomach. Elimination of all PI from soyabean and other legume seeds may be somewhat counter productive. Such seeds may have less resistance to insect and other pests and may have an inferior sulphur amino acid status of the proteins since certain PI (BBI family) represent rich sources of half-cystine.

[127] M. Friedman, O. K. Grosjean, and J. C. Zaunley, *J. Sci. Food Agric.*, 1982, **33**, 165.
[128] M. Friedman, M. R. Gumbmann, and O. K. Grosjean, *J. Nutr.*, 1984, **114**, 2241.
[129] J. L. Collins and G. G. Sanders, *J. Food Sci.*, 1976, **41**, 168.
[130] J. H. Orf and T. Hymowitz, *J. Am. Oil Chem. Soc.*, 1979, **56**, 722.
[131] T. Hymowitz, in 'Nutritional and Toxicological Significance of Enzyme Inhibitors in Foods', ed. M. Friedman, Plenum Press, New York, 1986, p. 291.

Fermentation

Prior to fermentation to produce tempeh or miso, chick peas, horse beans, and soyabeans are normally autoclaved (121 °C, 15 minutes) with almost complete destruction of trypsin inhibitor activity. Small residual amounts of trypsin inhibitor activity in the tempeh may be attributed to non-specific inhibition by free fatty acids produced during fermentation.[101]

Other procedures

Other industrial processes to inactivate trypsin inhibitors include dielectric heating and extrusion cooking.[101] Dielectric heating has a number of advantages over toasting including economy in energy use since much less water is required to inactivate the trypsin inhibitors, reduction in processing time and enhancement of functional properties. The process is also effective with intact seed (soyabean) at low moisture content. Extrusion cooking (cereal–legume mixes) with carefully selected temperatures and extrusion conditions is a highly effective means of inactivating trypsin inhibitors.

8 Nutritional, Metabolic, and Physiological Effects

General Effects of Dietary PI

An extensive literature exists on the nutritional significance of PI, much of it pertaining to soyabean in animal experiments.[6,132-135] The overall effects of PI when supplied in the diets of experimental animals as raw soyabean meals or flours, crude extracts containing PI activity or purified STI or BBI either alone or with inactivated soyabean meals can be summarized as follows:

1. Depress the growth rate of test animals (*e.g.* rats, mice, chicks). This effect can be prevented by inactivating the PI by heating, cooking or by methionine supplementation.
2. Marginally affects the protein digestibility of soyabean meal in rodents, *i.e.* heated compared with raw. PI depress the growth when added to diets containing predigested proteins or free amino acids. Proteolysis is depressed in chicks and dietary protein is excreted in the faeces.
3. Reduce nitrogen and sulphur absorption (raw soyabean meal) in both the rat and chick but different mechanisms were involved in the two species.
4. Induce pancreatic hypertrophy and hyperplasia and stimulate excessive enzyme secretion from this organ in rats, mice, and chicks.

[132] M. Friedman, 'Nutritional and Toxicological Aspects of Food Safety', ed. M. Friedman, Plenum Press, New York, 1984.
[133] M. Friedman, in 'Nutritional and Toxicological Significance of Enzyme Inhibitors in Foods', ed. M. Friedman, Plenum Press, New York, 1986.
[134] J. Huisman, T. F. B. van der Poel, and I. E. Liener, 'Recent Advances of Research in Antinutritional Factors in Legume Seeds', Pudoc, Wageningen, 1989.
[135] R. L. Ory, 'Antinutrients and Natural Toxicants in Foods', Food and Nutrition Press, Westport, 1981.

5 Stimulate trypsin and chymotrypsin synthesis in the pancreas thereby increasing the demand for methionine and cysteine.
6 Elicit endogenous losses of nitrogen and sulphur in rats and chickens.
7 Enhance the formation and release of humoral factors such as cholecystokinin which stimulate enzyme secretion from the pancreas.

Although mainly rats, mice, and chicks have been used to assess the growth response to trypsin inhibitors, guinea pigs, goslings, growing pigs, and calves respond similarly.[136,137]

The depression of growth of the animal appears to be moderated by factors such as age of the animal, the content and quality of protein in the diet, and the presence of other anti-nutritional factors. Adult animals appear to be less sensitive to the adverse effects of dietary PI. Increasing the content and nutritional quality of the dietary protein improves weight gain.

Sensitivity of Human Trypsins and Chymotrypsins to PI

Few studies on the sensitivity of human trypsins and chymotrypsins to plant PI had been undertaken prior to 1980. Several workers reported that a number of PI including STI and BBI were ineffective against human trypsins.[138,139] Subsequently STI was shown to inhibit Human trypsin I (Anionic) but was inactive towards Human trypsin II (Cationic) and chymotrypsin.[140] Contrary to these findings, STI, Lima bean inhibitor I (BBI family) and crude extracts from soyabean completely inhibited the trypsin and chymotrypsin activity in human pancreas juice *in vitro*.[141] These two enzymes accounted for between 40–50 % of the total proteolytic activity of the juice. Human and bovine pancreatic proteinases (crude extracts) varied in sensitivity to the PI from 10 legume seeds.[142] For example, Sword bean (*Canavalia ensiformis*) extracts completely inhibited bovine trypsin and chymotrypsin but had negligible activity towards the human enzymes while extracts from soyabeans, field bean (*Dolichos lablab*), kidney bean (*Phaseolus vulgaris*), and Bengal gram (*Cicer arietenum*) preferentially inhibitied the bovine enzymes. Cowpea (*Vigna unguiculata*) and Red gram (*Cajanus cajan*) extracts proved more effective in inhibiting human chymotrypsin. In similar studies in this laboratory using highly purified PI, human trypsin (crude pancreas extracts) and bovine trypsin were especially sensitive to cowpea and common bean inhibitors. Human trypsins were much less sensitive to the pea (*Pisum sativum*) and field bean (*Vicia faba*) inhibitors than the bovine enzymes. Cowpea and common bean inhibitors proved highly inhibitory towards human chymotrypsin but were less effective against the bovine

[136] D. Gallaher and B. O. Schneeman in 'Nutritional and Toxicological Aspects of Food Safety', ed. M. Friedman, Plenum Press, New York, 1984, p. 299.
[137] B. O. Schneeman and D. Gallaher, in 'Nutritional and Toxicological Significance of Enzyme Inhibitors in Foods', ed. M. Friedman, Plenum Press, New York, 1986, p. 185.
[138] R. E. Feeney, G. E. Means, and J. C. Bigler, *J. Biol. Chem.*, 1969, **244**, 1957.
[139] J. Travis and R. C. Roberts, *Biochemistry*, 1969, **8**, 2884.
[140] C. Figarella, G. A. Negri, and O. Guy, *Eur. J. Biochem.*, 1975, **53**, 457.
[141] A. Krogdahl and H. Holm, *J. Nutr.*, 1979, **109**, 551.
[142] K. S. Prabhu, K. Saldanha, and T. N. Pattabiraman, *J. Sci. Food Agric.*, 1984, **35**, 314.

enzymes. Human chymotrypsins were only weakly inhibited by winged bean (*Psophocarpus tetragonolobus*) and field bean inhibitors but were only moderately sensitive to those of the pea.

Recently the information on the potential of PI from grain legumes to inhibit human trypsins, chymotrypsin and other proteinases has been summarized.[143] In general, PI from the more important grain legume seeds were equally or slightly less inhibitory to human- than bovine-trypsins. In many cases as noted above some inhibitors were more active towards human chymotrypsin than to the bovine counterpart.

Fate of PI in the Gastrointestinal Tract

Stomach

Relatively little information exists on the effects of passage through the human stomach on PI. STI were inactivated when incubated with bovine pepsin at pH 1.5–2.0.[15] At pH 3.0, inactivation was negligible. In contrast, the BBI (AA) and other members of this family (Lima bean and Broad bean PI) are known to be stable to acidic pH and peptic digestion.[144,145] Krogdahl and Holm examined the sensitivity of STI, Lima bean PI and crude extracts of raw soyabean to low pH and human gastric juice *in vitro*.[146] STI was rapidly inactivated whereas the Lima bean inhibitor was unaffected over a prolonged incubation period. Between 30–40 % of the trypsin- and chymotrypsin-inhibitory activity of soyabean extracts was destroyed and this was thought to represent the contribution of STI to the total inhibitory activity. Thus BBI type inhibitors when ingested with food, would not be inactivated during passage through the stomach. The inactivation of STI by pepsin *in vivo* may be considerably less than that occurring at optimum pH *in vitro*. During digestion of a normal meal in the stomach, the pH is likely to be much higher than pH 2.0 for most of the time with commensurate effects on STI inactivation. Variable amounts of the active STI may also enter the duodenum following the ingestion of soyabean meals and products.

Intestinal Tract—Chickens

The fate of the BBI (AA) in the digestive tract of the chick has been investigated using radioimmunoassay procedures.[147] Inhibitors supplied in diets as raw soyabean meal or heated soyabean meal with added BBI were almost completely degraded during passage through the chick. The degradation of orally

---

[143] J. K. P. Weder, in 'Nutritional and Toxicological Significance of Enzyme Inhibitors in Foods', ed. M. Friedman, Plenum Press, New York, 1986, p. 239.
[144] Y. Birk and A. Gertler, *J. Nutr.*, 1961, **75**, 379.
[145] A. S. Warsy, G. Norton, and M. Stein, *Phytochemistry*, 1974, **13**, 2481.
[146] A. Krogdahl and H. Holm, *J. Nutr.*, 1981, **111**, 2045.
[147] Z. Madar, A. Gertler, and Y. Birk, *Comp. Biochem. Physiol.*, 1979, **62A**, 1057.

administered BBI was determined by following the disappearance of I^{125}-labelled BBI and BBI by radioimmunoassay after 3 hours. Only 10 % of the BBI administered was recovered, mainly from the faeces and intestinal tract, the remainder, although degraded, was equally distributed in the same fractions. Thus it was concluded that the BBI is degraded during passage through the intestine with the majority of the degradation products excreted in the faeces.

Intestinal Tract—Rats

Abbey and co-workers attempted to determine the precise nutritional and physiological significance of PI from field beans (BBI family) by incorporating the partially purified PI into synthetic diets devoid of all other legume seed proteins.[148,149] Even when incorporated at very high levels in the diet (20 times the endogenous levels found in field bean), no PI activity could be detected in the faeces using conventional assay procedures. The field-bean PI were shown to be inactivated in all sections of the gut, duodenum, upper- and lower-jejunum, and lower ileum. Crude enzyme preparations from the gut but notably the jejunum, inactivated and hydrolysed the inhibitors *in vitro*. From these studies it must be concluded that PI are inactivated by endogenous enzymes in the gut of the rat and chick. Small amounts of the PI appear to be excreted. The fate of the degradation products of the inhibitors in the rat is not known.

Short Term Physiological Responses to Dietary PI

Chicks, rats, mice, and many other species but not primates (monkeys), dogs, calves, pigs, and adult guinea pigs, when fed raw soyabean meal developed enlargement of the pancreas.[6,136,137,150–152] PI from various sources were shown to be responsible for the pancreatic enlargement which frequently occurred in the absence of growth effects generally with high protein diets.[153] Enlargement of the pancreas was due mainly to hypertrophy with some hyperplasia.[6,136,150] Pancreatic proteolytic enzyme secretion and synthesis was stimulated very quickly in the rat following the ingestion of a single meal (*e.g.* raw soyabean meal) containing PI. Pancreatic trypsin activity levels increased in monkeys fed raw soyabean meal.[152]

PI elicit pancreatic hypertrophy and increased enzyme secretions by interacting with the mechanism involved in the regulation of pancreatic secretion.[153] This regulation is mediated by a negative feedback stimulation of the pancreas

[148] B. W. Abbey, R. J. Neale, and G. Norton, *Brit. J. Nutr.*, 1979, **41**, 31.
[149] B. W. Abbey, G. Norton, and R. J. Neale, *Brit. J. Nutr.*, 1979, **41**, 39.
[150] D. Gallaher and B. O. Schneemann, in 'Nutritional and Toxicological Significance of Enzyme Inhibitors in Foods', ed. M. Friedman, Plenum Press, New York, 1986, p. 167.
[151] J. J. Rackis and M. R. Gumbmann, in 'Antinutrients and Natural Toxicants in Foods', ed. R. L. Ory, Food and Nutrition Press Inc., Westport, 1981, p. 203.
[152] B. J. Struthers, J. R. MacDonald, R. R. Dahlgren, and D. T. Hopkins, *J. Nutr.*, 1983, **113**, 86.
[153] M. R. Gumbmann, G. M. Dugan, W. L. Spangler, E. C. Baker, and J. J. Rackis, *J. Nutr.*, 1989, **119**, 1598.

by the hormone cholecystokinin (CCK). CCK release by the endocrine glands within the mucosa of the upper small intestine is repressed by pancreatic proteinases. This repression probably occurs indirectly via the breakdown of CCK releasing factors which are proteinase-labile polypeptides originating either in the pancreas or small intestine. These releasing factors would normally act on the endocrine sites promoting the release of CCK. Dietary proteins or PI reduce the level of free intestinal proteinases thereby increasing the levels of CCK releasing factors and as a consequence the blood CCK levels in the rat.[154,155] In turn, CCK initiated an increase in enzyme synthesis and secretion in the acinar pancreas. Eventually pancreatic hypertrophy and hyperplasia would occur from continued PI ingestion in the short term.[153] Dietary protein level modified the plasma CCK and pancreas responses to PI.[154] With adequate dietary protein levels, STI accentuated the transient response in plasma CCK levels obtained with protein alone. In turn pancreatic growth and proteinase synthesis was stimulated and the increased proteinase secretion suppressed CCK release, consequently plasma levels returned to normal. With adequate dietary protein, STI promoted large, increasing, and sustained plasma CCK levels. In this case pancreatic protein synthesis and growth was impaired due to inadequate amino acid supply. Particularly high demand is placed on the sulphur amino acids specifically cysteine. Supplementation of raw soyabean meal diets with cysteine or other S-amino acids, increased pancreas weight relative to the body weight of rats.[156] STI or potato PI in rat or mouse diets with sufficient protein to preclude growth inhibition stimulated pancreatic enlargement.[153] Pancreatic changes in the rat, mouse, chick, and other animals resulting from short term exposure to dietary PI can be regarded as reversible physiological adaptations to the digestive needs of the animal.

In man, raw but not heated soyabean flour in mixed meals stimulated CCK release specifically.[157] Duodenal infusions of raw soyabean meal extracts for 60 minutes, however, did not elicit an increase in plasma CCK.[158] While the raw soyabean extract completely inhibited the chymotrypsin in the duodenum, trypsin fell transiently and then increased to above pre-infusion levels. This apparent insensitivity to the raw soyabean extract was due to the presence of a previously unidentified inhibitor-resistant trypsin.[159]

Long Term Physiological and Pathological Responses to Dietary PI

The specific effects on the pancreas of feeding diets containing high levels of

---

[154] G. M. Green, V. H. Levan, and R. A. Liddle, in 'Nutritional and Toxicological Significance of Enzyme Inhibitors in Foods', ed. M. Friedman, Plenum Press, New York, 1986, p. 123.
[155] R. S. Temler and C. Mettraux, in 'Nutritional and Toxicological Significance of Enzyme Inhibitors in Foods', ed. M. Friedman, Plenum Press, New York, 1986, p. 133.
[156] M. R. Gumbmann and M. Friedman, *J. Nutr.*, 1987, **117**, 1018.
[157] J. Calam, J. C. Bojarski, and C. J. Springer, *Brit. J. Nutr.*, 1987, **58**, 175.
[158] H. Holm, L. E. Hanssen, A. Krogdahl, and J. Florholmen, *J. Nutr.*, 1988, **118**, 515.
[159] H. Holm, A. Krogdahl, and L. E. Hanssen, *J. Nutr.*, 1988, **118**, 521.

soyabean flour have been reviewed.[160] Raw full-fat soyabean flour enhanced pancreatic carcinogenesis when the diets were supplied together with or subsequent to the exposure of known pancreatic carcinogens. Raw full-fat soyabean supplied to rats over prolonged periods results in the formation of hyperplastic lesions (foci and nodules) in the pancreas including adenomas and adenocarcinomas. Pancreatic hypertrophy and hyperplasia developed rapidly in these animals but was completely reversible provided the exposure to full-fat soya was not too prolonged. The stimulatory effects of raw soyabean flour were greater for the growth of the carcinogen induced foci than for the normal pancreas. None of the above effects developed when heated soya flour was used and the conclusion was drawn that PI were responsible for these effects.

In a long-term investigation (2 years) involving rats fed soyabean-flour and -isolates, a high incidence of pancreatic nodular hyperplasia often visible by six months and the benign neoplastic lesions, acinar adenoma, developed even at moderate dietary PI levels.[161,162] The control animals in this experiment (casein diets) also developed an unusual incidence of nodular hyperplasia indicating that other factors may be involved in precancerous changes in the pancreas.

Soya protein isolates in long term feeding experiments (18 months) with rats promoted an increase in the activity, weight, and hyperplasia of the pancreas, but did not stimulate the development of lesions.[163] Raw soyabean flours (defatted) did not elicit pancreatic enlargement or the development of pancreatic lesions in hamsters in long term feeding trials over 15 months.[164] The pancreatic response of rats and mice to PI from soya bean and potato in isonitrogenous casein diets has been compared.[153] Rats developed pancreatic nodular hyperplasia and acinar adenoma in a dose related manner to both soyabean and potato inhibitors. Mice, however, did not develop pancreatic lesions in the long term despite responding identically to rats in the short term (pancreatic hypertrophy and hyperplasia).

Long term exposure (up to 5 years) of primates (Cebus monkeys) to semisynthetic diets containing low levels of PI did not cause any adverse effects in the pancreas such as hypertrophy, hyperplasia, pathological, or biochemical changes.[165,166]

Different species or even strains, *e.g.* Wistar rats exhibit contrasting propen-

[160] E. E. McGuinness, R. G. H. Morgan, and K. G. Wormsley, *Environ. Health Perspect.*, 1984, **56**, 205.
[161] M. R. Gumbmann, W. L. Spangler, G. M. Rackis, and I. E. Liener, *Qual. Plant. Plant Foods Hum. Nutr.*, 1985, **35**, 275.
[162] M. R. Gumbmann, W. L. Spangler, G. M. Dugan, and J. J. Rackis, in 'Nutritional and Toxicological Significance of Enzyme Inhibitors in Foods', ed. M. Friedman, Plenum Press, New York, 1986, p. 33.
[163] B. D. Richter and B. O. Schneeman, *J. Nutr.*, 1987, **117**, 247.
[164] I. E. Liener and A. Hasdai, in 'Nutritional and Toxicological Significance of Enzyme Inhibitors in Foods', ed. M. Friedman, Plenum Press, New York, 1986, p. 189.
[165] J. P. Harwood, L. M. Ausman, N. W. King, P. K. Sehgal, R. J. Nicolosi, I. E. Liener, D. Donatucci, and J. Tarcza, in 'Nutritional and Toxicological Significance of Enzyme Inhibitors in Foods', ed. M. Friedman, Plenum Press, New York, 1986, p. 223.
[166] J. P. Harwood, B. A. Jackson, and N. McCabe, in 'Recent Advances of Research in Antinutritional Factors in Legume Seeds', ed. J. Huisman, T. F. B. van der Poel, and I. E. Liener, Pudoc, Wageningen, 1989, p. 95.

sities to pre-neoplastic and neoplastic lesions in the pancreas. Such responses cannot be predicted by short term hypertrophic and hyperplastic response of the pancreas to PI.

Pancreatic cancer is the fifth most common cause of death due to cancer in the USA and its incidence appears to be rising.[167] While the epidemiology of the disease is not known, the studies with rats in which high soya diets containing PI induced pancreatic carcinogenesis have assumed particular if somewhat premature importance. From the evidence presented above it would appear that the rat may not be an accurate model to predict the response of man to dietary PI. At this time there is not compelling evidence that pancreatic cancer is caused by PI.[168]

Therapeutic Applications of PI

Troll and co-workers have proposed an anti-carcinogenic role for dietary PI including those from non-plant sources.[168] Diets containing purified PI or raw soya products aided in the prevention of breast, colon, and prostrate cancer in humans and mammary gland, colon, and skin tumours in rodents. PI may be active in cancer prevention by blocking the formation of active oxygen species by stimulated neutrophils, inhibiting poly-ADP ribosylation, and impeding the digestion of proteins and thereby depriving rapidly growing cancer cells of essential nutrients. The anti-carcinogenic activity of BBI was attributed to the inhibition of chymotrypsin. Raw soya flour in diets prevented the development of pancreatic cancer in hamsters injected with N-nitroso-bis(2-oxypropyl)-amine.[164] Neither raw soya flour diets nor azaserine (a pancreatic carcinogen in rats) affected atypical acinar cell nodule incidence in long term experiments with mice. PI arrested oncogenic transformation and promotion in tissue cultures of mouse cells. STI were without effect on the X-ray induced transformation of these cells but suppressed the promotional effects of 12-O-tetradecanoyl phorbol-13-acetate (TPA) on transformed cells.[169] Nanogram quantities of BBI irreversibly blocked the X-ray induced transformation of mouse cells but had no effect on the promotion of transformed cells by TPA.[170]

Raw soyabean flour has proved to be an effective treatment of chronic pancreatitis with moderate pancreatic insufficiency.[121] Peanut trypsin inhibitors possess anti-plasmin activity and have proved to be an haemostatic agent in the treatment of haemophiliacs.[6]

A number of sequence variants of the squash seed CMTI-III were found to inhibit human leucocyte elastase or cathepsin G or both.[91] These novel inhibitors were considered possible agents for controlling various inflammatory proteolytic enzymes. A potent inhibitor of Hageman Factor Fragment (HG$_{FI}\beta$-Factor XIIa) and bovine trypsin has been isolated from pumpkin

---

[167] B. D. Roebuck, *J. Nutr.*, 1987, **117**, 398.
[168] W. Troll, K. Frenkel, and R. Wiesner, in 'Nutritional and Toxicological Significance of Enzyme Inhibitors in Foods', ed. M. Friedman, Plenum Press, New York, 1986, p. 153.
[169] A. R. Kennedy and J. B. Little, *Cancer Res.*, 1981, **41**, 2103.
[170] J. Yavelow, M. Collins, Y. Birk, W. Troll, and A. R. Kennedy, *Proc. Natl. Acad. Sci., (USA)*, 1985, **82**, 5395.

seed.[171] This small polypeptide inhibitor was inactive towards a range of human serine proteinases and may prove to be useful in physiological investigations and therapies. A similar but much larger inhibitor has been isolated from maize.[86]

9 Functions of PI in the Plant

Regulation of Endogenous Proteinases

Certain PI have a regulatory function. Some plant proteinases are insensitive to endogenous inhibitors but many are not.[5,20] In germinating mung bean seeds the proteinase inhibitor specific for the major endopeptidase was shown not to be involved in the regulation of protein turnover.[8] This conclusion was based on compartmentalization of the PI and the endopeptidase in the cytosol and protein bodies respectively. The most likely function of the PI in the cell was to protect the cytosol against the proteinases released by accidental rupture of the protein bodies.

Storage or Reserve Protein

PI may serve as storage proteins in seeds and tubers which are mobilized during germination and sprouting. Some evidence has been obtained supporting such a role in potato tubers,[172] but that for seeds is equivocal.[5]

Defence Mechanism

PI may afford protection to the various parts of the plant (seed, leaf, tuber) against insects and micro-organisms. The most significant evidence in support of this hypothesis was the demonstration that wounding the leaves of potato and tomato plants by insects (larvae or adult Colorado beetle) induced a rapid accumulation of PI-I and -II throughout the aerial parts of the plant.[173] Any stimulus, whether mechanical, microbial, or fungal infection elicited this response which was mediated by a PI inducing factor (PIIF) or wound hormone. This factor, identified as an oligomer of α-(1–4) linked D-galactopyranosyluronic acid moieties, was produced from the cell wall pectins by endopolygalacturonases released from the damaged tissue.[174] PIIF was transported mainly upwards to the undamaged leaves where it initiated the synthesis and vacuolar accumulation of the inhibitors within 4–6 hours of the initial stimulus. Within 2–3 days, PI-I and II could account for in excess of 1% of the total leaf protein. Since both inhibitors persist in the leaves for prolonged periods they constitute a defence mechanism against further attack.

There are numerous reports in the literature in which the resistance to insect

---

[171] Y. Hosima, J. V. Pierce, and J. J. Pisano, *Biochemistry*, 1982, **21**, 3741.
[172] C. A. Ryan, *Ann. Rev. Plant Physiol.*, 1973, **24**, 173.
[173] T. R. Green and C. A. Ryan, *Science*, 1973, **175**, 776.
[174] P. Bishop, G. Pearce, J. E. Bryant, and C. A. Ryan, *J. Biol. Chem.*, 1984, **259**, 13172.

predation has been attributed to PI.[5,20,175,176] Gatehouse and co-workers showed that a cultivar of cowpea (*Vigna unguiculata*) TVu 2027 (out of 11 000 accessions) contained high levels of PI in the seed and this conferred resistance against the Bruchid beetle larva (*Callobruchus maculatus*).[177] Recent work has confirmed that cowpea weevil larvae were sensitive to the Nigerian cultivar TVu 2027 and its progeny (IT81D-1045 and -1064) but not the Brazilian cultivars CE-31, -11, and -524.[176] No correlation was found between proteinase (trypsin, chymotrypsin, subtilisin BPN', and papain) inhibitor content and insect resistance as shown by the Nigerian cultivars. Cultivars TVu 2027 and CE-11 and -524 all contained high levels of PI whereas IT81D-1045 and -1064 contained low levels of trypsin and chymotrypsin inhibitors. CE-31 contained high levels of trypsin inhibitor activity only. No correlation was apparent between lectin and tannin levels and insect resistance. Clearly factors other than PI and other anti-nutritional factors conferred insect resistant on the seed. This conclusion is consistent with the fact that the larvae of seed-eating bruchids utilize cysteine proteinases and not serine proteinases for protein digestion. Cowpea cultivars, Bruchid resistant (*C. maculatus*), IT81D-1045 and -1064, supported the growth of *Zabrotes subfasciatus*, a bruchid that infests the seeds of *Phaseolus vulgaris* and *Vigna unguiculata*.

Leaves of tobacco plants transformed with the cowpea trypsin inhibitor gene inhibited the growth and development of the tobacco budworm larvae.[120] These lepidopterous larvae use serine proteinases for the digestion of protein.

10 Conclusions

Protein PI particularly those inhibiting serine proteinases occur in a wide range of plant tissues but despite the use of sensitive immunological procedures, their precise location within the cell remains equivocal.

All PI inhibit their respective proteinases by the standard mechanism. Most have M_r between 3–22 kd. Many inhibitors have been classified into 6 inhibitor families on the basis of M_r, homology of primary amino acid sequences, and reactive site characteristics. Others still have to be classified.

The most widely used assay procedures for PI are those based on the inhibition of proteinase activity using natural or synthetic substrates. A recently introduced enzyme linked immunosorbent assay procedure (ELISA) affords a highly sensitive and specific technique for specialized uses.

PI are synthesized as either pre- or prepro-proteins which are processed *in vivo* to produce the native inhibitor. The genes coding for BBI (soyabean and cowpea) and Potato Inhibitors I and II have been isolated. mRNA for the wound inducible Inhibitors I and II in tomato and potato appeared 2–4 hours after the stimulus and 2 hours before the inhibitors were detected.

[175] A. M. R. Gatehouse, in 'Developments in Food Proteins', ed. B. J. F. Hudson, Elsevier, London and New York, 1984, p. 245.
[176] J. Xavier-Filho, F. A. P. Campos, M. B. Ary, C. Peres Silva, M. M. M. Carvalho, M. L. R. Macedo, F. J. A. Lemos, and G. Grant, *J. Agric. Food Chem.*, 1989, **37**, 1139.
[177] A. M. R. Gatehouse, J. A. Gatehouse, P. Dobie, A. M. Kilminster, and D. Boulter, *J. Sci. Food Agric.*, 1979, **30**, 948.

PI in meals and flours can be inactivated by moist heat (live steam) whereas the purified inhibitors are generally more heat stable. Complete inactivation of PI in feeds and meals is neither necessary nor desirable. Newer techniques that would prove useful for PI inactivation include the use of reducing agents (sulphite or cysteine) with mild heat and dielectric heating.

The nutritional significance of dietary PI in rats, mice, and chicks is well documented. PI only marginally reduce protein digestibility but depress growth. This growth depression is largely due to methionine stress induced by the PI stimulated synthesis and secretion of methionine-rich trypsin and chymotrypsin, increased endogenous losses and lowered uptake of N and S. PI inactivation or methionine supplementation prevented growth inhibition. Trypsin and chymotrypsin from all species including man are sensitive to PI to varying degrees. While pepsin inactivates STI, but not BBI types *in vitro*, the effect *in vivo* is likely to be small. Most PI appear to be degraded in the intestine.

In the short term, PI elicited a rapid increase in pancreatic trypsin and chymotrypsin synthesis and secretion in most animals. PI by reducing the levels of active trypsin and chymotrypsin, activate a negative feedback stimulation of the pancreas mediated by CCK. Evidence for PI stimulating CCK release in humans is equivocal. Increased pancreatic activity in rats, mice and chicks is associated with hypertrophy and hyperplasia. These short-term effects are reversible. Long term ingestion of PI (Wistar rats only) may lead to pancreatic nodular hyperplasia, acinar adenomas and adenocarcinomas depending on intake and dietary protein levels. PI most likely act as co-carcinogens with other initiators. From the evidence available at this time it would appear that the low residual levels of PI in processed and cooked foods do not present a significant health risk to humans.

PI have potential in certain therapeutic applications. In the plant, PI play a defensive role against insect and fungal predators.

CHAPTER 5

Antigenic Proteins

J. P. FELIX D'MELLO

1 Introduction

The storage proteins of leguminous and cereal grains are primary sources of nutrients for farm livestock and man. However, these proteins are deficient in certain indispensable amino acids, although this limitation may be overcome by dietary supplementation with protein from animal by-products or with crystalline amino acids.

It is well established that the terminal stages of protein digestion in the alimentary canal occur intracellularly,[1] after transport of peptides into the absorptive cells of the intestinal mucosa. However, there is now unequivocal evidence[2] that intact proteins are also capable of crossing the epithelial barrier of the intestinal mucosa. Furthermore, this absorption of intact protein occurs well after 'closure' to immunoglobulin transfer has taken place. In the normal course of events, absorption of intact protein proceeds without evoking deleterious responses in most animals and in human subjects. However, in certain circumstances, gut uptake of intact proteins is attended by pathological and immunological disorders[3-6] leading to malabsorption and hypersensitivity in young farm animals. In humans, coeliac disease or gluten-sensitive enteropathy is the prime example of a food allergy activated by the absorption of certain

---

[1] D. B. A. Silk, G. K. Grimble, and R. G. Rees, *Proc. Nutr. Soc.*, 1985, **44**, 63.
[2] M. L. G. Gardner, *Ann. Rev. Nutr.*, 1988, **8**, 329.
[3] J. W. Sissons, *Proc. Nutr. Soc.*, 1982, **41**, 53.
[4] T. J. Newby, B. Miller, C. R. Stokes, D. Hampson, and F. J. Bourne, in 'Recent Advances in Animal Nutrition—1984', ed. W. Haresign and D. J. A. Cole, Butterworths, London, 1984, p. 49.
[5] P. Porter and M. E. J. Barratt, in 'Recent Advances in Animal Nutrition—1987', ed. W. Haresign and D. J. A. Cole, Butterworths, London, 1987, p. 107.
[6] J. W. Sissons, in 'Recent Advances in Animal Nutrition—1989', ed. W. Haresign and D. J. A. Cole, Butterworths, London, 1989, p. 261.

cereal proteins.[7-12]

The components of leguminous and cereal grains which elicit antigenic effects in animals and man are certain fractions of storage proteins. In the case of soyabean (*Glycine max*), the antigenic proteins have been identified as glycinin and β-conglycinin.[6,13,14] In wheat (*Triticum aestivum*) it is the alcohol-soluble gliadin component of the gluten fraction which evokes intestinal damage in coeliacs.[7-12] Although the storage proteins of maize appear to be free of antigenic properties and do not, for example, activate or perpetuate coeliac disease in man, it has recently been observed[15] that diets containing high proportions of maize gluten meal depress utilization of lysine in broiler chicks. D'Mello[16] has concluded that this depression cannot be explained solely on the basis of the poor digestibility of maize gluten meal. A possible antigenic role of maize gluten cannot, therefore, be ruled out.

2 Properties of Storage Proteins

Two major classes of globulin storage proteins have been identified[17] in legumes on the basis of sedimentation characteristics. Proteins with sedimentation coefficients of 11–12 S are generally known as legumins while those with coefficients of 7–8 S are termed vicilins. In soyabean seeds, legumins and vicilins are accorded the trivial names glycinin and conglycinin, respectively. The conglycinin fraction is further sub-divided into α, β, and γ isotypes. Glycinin and β-conglycinin predominate, representing almost 0.7 of the total protein in the seed. The ratio of glycinin to β-conglycinin varies, according to cultivar, from 1:1 to 3:1. Both fractions are characteristically deficient in the sulphur containing amino acids, methionine and cystine.

The storage proteins in cereals occur primarily in the form of prolamines, representing 0.5 to 0.6 of the total protein in the endosperm.[17] Although the major cereal prolamines comprise of glutenins and gliadins, only the latter fraction has been implicated in coeliac disease.[10] The gliadins are single chain polypeptides rich in glutamine and proline but low in lysine and tryptophan, features which are largely responsible for the low nutritional value of cereals. The gliadin components of wheat comprise four major electrophoretic frac-

---

[7] M. F. Kagnoff, in 'Gastrointestinal Disease', ed. M. H. Sleisenger and J. S. Fordtran, W. B. Saunders Company, Philadelphia, 1983, p. 20.
[8] M. L. Clark, in 'Textbook of Gastroenterology', ed. I. A. D. Bouchier, R. N. Allan, H. J. F. Hodgson, and M. R. B. Keighley, Bailliere Tindall, London, 1984, p. 448.
[9] W. T. Cooke and G. K. T. Holmes, 'Coeliac Disease', Churchill Livingstone, Edinburgh, 1984.
[10] S. G. Cole and M. F. Kagnoff, *Ann. Rev. Nutr.*, 1985, **5**, 241.
[11] B. McNicholl, B. Egan-Mitchell, F. M. Stevens, P. F. Fottrell, and C. F. McCarthy, Proceedings of the XIIIth International Congress of Nutrition, Brighton, 1985, p. 752.
[12] M. D. Hellier, in 'Diseases of the Gut and Pancreas', ed. J. J. Misiewicz, R. E. Pounder, and C. W. Venables, Blackwell Scientific Publications, Oxford, 1987, p. 619.
[13] J. W. Sissons, H. E. Pedersen, and S. M. Thurston, *Proc. Nutr. Soc.*, 1984, **43**, 115A.
[14] L. M. J. Heppell, J. W. Sissons, and H. E. Pedersen, *Br. J. Nutr.*, 1987, **58**, 393.
[15] J. P. F. D'Mello, *World's Poult. Sci. J.*, 1988, **44**, 92.
[16] J. P. F. D'Mello, Proceedings of VIIIth European Poultry Conference, Barcelona, 1990, p. 302.
[17] M. A. Shotwell and B. A. Larkins, in 'The Biochemistry of Plants', ed. P. K. Stumpf and E. E. Conn, Academic Press, San Diego, 1989, p. 297.

tions: α-, β-, γ-, and ω-gliadins. There is considerable controversy over which of these fractions precipitates coeliac disease, although it is generally conceded that the α-gliadins are the primary activators.

Glycinin, β-conglycinin, and the gliadins are characterized by their resistance to denaturation by the conventional thermal methods employed in the processing of leguminous and cereal grains. It is also established that these proteins are resistant to digestive attack in the alimentary canal.[13]

3 Defence Mechanisms in the Gut

The major function of the gastrointestinal tract is to provide a site for the absorption of nutrients derived from the digestion of food. In carrying out these activities, the gut is exposed to a diverse array of ingested antigens, some of which inevitably escape denaturation to be absorbed and, subsequently, to evoke deleterious effects in susceptible animals. Consequently, the gastrointestinal system has evolved a second function, that of providing a defence mechanism against micro-organisms and dietary antigens.[18–20]

A diverse range of non-immune factors contributes to the defence system in the gut,[5,21] including digestive secretions, particularly acid produced in the stomach and the proteolytic activity of several enzymes. In addition, secretion of mucous by the goblet cells lining the intestinal tract exerts a protective role by inhibiting the attachment of antigens to the mucosal epithelium (Figure 1). Other factors contributing to the non-immune defence system are motility and constant renewal of the mucosal lining of the intestinal tract.

The principal defence mechanisms of the gut, however, reside in the extensive lymphoid tissue ramifying through the lamina propria of the intestine.[5,7,9,22] The lymphoid tissue, particularly that located in the Peyer's patches, is responsible for the initiation and expression of the humoral immune response which comprises the local production of two classes of immunoglobulins, IgA and IgM, derived from B cells (Figure 1). These antibodies are secreted into the lumen where they act to prevent absorption of food-borne and other antigens. Those antigens which penetrate the first lines of defence are capable, on absorption, of initiating an inflammatory reaction which eventually leads to a disruption of the cellular integrity and architecture of the intestinal mucosa.[5,7,10,21] The inflammatory response is induced under the agency of the cytophilic immunoglobulin, IgE which reacts with antigen causing degranulation of the mast cells to which it is attached with the release of vasoactive amines. In addition, IgG complement induced reactions may also be initiated leading to cellular damage in the intestinal mucosa. Finally, the arrival of antigen may activate the cell-mediated arm of the immune system by stimulating the proliferation of T lymphocytes[5] with the generation of cytotoxic T cells

---

[18] E. Bleumink, *Proc. Nutr. Soc.*, 1983, **42**, 219.
[19] A. Ferguson, *Proc. Nutr. Soc.*, 1985, **44**, 73.
[20] R. J. Levinsky, *Proc. Nutr. Soc.*, 1985, **44**, 81.
[21] F. M. Atkins and D. D. Metcalfe, *Ann. Rev. Nutr.*, 1984, **4**, 233.
[22] M. H. Lessof and P. D. Buisseret, in 'Immunological and Clinical Aspects of Allergy', ed. M. H. Lessof, M.T.P. Press, Lancaster, p. 141.

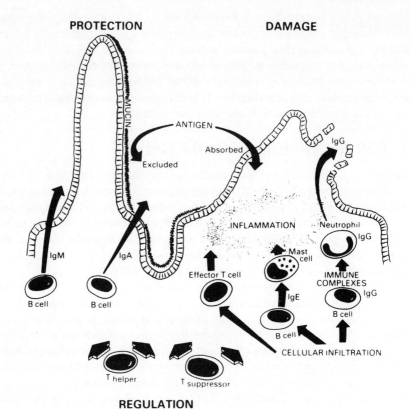

Figure 1 *Immune reactions in the intestine*
(Reproduced by permission of the publishers, Butterworth–Heinemann Ltd. ©, from P. Porter and M. E. J. Barratt, in 'Recent Advances in Animal Nutrition—1987', ed. W. Haresign and D. J. A. Cole, Butterworths, London, 1987, p. 107.)[5]

which damage or kill their targets in a highly selective manner, without antibody involvement, but in doing so exacerbate the IgE induced inflammation. Other T cells modulate the immune system by the induction of B cells to antibody-forming plasma cells or, alternatively, by suppressing humoral and cell-mediated immune reactions (Figure 1).

4 Adverse Effects of Food Antigens

There is now unequivocal evidence that food antigens precipitate extensive intestinal lesions and activate the immune system to such a degree that both growth and survival may be compromised (Table 1). The changes in intestinal morphology will be considered first since they provide supporting evidence for an immunological response to the absorption of antigenic proteins.

Intestinal Lesions

Striking changes occur in mucosal morphology on administration of antigenic proteins to sensitized animals and to patients with active coeliac disease. There is a marked degree of uniformity in these lesions (Table 1), irrespective of animal species or of the source of antigenic agent, although some differences are apparent. These similarities in the morphological response to dietary antigens have been interpreted by some authors[5,7,23] to signify a common underlying immunological aetiology.

Detailed analysis of the morphological changes show that while normal villi appear long and slender (Figures 2 and 3) with tall columnar epithelial cells,[23] those of sensitized calves[23-25] and pigs[26,27] are substantially shorter and broader with some evidence of disorganization of enterocyte architecture (Figures 4 and 5). For example, in one study,[25] reductions in mean villous heights of up to 517 μm were observed in the mid-jejunum of calves sensitized with soyabean antigens. Furthermore, villous atrophy was more prevalent in the mid-gut, a region where villous heights are normally at their greatest. Earlier studies[24] had shown that tips of villi may be damaged after sensitization of calves with heated soyabean flour, with extrusion of enterocytes and the discharge of red and white blood cells into the lumen. The surface of villi may be coated with a mesh of fibrin incorporating blood cells, bacteria, and undigested food residues. Coeliacs are often totally devoid of intestinal villi (Figure 6) with the epithelial cells assuming a cuboidal rather than columnar shape.[9,11] The brush border may be poorly developed or absent altogether in untreated patients, while the microvilli are generally stunted and several may be joined at their bases to present a branched appearance.

A consistent feature accompanying villous atrophy is the marked increase in crypt depth following antigen stimulation. For example, in sensitized calves[25] duodenal crypt depth was increased by 251 μm and similar effects were observed elsewhere in the small intestine and in the early-weaned piglet.[26] In patients with untreated coeliac disease, crypts in the jejunum also show marked hypertrophy and many appear to be branched.[9]

In normal villi, some lymphocytes are usually observed between cells of the mucosal epithelium, and furthermore, lymphocytes, plasma cells, and eosinophils may occur in moderate numbers in the lamina propria.[23] In contrast, villi from sensitized animals are distinguished by prominent infiltration with lymphocytes which may extend into the lamina propria.[23] Increased numbers of mast cells in the lamina propria of calves challenged with soyabean antigens are occasionally seen.[25]

Soyabean protein antigens may also elicit marked changes in jejunal histo-

[23] M. E. J. Barratt, P. J. Strachan, and P. Porter, *Clin. Exp. Immunol.*, 1978, **31**, 305.
[24] J. W. Sissons, H. E. Pedersen, and K. Wells, *Proc. Nutr. Soc.*, 1984, **43**, 113A.
[25] H. E. Pedersen, J. W. Sissons, A. Turvey, and I. Sondergaard, *Proc. Nutr. Soc.*, 1984, **43**, 114A.
[26] B. Ratcliffe, M. W. Smith, B. G. Miller, P. S. James, and F. J. Bourne, *J. Agric. Sci.*, 1989, **112**, 123.
[27] D. F. Li, J. L. Nelssen, P. G. Reddy, F. Blecha, J. D. Hancock, G. L. Allee, R. D. Goodband, and R. D. Klemm, *J. Anim. Sci.*, 1990, **68**, 1790.

Table 1 Adverse Effects of Soyabean and Gluten Antigens

Adverse effects	Soyabean hypersensitivity			Coeliac disease (Gluten-sensitive enteropathy) Human[h,i]
	Calf[a,b,c]	Pig[d,e,f]	Human[g]	
Intestinal lesions	Villi short and broad with extrusion of enterocytes	Atrophy of villus	Villi absent	Villi absent; epithelial cells cuboidal in shape; microvilli stunted with several joined at their bases
	Crypt depth increased	Crypt hyperplasia	Crypt hypertrophy	Crypt hypertrophy
	Lymphocyte infiltration into villi and lamina propria		Oedema and polymorphonuclear leucocyte infiltration of lamina propria	Increased number of intra-epithelial lymphocytes in jejunal mucosa
	Increased number of mast cells in lamina propria			Increased plasma cells in lamina propria
				Enlarged mesenteric lymph nodes
Immunological reactions	Elevated serum IgG and IgE concentrations	Synthesis of anti-soyabean antibody (IgG)		Increased synthesis of IgG (anti-gliadin) antibodies, IgA and IgM
	Cell-mediated reactions	Local production of IgA		Delayed hypersensitivity to recall antigens; impaired lymphocyte transformation; inhibition of leucocyte migration
		Cutaneous inflammation		

Digestive abnormalities	Impaired xylose uptake Decreased net absorption of nitrogen	Transient depression of xylose uptake Decreased digestibility of soyabean protein Reduced lactase activities		Impaired xylose uptake Persistently low lactase levels
Whole-body response	Diarrhoea Reduced growth Increased mortality	Diarrhoea (transient) Reduced growth Increased mortality	Diarrhoea	Diarrhoea Growth retardation Mortality exacerbated by malignant disease of intestine

[a] J. W. Sissons, H. E. Pedersen, and K. Wells, *Proc. Nutr. Soc.*, 1984, **43**, 112A.
[b] M. E. J. Barratt, P. J. Strachan, and P. Porter, *Clin. Exp. Immunol.*, 1978, **31**, 305.
[c] F. J. Seegraber and J. L. Morrill, *J. Dairy Sci.*, 1979, **62**, 972.
[d] D. F. Li, J. L. Nelssen, P. G. Reddy, F. Blecha, J. D. Hancock, G. L. Allee, R. D. Goodband, and R. D. Klemm, *J. Anim. Sci.*, 1990, **68**, 1790.
[e] A. D. Wilson, C. R. Stokes, and F. J. Bourne, *Res. Vet. Sci.*, 1989, **46**, 180.
[f] B. G. Miller, A. D. Phillips, T. J. Newby, C. R. Stokes, and F. J. Bourne, *Proc. Nutr. Soc.*, 1984, **43**, 116A.
[g] M. E. Ament and C. E. Rubin, *Gastroenterology*, 1972, **62**, 227.
[h] W. T. Cooke and G. K. T. Holmes, 'Coeliac Disease', Churchill Livingstone, Edinburgh, 1984.
[i] B. McNicholl, B. Egan-Mitchell, F. M. Stevens, P. F. Fottrell, and C. F. McCarthy, Proceedings of the XIIIth International Congress of Nutrition, Brighton, 1985, p. 752.

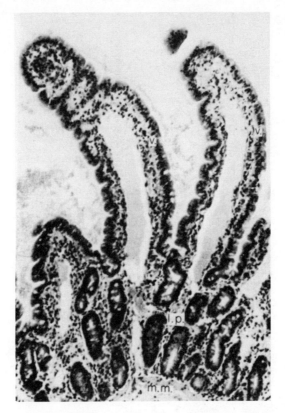

Figure 2 *Section of normal ileum of calf showing villi (v), lamina propria (l.p.), and muscularis mucosae (m.m.)*
(Reproduced by permission from *Clin. Exp. Immunol.*, 1978, **31**, 305.)[23]

logy in susceptible human subjects.[28] In such cases, oedema and polymorphonuclear leucocyte infiltration of the lamina propria occurs within 12 hours of antigenic challenge. Although some remission occurs 24 hours after challenge, the infiltration of the lamina propria with plasma cells and lymphocytes is still conspicuous.

Broadly similar changes are observed in human subjects with active coeliac disease.[7,9] Thus, there is evidence of lymphocyte infiltration of intestinal epithelium and increased cellularity of the lamina propria, attributable mainly to the proliferation of plasma cells.

Immunological Derangement

Hypersensitivity in the Calf

Early indications of the involvement of the immune system in antigen challenge arose from observations in the calf that the adverse effects occurred only after

[28] M. E. Ament and C. E. Rubin, *Gastroenterology*, 1972, **62**, 227.

Figure 3 *Scanning electron micrograph of normal duodenum of milk-fed piglet*
(Reproduced by permission from © *J. Anim. Soc.*, 1990, **68**, 1790.)[27]

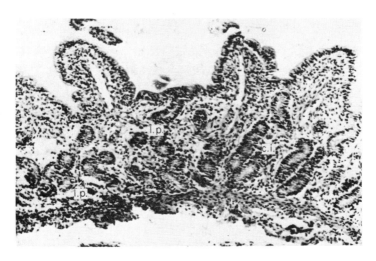

Figure 4 *Lesions in ileum of calf fed a diet containing heated, ethanol-extracted soyabean meal showing villous atrophy and cellular infiltration (c.i.) in lamina propria (l.p.)*
(Reproduced by permission from *Clin. Exp. Immunol.*, 1978, **31**, 305.)[23]

Figure 5 *Scanning electron micrograph of duodenum of soyabean-fed piglet showing damage to villi*
(Reproduced by permission from © *J. Anim. Sci.*, 1990, **68**, 1790.)[27]

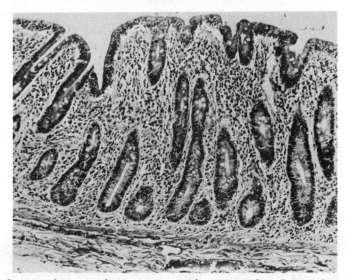

Figure 6 *Section of jejenum from a patient with active coeliac disease showing crypt hypertrophy and absence of villi*
(Reproduced by permission from W. T. Cooke and G. K. T. Holmes, 'Coeliac Disease', Churchill Livingstone, Edinburgh, 1984.)[9]

administration of successive feeds containing defatted soyabean products.[6] This implied that calves were sensitized by the initial feeds with the onset of overt digestive disturbances ensuing after subsequent feeds. Such a response is consistent with the classical pattern of antibody synthesis. Following first administration of a novel antigen, there is invariably a lag phase after which time antibodies may be synthesized in concentrations which increase with time up to a maximal level before receding. Presentation of the antigen on a subsequent occasion elicits a secondary response which is characterized by a much reduced lag phase, followed by a significantly enhanced rate of antibody production compared with the primary response. Furthermore, the duration of antibody synthesis is more prolonged in the secondary phase.

Confirmation of the involvement of the immune system in soyabean allergy in the calf arose from direct observations linking elevated serum IgG concentrations with the specific antigens, namely glycinin and β-conglycinin.[29] It has been suggested that local immuno-deficiency of the digestive trace predisposes the calf to allergic reactions.

Although the evidence for the participation of the immune system in calf hypersensitivity is now unequivocal, it is still not clear which of the immune reactions precipitates the gut lesions. Barratt et al.[23] proposed the intervention of reactions which involve the formation of complexes between soyabean antigens and complement fixing IgG antibodies resulting in increased capillary permeability, release of lysosomal enzymes, and tissue damage. However, anaphylactic reactions may also occur whereby antigen combines with reaginic (IgE) antibodies at the surface of mast cells, releasing an array of substances such as histamine, kinins, leukotrienes, and prostaglandins which evoke the inflammatory response.[6] This proposition is supported by the detection of IgE antibodies in the serum of some calves sensitive to soyabean antigens.[29] Finally, there is limited evidence to indicate the intervention of cell-mediated immune response to soyabean antigens in the calf. This is based on observations of delayed hypersensitive reactions in sensitized animals.[30]

Transient Hypersensitivity in Piglets

Early-weaned piglets also display allergic responses to soyabean antigens.[31] However, it is claimed that this intolerance is ephemeral and adverse effects do not persist providing that the period of hypersensitivity is not associated with infection by opportunistic micro-organisms. As in the calf, initial exposure of the piglet to soyabean antigens in the creep-feeding phase primes the animal such that subsequent feeding of soyabean at the weaning stage provokes a series of intestinal lesions and disorders.[4] Typical manifestations include morphological aberration of mucosal architecture, impaired nutrient absorption, and predisposition to diarrhoea.[26,31] These events implicate the immune system in

---

[29] P. J. Kilshaw and J. W. Sissons, Res. Vet. Sci., 1979, 27, 361.
[30] P. J. Kilshaw and H. Slade, Res. Vet. Sci., 1982, 33, 305.
[31] B. G. Miller, A. D. Phillips, T. J. Newby, C. R. Stokes, and F. J. Bourne, Proc. Nutr. Soc., 1984, 43, 116A.

piglets, as in calves, and there is direct evidence in support of this contention. Thus, piglets orally sensitized with soyabean protein show higher serum IgG titres than milk-fed controls.[27] In addition, an antigen–antibody association has recently been observed in young pigs.[32] Animals weaned at three weeks of age absorb glycinin from ingested soyabean, the quantity absorbed declining with time. This fall is associated with increasing serum titres of antibody specific to the soyabean antigens. The decreased absorption of antigen has been attributed to local production of IgA in the lamina propria which occurred only in pigs fed soyabean diets. In pigs weaned at ten weeks of age, antigen absorption and antibody production were again demonstrated on feeding soyabean flour. As before, absorption of antigen declined with time, but this reduction was specific for soyabean antigen as these pigs were still capable of absorbing another novel protein in the form of ovalbumin. At six months of age pigs no longer absorbed soyabean antigens when transferred from a soyabean-free diet to one containing high proportions of this protein source.

The immune response is no less striking in the gnotobiotic piglet[26] with one report of a six-fold increase in anti-soyabean IgG concentrations in serum on feeding soyabean protein to sensitized animals. Studies by Thiry–Vella loop perfusion in piglets indicated that soyabean protein solutions induced detectable quantities of antibody in the intestinal secretions six days after local antigenic stimulation.[23] The accompanying adverse effects on flow rates through the intestinal loop occurred only after prior sensitization with soyabean antigens.

The intervention of cell-mediated immune reactions has also been implicated[31] in soyabean-induced hypersensitivity in piglets. Marked increases in ear thickness have been recorded on intradermal administration of a soyabean protein extract, coinciding with a period of malabsorption. The cutaneous inflammation implies that intestinal responses to soyabean may originate, at least in part, from delayed cell-mediated reactions to antigens present in the legume.

Soyabean Intolerance in Other Species

Hypersensitivity to soyabean antigens is not restricted to calves and piglets. There are cases of soyabean allergies with associated immunological dysfunction in human subjects.[28] Even avian species show increased immunoglobulin titres and delayed hypersensitivity when challenged with isolated soyabean protein at four and eight days of age and subsequently placed on diets containing soyabean meal.[33] Furthermore, the immune response to initial sensitization persisted even in chicks subsequently fed on a control diet devoid of soyabean antigens.

---

[32] A. D. Wilson, C. R. Stokes, and F. J. Bourne, *Res. Vet. Sci.*, 1989, **46**, 180.
[33] K. C. Klasing, P. M. Maynard, and D. E. Laurin, *Poult. Sci.*, 1988, **67**, 104.

Coeliac Disease

It is widely accepted that the intestinal response in coeliac disease is associated with immune reactions to the gliadin fractions of cereals rather than a direct toxic effect of these proteins.[7-12] Evidence for the pre-eminent role of the immune system in coeliac disease emanates from both indirect and direct assessments. The enlarged mesenteric lymph nodes seen in patients with the active form of the disease is usually interpreted as a local immunological reaction to gluten fractions since peripheral lymph nodes are unaffected.[12] In untreated subjects, the increased number of immunocompetent cells in the form of intraepithelial lymphocytes in jejunal mucosa and their subsequent reduction to near-normal values following the imposition of a gluten-free dietary regime is again indicative of immunoreactive intervention in coeliac disease. Furthermore, increased proliferation of these cells can be induced in coeliacs maintained on a gluten-free diet, by a single gluten challenge, or by more protracted exposure to gluten antigens. Increased numbers of plasma cells in the lamina propria of jejunal mucosa of coeliacs has also been observed. However on imposition of a gluten-free dietary regime, plasma cell counts decline, but not to normal values.[9]

Direct assessments indicate complicity of both the humoral and cellular arms of the immune system in the activation of coeliac disease. Thus increased mucosal antibody synthesis coupled with enhanced antibody titres have been observed in patients with the active disease.[7-12] The IgG anti-gliadin antibody response may be a reaction to any one or more of the four fractions of gliadin. Antibodies of the IgA and IgM classes are also enhanced in coeliacs.[12] These antibodies tend to decline after the imposition of a gluten-free regime, although there is evidence that anti-gliadin IgG and to a lesser extent IgA antibodies can persist for protracted periods of time long after overt clinical manifestations have ceased. While the overwhelming weight of evidence points to a humoral involvement in the pathogenesis of coeliac disease, the intervention of cell-mediated immune reactions should not be ignored. Indeed, it has been postulated that a cell-mediated cytotoxic mechanism may contribute to the intestinal lesions in coeliac disease. Moreover, delayed hypersensitivity to recall antigens, impaired lymphocyte transformation, and inhibition of leucocyte migration by lymphokines are all suggested as being indicative of cell-mediated immune reactions.[12]

Dysfunction of the Digestive System

Malabsorption

The extensive morphological lesions which occur in the small intestine of sensitized animals and coeliacs significantly impair digestive and absorptive function by virtue of markedly reduced surface area in the gut, through abnormalities in the maturation of enterocytes and through reductions in the secretion of key enzymes such as lactase.

Absorptive competence in abnormal states is widely assessed in animals[31,34] and man[35] by xylose uptake measurements. In one such study, xylose absorption, as reflected in plasma xylose concentrations, was markedly higher in milk-fed calves than in their soyabean-fed counterparts with these differences diverging with age.[34] Xylose uptake increased in milk-fed calves over a five-week period but not in animals fed soyabean products.

The effects of soyabean flour on xylose uptake by piglets[31] is no less striking with a marked depression in plasma xylose concentration occurring five days post-weaning. However, in contrast to the response in calves, plasma xylose concentration in piglets is restored to normal by the eleventh day post-weaning. Thus in piglets the immune hypersensitivity to soyabean antigens is associated with a transient period of malabsorption.

Direct measurements of nutrient uptake confirm the abnormalities implied by the xylose tests. For example, calves orally sensitized with soyabean flour show a precipitous decline in net nitrogen absorption efficiency to 0.25 as compared with 0.57 for unsensitized and 0.85 for casein-fed calves.[6] It has long been recognized that the protein fraction of soyabean meal is consistently less digestible by the young pig in comparison with milk proteins.[36] While this difference may be attributed to limited, age-dependent production of certain digestive secretions,[37,38] the adverse effects of soyabean antigens on nutrient absorption should not be discounted.

Weaning on to soyabean diets elicits striking changes in lactase activities of porcine enterocytes.[26] In the normal course of events in unweaned piglets or in those weaned on to milk-based diets, lactase activity in enterocytes emerging from the intestinal crypt is low, but activity increases as enterocytes migrate along the length of the villus with a decline as migration is completed. In piglets weaned on to soyabean protein diets, lactase activities follow a similar pattern, but absolute activities are markedly lower. These reductions, which have been observed in conventionally reared[39] and gnotobiotic[26] piglets, are attributed to soyabean antigens through their effects on the intestinal epithelium.

Changes in brush-border hydrolase activities are also seen in children with coeliac disease. Mucosal disaccharidase activities of untreated subjects are invariably reduced, but the imposition of a gluten-free diet restores enzyme levels to normal values with the exception of lactase.[9]

Gut Motility

Calves given a series of liquid feeds containing heated soyabean flour readily develop abnormalities in movement of digesta.[40] An early manifestation of such disorders is an inhibition of the rate of abomasal evacuation which develops

[34] F. J. Seegraber and J. L. Morrill, *J. Dairy Sci.*, 1979, **62**, 972.
[35] P. A. Christiansen, J. B. Kirsner, and J. Ablaza, *Amer. J. Med.*, 1959, **27**, 443.
[36] G. E. Combs, F. L. Osegueda, H. D. Wallace, and C. B. Ammerman, *J. Anim. Sci.*, 1963, **22**, 396.
[37] D. J. Hampson, *Res. Vet. Sci.*, 1986, **40**, 313.
[38] D. Kelly, J. A. Smyth, and K. J. McCracken, *Res. Vet. Sci.*, 1990, **48**, 350.
[39] B. G. Miller, P. S. James, M. W. Smith, and F. J. Bourne, *J. Agric. Sci.*, 1986, **107**, 579.
[40] J. W. Sissons and R. H. Smith, *Br. J. Nutr.*, 1976, **36**, 421.

progressively with successive feeds containing soyabean antigens. Accompanying these changes are decreased transit time of digesta through the small intestine, increased flow of water and certain mineral ions in the terminal ileum, and decreased absorption of nitrogen. Such conditions predispose the calf to diarrhoea by virtue of osmotic effects exerted by unabsorbed nutrients in the intestinal lumen. The inflammatory reactions in the mucosal tissue are also contributory factors in the aetiology of diarrhoea.[6]

Piglets fed diets containing heated soyabean flour at weaning also readily develop enteritis but this is of a transient nature and, furthermore, may be prevented by ensuring that piglets consume sufficient antigen in the creep feed to induce 'tolerance' rather than sensitization.[4,31]

Some of the classical symptoms of coeliac disease include the incidence of diarrhoea[8,11] in 70 to 80 % of patients which may be accompanied by abdominal distention and discomfort. Diarrhoea may be transient, occasionally alternating with constipation.

Whole-body Responses

The incidence of diarrhoea in animals[31,41] and human subjects[11] sensitive to food antigens is invariably attended with some mortality. In coeliacs, an alarming feature is the increased mortality arising from malignant diseases of the intestine,[11] for which some remission may be secured by the imposition of, and strict adherence to, a gluten-free regime.

Although the transient nature of immune hypersensitivity in pigs has been emphasized in some studies,[4,31] it is salutary to note that the effects on growth and efficiency of food utilization can be more permanent. Indeed, it is now universally accepted[42,43] that heated soyabean flour and conventionally processed soyabean meal elicit disappointing growth rates and efficiency of food conversion unless the products undergo further treatments to inactivate or reduce the offending antigens.

5 Inactivation of Antigenic Proteins

Analytical Methods

As a prerequisite for the development of any treatments to inactivate antigenic proteins it is necessary to devise assays for the quantitative analysis of these anti-nutritional factors. The concentrations of glycinin and β-conglycinin present in saline extracts of several types of soyabean products have been determined by haemagglutination inhibition tests and by immunoelectrophoresis.[44] Levels of antigens determined by these methods correlate well with the

[41] A. L. Grant, J. W. Thomas, K. J. King, and J. S. Liesman, *J. Anim. Sci.*, 1990, **68**, 363.
[42] A. M. Lennon, H. A. Ramsey, W. L. Alsmeyer, A. J. Clawson, and E. R. Barrick, *J. Anim. Sci.*, 1971, **33**, 514.
[43] W. Eeckhout and M. De Paepe, *Anim. Feed Sci. Technol.*, 1989, **24**, 1.
[44] J. W. Sissons, A. Nyrup, P. J. Kilshaw, and R. H. Smith, *J. Sci. Food Agric.*, 1982, **33**, 706.

incidence of gastrointestinal disorders caused by feeding the different soyabean products to calves. More recently,[45] an enzyme-linked immunosorbent assay (ELISA) has been used for the quantitative measurement of soyabean proteins in processed food products.

The gliadins of wheat are generally recognized as being soluble in ethanol at room temperature whereas the glutinins are insoluble. The soluble components may then be subjected to gel electrophoresis to separate the four gliadin fractions.[17] Alternatively, enzyme immunoassay may be used for the determination of gliadin in food. In a recent study[46] employing this methodology, gliadin concentrations for wheat flour, corn flour, and rice were found to be 57, 0.47, and 0.46 g kg^{-1}, respectively, while two samples of gluten-free flour[11] yielded values of 0.35 and 0.50 g kg^{-1}.

Denaturation of Soyabean Antigens

Conventional heat processing techniques, while adequate to inactivate the thermolabile protease inhibitors present in soyabean, are patently ineffective in reducing the antigenicity of glycinin and β-conglycinin. Consequently, additional treatments are necesssary to minimize the risk of hypersensitive reactions and diarrhoea in calves and pigs.

Inactivation of antigens by the use of hot aqueous ethanol extraction has proved successful as a means of detoxifying soyabean products for the calf. [44,47] It is claimed that this treatment achieves detoxification by altering protein structure rather than by the removal of osmotically active components such as sucrose and the oligosaccharides. Denaturation of soyabean antigens is crucially dependent upon alcohol concentrations and temperature. Maximum efficacy is achieved with proportions of 0.65 to 0.70 ethanol at 78 °C. Lack of adherence to these factors may explain why some batches of alcohol extracted soyabean protein concentrate still retain their antigenic properties.[23] Although this extraction removes antigenicity and prevents gut inflammatory reactions in previously sensitized calves, treated soyabean products are still inferior to milk proteins.[48] This discrepancy has been explained on the basis of exposure of novel immunoreactive protein structures following treatment with ethanol.[6] However, it is equally likely that differences in clotting properties and in amino acid composition and availability may contribute to the relatively poor nutritional value of treated soyabean products.

Ethanol extraction is not an effective treatment for enhancing the nutritive value of soyabean protein for early weaning of piglets. Thus in one study[49] the incidence of diarrhoea and mortality was increased markedly on replacing dietary skim-milk with an ethanol-extracted soyabean concentrate.

Alkali treatment offers an alternative strategy for the denaturation of

[45] K. Yasumoto, M. Sudo, and T. Suzuki, *J. Sci. Food Agric.*, 1990, **50**, 377.
[46] D. F. McKillop, J. P. Gosling, F. M. Stevens, and P. F. Fottrell, *Biochem. Soc. Trans.*, 1985, **13**, 486.
[47] J. W. Sissons, R. H. Smith, D. Hewitt, and A. Nyrup, *Br. J. Nutr.*, 1982, **47**, 311.
[48] I. J. F. Stobo, P. Ganderton, and H. Connors, *Anim. Prod.*, 1983, **36**, 512.
[49] M. J. Newport and H. D. Keal, *Br. J. Nutr.*, 1982, **48**, 89.

antigens in soyabean products destined for piglet diets. It has been known for at least 20 years[42] that the nutritive value of soyabean for early-weaned pigs is greatly enhanced by treatment for 0.5 h with sodium hydroxide at pH 10.6 and at a temperature of 37 °C. For example, in one experiment daily weight gains and efficiency of food utilization were as low as 0.15 kg and 0.48 respectively for piglets fed conventionally processed soyabean meal. However, on treatment of the soyabean meal with alkali, values of 0.24 kg and 0.65 respectively were recorded.[42] More recent data[43] serve to corroborate these findings and further demonstrate that alkali treatment improves retention of nitrogen by early-weaned piglets. In the case of young calves, prior treatment of soyabean flour with alkali or with acid promotes higher weight gain than diets containing untreated flour.[50] Young chicks also respond favourably to diets containing alkali-treated seeds of *Canavalia ensiformis*.[51] Whether the effects of alkali or acid are mediated through inactivation of the antigenic proteins in soyabean and *C. ensiformis*[52] remains to be established, but this seems to be the most appropriate explanation at the present time.

Reductions of soyabean antigens have been achieved by other means,[53] including fermentation with rumen fluid and moist extrusion at elevated temperatures. The effectiveness of these treated products as components of milk replacers and early-weaning diets remains to be established.

The treatments discussed thus far are all essentially of an experimental nature. However, at least two soyabean products are now available for commercial use in calf and pig diets.[54] The products are guaranteed free of antigens, although the exact extraction technique used in the processing procedure remains undisclosed.

6 Dietary Management

A variety of dietary measures are available to mitigate hypersensitivity in animals and man. The reduced mucosal growth observed in calves[55] and piglets[41] sensitized with soyabean proteins can be restored to normal by dietary additions of the polyamines, putrescine and ethylamine. Xylose absorption is also improved but not to that observed in milk-fed animals. These responses are consistent with the putative role of the polyamines in mucosal growth, development, and maturation. However, since there was some evidence of diarrhoea in piglets, the value of such supplements should be considered further to include an economic appraisal of the potential benefits.

An alternative strategy may be to ensure adequate intakes of food in the creep feeding phase to induce 'tolerance' rather than sensitization in piglets.[4] A more

---

[50] B. M. Colvin and H. A. Ramsey, *J. Dairy Sci.*, 1969, **52**, 270.
[51] J. P. F. D'Mello and A. G. Walker, *Anim. Feed Sci. Technol.*, 1991, **33**, 117.
[52] S. C. Smith, S. Johnson, J. Andrews, and A. McPherson, *Plant Physiol.*, 1982, **70**, 1199.
[53] P. S. Mir, J. H. Burton, B. N. Wilkie, and F. R. van De Voort, *Can. J. Anim. Sci.*, 1989, **69**, 727.
[54] Technical Brochure on Soycomil®, The Antigen Story, Loders Croklaan B.V., P.O. Box 4, 1520 AA Wormerveer, Holland.
[55] A. L. Grant, R. E. Holland, J. W. Thomas, K. J. King, and J. S. Liesman, *J. Nutr.*, 1989, **119**, 1034.

viable approach might be to reduce intake of antigens by lowering overall protein levels of diets fed at weaning in conjunction with increased use of crystalline amino acids. This course of action would achieve a further objective, that of more efficient utilization of both protein and added amino acids.[15,16]

With regard to the treatment of patients with coeliac disease, the primary, if not only, strategy involves the imposition of a gluten-free dietary regime.[9] This procedure, more often than not, leads to clinical and mucosal remission within a relatively short period of time. The foods responsible for the activation of coeliac disease are now well recognized[46] but adherence to a stringent gluten-free regime is made more difficult by the ubiquitous distribution of cereal antigens in processed foods.

7 Conclusions

There is considerable evidence linking the ingestion of certain antigenic proteins with the induction of immune hypersensitive reactions in mammals. Soyabean products, whether raw or processed by conventional techniques, contain two antigens in the form of glycinin and β-conglycinin. By virtue of their refractory nature, these proteins are capable of surviving the digestive processes in the gut and, on absorption, of provoking extensive local and systemic immunological reactions together with severe intestinal lesions. Similar disorders are seen in coeliac disease (gluten-sensitive enteropathy) of humans when susceptible individuals consume wheat proteins of the gliadin fraction. The similarities among antigen-induced disorders of animals and man are striking (Table 1). However, considerable variation is often observed in these manifestations, with some reports[38,56] of animals being unaffected by the intake of soyabean and other food antigens. Lack of effect may be attributed to previous immunological and nutritional status of both neonate and mother. In the case of coeliac disease there is compelling evidence that genetic factors[10,11] may exert a significant role in determining the aetiology of the disease.

An outstanding issue requiring resolution remains that of establishing cause and effect in hypersensitivity responses to specific food antigens. The diverse humoral and cell-mediated immune reactions which occur in sensitized neonates and in human coeliacs, suggest an immunological aetiology in these disorders. Indeed, it is widely held that the effects on intestinal morphology and function are secondary manifestations of an underlying immunological disorder.[5,7,23] Nevertheless, there is a groundswell of opinion[9,11,12] supporting the view that the immunological reactions merely reflect the entry of antigens through a perforated mucosal barrier thereby securing access to and activating the immune system. For example, Cooke and Holmes[9] maintain that, in coeliac disease, there is no evidence that immunological reactions precipitate mucosal damage, but once initiated, they may well sustain the lesion. Thus the primary mechanism whereby food antigens induce intestinal damage remains an issue of considerable debate.

Although remission of adverse effects on mucosal morphology of the

[56] V. R. Fowler and D. Fraser, *Anim. Prod.*, 1985, **40**, 547.

intestine occurs promptly on withdrawal of antigen, a disquieting feature appears to be the persistence of residual effects in animals and man. It is believed[4] that following antigen challenge, the intestine of the neonate never returns to its pre-weaning condition. In coeliacs, the imposition of a gluten-free regime fails to restore all features of intestinal histology.[9] Thus the number of plasma cells in the lamina propria of jejunal mucosa remains above the normal range even after protracted abstention from gluten.

Experimentally, the inactivation of antigenic proteins may be accomplished by a variety of procedures[6] including treatment with alcohol, alkali, or acid. However, there are species differences in their responses to some treated products. In addition, it is possible that alcohol treatment may yield novel immunoreactive protein structures and that severe treatments of protein with alkali can lead to the synthesis of unusual toxic amino acids such as lysinoalanine.[57] While these procedures remain essentially, but not entirely, experimental, it is likely that dietary manipulation offers the best prospect for prevention of hypersensitivity in neonatal animals. Thus it has been advocated that adequate intakes of food, and therefore antigens, during the pre-weaning phase should induce tolerance rather than sensitization in piglets.[4] Alternatively, antigen intake may be reduced by minimizing dietary protein concentrations through increased use of crystalline amino acids. This approach would confer the added advantage of enhancing overall efficiency of protein utilization.

---

[57] P. A. Finot, *Nutr. Abstr. Rev.*, 1983, **53A**, 67.

CHAPTER 6

Glucosinolates

ALAN J. DUNCAN

1 Introduction

Glucosinolates are a group of plant thioglucosides found principally among members of the Cruciferae. Their presence in agricultural crop plants, such as oilseed rape and brassica forages, is important because of the potentially harmful effects of their breakdown products to livestock consuming such crops. Despite extensive research on the chemistry, biosynthesis, and taxonomic distribution of glucosinolates, studies on their toxicology have tended to lack physiological understanding by focusing primarily on their effects on agricultural production variables, such as growth and egg production. In consequence, knowledge of the physiological effects of specific breakdown products is fragmentary and few attempts have been made to synthesize findings. This review will concentrate therefore on work relating to the physiological effects and metabolic fate of glucosinolates and their breakdown products in animals. To set such a review in context, brief consideration will be given initially to the chemistry of glucosinolates, their distribution and biosynthesis and to their possible natural functions within the plant. Methods of glucosinolate analysis will also be briefly reviewed.

2 Structure, Distribution, and Adaptive Significance of Glucosinolates

The general structure of glucosinolates was elucidated by Ettlinger and Lundeen[1] as consisting of a thioglucose group linked, via a sulphonated oxime group, to an R-group (Scheme 1). A range of R-groups, approaching a hundred in number, have since been described. Among the aromatic representatives, indole groups are common, while among aliphatic glucosinolates a terminal unsaturated bond is a typical feature. The presence of methyl, thiol, and hydroxyl groups further increase the diversity of glucosinolates.

---

[1] M. G. Ettlinger and A. J. Lundeen, *J. Am. Chem. Soc.*, 1956, **78**, 4172.

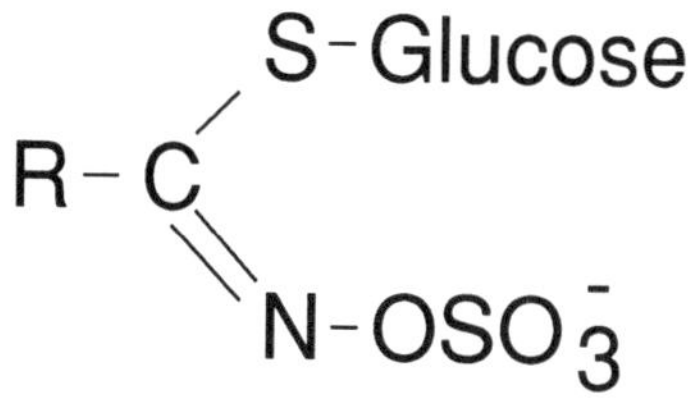

$$R-C\overset{\displaystyle S-Glucose}{\underset{\displaystyle N-OSO_3^-}{}}$$

Scheme 1 *Structure of glucosinolates*

Glucosinolates occur mainly in the order Capparales and principally in the families, Cruciferae, Resedaceae, and Capparidaceae, although their presence in other families has been reported.[2] Their agricultural importance lies in their occurrence among members of the genus *Brassica*, which have widespread use in both human and animal nutrition. Glucosinolate concentrations vary throughout the plant with seeds containing concentrations of glucosinolates approximately an order of magnitude higher than those in the vegetative tissues. *Brassica napus*, for example, may typically contain 100 mmol kg^{-1} DM in the seeds, while leaf concentrations of 10–20 mmol kg^{-1} DM are common. Different plant parts contain different glucosinolate concentrations and also have different profiles of individual glucosinolates.[3] The adaptive significance of such differences is unclear, although high glucosinolate concentrations tend to occur in rapidly growing plant parts such as shoot and root tips[4] and may be associated with protection against damage by herbivores.

The presence of glucosinolates has been used as a chemotaxonomic parameter[5] and in cultivar identification within species,[6] indicating the importance of genotype in determining glucosinolate profiles. A number of environmental factors may also affect glucosinolate profiles and concentrations; for example, increased concentrations of sulphate in soil solution increase glucosinolate concentrations in both leaves[7] and seeds[8,9] while increased soil nitrogen status has the opposite effect of reducing glucosinolate concentrations.[7,8,10] The effects of water availability on glucosinolate concentrations in crucifers are unclear with some studies indicating a negative correlation between water supply and glucosinolate concentration[11] and others showing no response.[10,12,13] Decreased light supply has also been shown to increase glucosinolate concentrations in the plant. Finally, herbivore damage may increase

---

[2] P. O. Larsen, in 'The Biochemistry of Plants, Vol. 7.' ed. P. K. Stumpf and E. E. Conn, Academic Press, London, 1981, p. 501.
[3] J. P. Sang, I. R. Minchinton, P. K. Johnstone, and R. J. W. Truscott, *Can. J. Plant Sci.*, 1983, **64**, 77.
[4] B. Uppstrom, *Sver. Utsaedesfoeren. Tidskr.*, 1983, **93**, 331.
[5] J. E. Rodman, *Phytochem. Bull.*, 1978, **11**, 6.
[6] H. Adams and J. G. Vaughan, *J. Sci. Food Agric.*, 1989, **46**, 319.
[7] J. L. Wolfson, *Environ. Entomol.*, 1982, **11**, 207.
[8] E. Joseffson, *J. Sci. Food Agric.*, 1970, **21**, 98.
[9] R. J. Mailer, *Aust. J. Agric. Res.*, 1989, **40**, 617.
[10] R. K. Heaney, E. A. Spinks, and G. R. Fenwick, *Z. Pflanzenzuecht.*, 1983, **91**, 219.
[11] R. J. Mailer and P. S. Cornish, *Aust. J. Agric. Res.*, 1987, **27**, 705.
[12] S. M. Louda and J. E. Rodman, *Biochem. Syst. Ecol.*, 1983, **11**, 199.
[13] M. J. Blua, Z. Hanscom, and B. D. Collier, *J. Chem. Ecol.*, 1988, **14**, 623.

glucosinolate concentrations presumably as an adaptive measure to prevent further herbivory.[14,15]

The function of glucosinolates in plant metabolism is not entirely clear. Their rapid turnover within the plant tissues[2] with the associated metabolic costs involved suggests an important adaptive function and there is increasing evidence that they serve to protect the host plant against herbivore damage in a manner analogous to other groups of plant secondary compounds.[16] Research has centred on the study of protection against insect herbivory in which the glucosinolate content of various plant species has been shown to affect larval development[7,17] and pupation,[7] and to influence the degree of herbivory of phytophagous insects.[12,18] Some insect species have evolved means of detoxifying glucosinolate breakdown products[19] and have become 'Brassica-specialists'. In some cases it has been found that glucosinolates are used by insects to locate appropriate food sources[20] and glucosinolates may also act as phago-stimulants.[21-23] Glucosinolates breakdown products have also been shown to have anti-fungal[24,25] and anti-microbial properties[26] and may thus have a role in pathogen resistance.

3 Biosynthesis

Glucosinolates are synthesized in the plant from amino acid precursors according to Scheme 2. The number of amino acids identified as precursors in glucosinolates biosynthesis is relatively small. For example, indole glucosinolates are produced from D,L-tryptophan[27] while benzyl glucosinolate and *p*-hydroxybenzyl glucosinolate are produced from phenyl alanine and tyrosine respectively.[28] Many of the aliphatic glucosinolates have methionine as their original precursor.[29] The diversity in the R-group of the glucosinolates is generated by two processes in the biosynthesis of glucosinolates. Firstly, the amino acid precursors may be modified prior to the initial hydroxylation step of glucosinolate biosynthesis by a process of chain lengthening analogous to that seen in the formation of leucine from valine. For example, prop-2-enyl glucosinolate is produced following chain-lengthening of methionine to homo-

[14] J. Lammerlink, D. B. MacGibbon, and A. R. Wallace, *N.Z. J. Agric. Res.*, 1984, **27**, 89.
[15] A. N. E. Birch, D. W. Griffiths, and W. H. MacFarlane Smith, *J. Sci. Food Agric.*, 1990, **51**, 309.
[16] W. J. Freeland and D. H. Janzen, *Am. Nat.*, 1974, **108**, 269.
[17] P. A. Blau, P. Feeny, L. Contardo, and D. S. Robson, *Science*, 1978, **200**, 1296.
[18] F. Klingauf, C. Sengonca, and H. Bennewitz, *Oecologia*, 1972, **9**, 53.
[19] E. W. Wadleigh and S. J. Yu, *J. Chem. Ecol.*, 1988, **14**, 1279.
[20] S. Finch, *Entomol. Exp. Appl.*, 1978, **24**, 350.
[21] L. R. Nault and W. E. Styer, *Entomol. Exp. Appl.*, 1972, **15**, 423.
[22] J. K. Nielsen, L. Dalgaard, L. M. Larsen, and H. Sorenssen, *Entomol. Exp. Appl.*, 1979, **25**, 227.
[23] J. K. Nielsen, *Entomol. Exp. Appl.*, 1989, **51**, 249.
[24] L. Drobnica, in 'Mechanisms of Action of Fungicides and Antibiotics', ed. M. Girbardt, Academie-Verlag, Berlin, 1977, p. 131.
[25] R. F. Mithen, B. G. Lewis, and G. R. Fenwick, *Trans. Br. Mycol. Soc.*, 1986, **87**, 433.
[26] T. Zsolnai, *Arzneim. Forsch.*, 1966, **16**, 870.
[27] M. Kutacek, Z. Prochazka, and K. Veres, *Nature (London)*, 1962, **194**, 393.
[28] E. W. Underhill, M. D. Chisholm, and L. R. Wetter, *Can. J. Biochem. Physiol.*, 1962, **14**, 1505.
[29] J. R. Glover, C. C. S. Chapple, S. Rothwell, I. Tober, and B. E. Ellis, *Phytochemistry*, 1988, **27**, 1345.

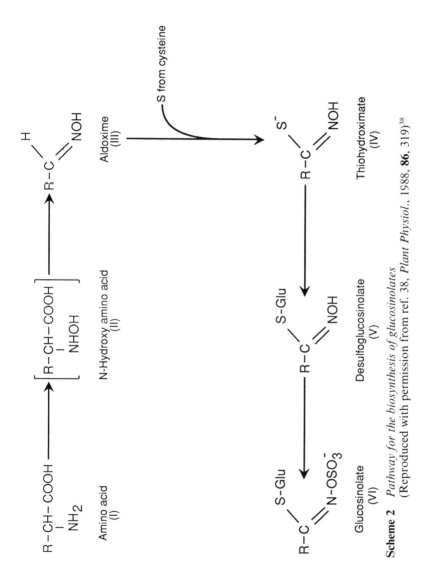

Scheme 2 *Pathway for the biosynthesis of glucosinolates* (Reproduced with permission from ref. 38, *Plant Physiol.*, 1988, **86**, 319)[38]

methionine.[30] Secondly, side chain modification may occur following the last stage of glucosinolate biosynthesis. For example, elimination of the methylsulphonyl group of methyl-sulphonyl-propyl glucosinolate yields prop-2-enyl glucosinolate.[31]

The intermediates of glucosinolate biosynthesis have been elucidated in tracer studies. Initial hydroxylation of the amino acid precursor[32] is followed by formation of an aldoxime intermediate.[33] Incorporation of a sulphur atom from cysteine results in formation of a thio-hydroxymate[34] which is then glycosylated to yield the desulpho-glucosinolate.[34] Sulphonation of the oxime group is the final step in glucosinolate biosynthesis. Details of the elucidation of glucosinolate biosynthesis are contained in specialist reviews.[35,36]

The control of glucosinolate biosynthesis is not clearly understood, although the enzymes catalysing a number of the steps have been characterized and shown to have activity in *in vitro* systems. A thiohydroximate glucosyltransferase enzyme, catalysing the glycosylation of phenyl-aceto-thiohydroximate to desulphobenzyl glucosinolate, has been purified.[37] The activity of the sulphotransferase catalysing the final step in benzyl glucosinolate biosynthesis has been measured in *Lepidium sativum* seedlings and shown to require 3'-phosphoadenosine-5'-phosphosulphate as the sulphur donor.[38] The distribution of putative biosynthetic enzymes has been suggested as having importance in determining glucosinolate profiles in different plant parts. Thus, the activity of an aminotransferase was found to be highly correlated with prop-2-enyl glucosinolate concentration in different parts of *Brassica carinata*, suggesting its involvement in chain lengthening of methionine at an early stage in prop-2-enyl glucosinolate biosynthesis.[29] Similarly, the distribution of enzymes catalysing the final glycosylation and sulphation steps in glucosinolate biosynthesis in *Brassica juncea* has been determined; the correlation of enzyme activity with ultimate glucosinolate concentrations implicated these enzymes in the control of glucosinolate biosynthesis.[39] Despite these recent developments, work on the control of glucosinolate biosynthesis is at an early stage and a clearer understanding of the mechanisms of control may facilitate artificial manipulation of glucosinolate concentrations. The potential for manipulating glucosinolate concentrations by artifical means has recently been demonstrated in a study which showed sulphur availability to be an important factor limiting glucosinolate biosynthesis. Infection of *Brassica campestris* plants with a cauliflower mosaic virus harbouring the gene for mammalian metallothionein synthesis suppressed glucosinolate synthesis; expression of the metallothionein

[30] E. W. Underhill, *Can. J. Biochem.*, 1965, **43**, 179.
[31] M. D. Chisolm and M. Matsuo, *Phytochemistry*, 1972, **11**, 203.
[32] H. Kindl and E. W. Underhill, *Phytochemistry*, 1968, **7**, 745.
[33] E. W. Underhill, *Eur. J. Biochem.*, 1967, **2**, 61.
[34] E. W. Underhill and L. R. Wetter, *Plant Physiol.*, 1969, **44**, 584.
[35] E. W. Underhill, L. R. Wetter, and M. D. Chisolm, *Biochem. Soc. Symp.*, 1973, **38**, 303.
[36] P. M. Dewick, *Nat. Prod. Rep.*, 1984, **1**, 545.
[37] M. Matsuo and E. W. Underhill, *Phytochemistry*, 1971, **10**, 2279.
[38] T. M. Glendening and J. E. Poulton, *Plant Physiol.*, 1988, **86**, 319.
[39] J. C. Jain, M. R. Michayluk, J. W. D. Grootwassink, and E. W. Underhill, *Plant Sci.*, 1989, **64**, 25.

gene created a 'sulphur sink' in the plant tissues and thus reduced sulphur availability for glucosinolate biosynthesis.[40]

4 Chemical Analysis

The analysis of glucosinolates has been recently reviewed[41,42] and only a broad overview of methods will be presented here to indicate the range of methods currently employed. The wide range of analytical methods applied to the glucosinolates reflects the diversity of the compounds themselves, especially following hydrolysis, as well as the diversity in the requirements of the analyst with respect to specificity and accuracy.

Glucosinolate analyses can be divided into those which distinguish between individual glucosinolates and those which measure total concentrations of glucosinolates, regardless of chemical identity.

Analysis of Individual Glucosinolates

Determination of the concentration of individual glucosinolates relies almost exclusively on chromatographic methods and advances in this field have led to huge improvements in resolution and quantification in the last thirty years. Early methods included paper chromatography and thin-layer chromatography of intact glucosinolates[43] and of the thiourea derivatives of isothiocyanates. Paper chromatography of the thiourea derivatives of isothiocyanates was used extensively in the chemo-taxonomic studies of Rodman.[5] Such methods are useful for identification of compounds present but are of limited value in quantitative analysis.

Gas–liquid chromatography (GLC) is a more useful approach to the quantitative analysis of glucosinolates. Original methods were based on the separation of hydrolysis products following either autolysis or controlled hydrolysis with exogenous myrosinase.[44–46] Gas–liquid chromatography is well suited to the separation of these compounds because of their volatile nature and capillary GLC studies have demonstrated the impressive resolution that is possible.[47] However the fact that such methods rely on measurement, not of the parent compounds themselves, but of their enzymic breakdown products, makes such techniques rather tedious and potentially confusing because of the many factors which may influence the course of enzymic hydrolysis (see following section). More recent methods have therefore concentrated on the analysis of the parent glucosinolates themselves, either by GLC or by HPLC. Gas–liquid chromatography methods for glucosinolate separation rely on prior derivatization with

[40] D. D. Lefebvre, *Plant Physiol.*, 1990, **93**, 522.
[41] O. Olsen and H. Sorensen, *J. Am. Oil Chem. Soc.*, 1981, **58**, 857.
[42] D. I. McGregor, W. J. Mullin, and G. R. Fenwick, *J. Assoc. Off. Anal. Chem.*, 1983, **66**, 825.
[43] A. Kjaer, *Fortschr. Chem. Org. Naturst.*, 1960, **18**, 122.
[44] M. E. Daxenbichler, G. F. Spencer, R. Kleiman, C. H. VanEtten, and I. A. Wolff, *Anal. Biochem.*, 1970, **38**, 373.
[45] M. E. Daxenbichler and C. H. VanEtten, *J. Assoc. Off. Anal. Chem.*, 1977, **60**, 950.
[46] H. A. Macleod, G. Benns, D. Lewis, and J. F. Lawrence, *J. Chromatogr.*, 1978, **157**, 285.
[47] K. Grob Jr. and P. Matile, *Phytochemistry*, 1980, **19**, 1789.

tri-methylsilane (TMS) to increase the volatility of the compounds[48] and unless the derivatization conditions are carefully controlled, certain indole glucosinolates may not be detected.[49,50] High performance liquid chromatography is increasingly the method of choice for determination of parent glucosinolates. Methods for direct separation of intact glucosinolates have been described[51,52] but HPLC separation of enzymically desulphated glucosinolates is a simpler and more robust technique.[53,54]

A recently reported enzyme-linked immuno-sorbent assay (ELISA) for determination of sinigrin[55] combines the specificity and accuracy of slower chromatographic techniques with the speed required for the screening of large numbers of samples and this may prove valuable in the context of plant breeding.

Total Glucosinolate Analyses

If quantification of individual glucosinolates is not required, simpler 'total glucosinolate' methods of glucosinolate analysis may be employed. These rely generally on detection of enzymically released glucose, although release of sulphate has also been used.[56] Glucose-release methods can be applied to both vegetative and seed material[57-60] by retention of glucosinolates on an ion-exchange resin, followed by washing and on-column hydrolysis. Glucose may then be detected in the eluent using a colorimetric method. Various methods for the direct determination of total glucosinolates without myrosinase hydrolysis have been described, including a tetra-chloropalladate complexing method[61] and a method based on the reaction of glucosinolates with thymol.[62] In addition, recent 'dry' methods of glucosinolate determination using X-ray fluorescence[63] and near infrared reflectance spectroscopy techniques[64] have been developed. These rely on the correlation between glucosinolate content and the total sulphur content of cruciferous material.

Clearly, techniques for the determination of glucosinolates have reached high levels of sophistication but considering their importance as toxic substances in food and livestock feeds, methods for detection of their hydrolysis products

---

[48] W. Thies, *Fette, Seifen, Anstrichm.*, 1976, **78**, 231.
[49] R. K. Heaney and G. R. Fenwick, *J. Sci. Food Agric.*, 1980, **31**, 593.
[50] R. K. Heaney and G. R. Fenwick, *J. Sci. Food Agric.*, 1982, **33**, 68.
[51] P. Helboe, O. Olsen, and H. Sorensen, *J. Chromatogr.*, 1980, **197**, 199.
[52] B. Bjorkvist and A. Hase, *J. Chromatogr.*, 1988, **435**, 501.
[53] I. Minchington, J. Sang, D. Burke, and R. J. W. Truscott, *J. Chromatogr.*, 1982, **247**, 141.
[54] E. A. Spinks, K. Sones, and G. R. Fenwick, *Fette, Seifen, Anstrichm.*, 1984, **86**, 228.
[55] F. Hassan, N. E. Rothnie, S. P. Yeung, and M. V. Palmer, *J. Agric. Food Chem.*, 1988, **36**, 398.
[56] E. Joseffson and L. A. Appelqvist, *J. Sci. Food Agric.*, 1968, **19**, 564.
[57] R. K. Heaney and G. R. Fenwick, *Z. Pflanzenzuecht.*, 1981, **87**, 89.
[58] R. K. Heaney, E. A. Spinks, and G. R. Fenwick, *Analyst (London)*, 1988, **113**, 1515.
[59] C. H. VanEtten, C. E. McGraw, and M. E. Daxenbichler, *J. Agric. Food Chem.*, 1974, **22**, 452.
[60] C. H. VanEtten and M. E. Daxenbichler *J. Assoc. Off. Anal. Chem.*, 1977, **60**, 946.
[61] W. Thies, *Fette, Seifen, Anstrichm.*, 1982, **84**, 338.
[62] J. T. Tholen, S. Shifeng, and R. J. W. Truscott, *J. Sci. Food Agric.*, 1989, **49**, 157.
[63] E. Schnug and S. Haneklaus, *J. Sci. Food Agric.*, 1988, **45**, 243.
[64] Z.-H. Yang, J.-H. Xin, Y.-M. Zhu, and Shi Xi-Kui, *Analyst (London)*, 1988, **113**, 355.

and their excretory metabolites in body fluids and tissues are scarce. In particular, measurements of glucosinolate hydrolytic products in digestive fluids are lacking and this is an important area for future technique development to facilitate a clearer understanding of the toxicity and metabolic fate of glucosinolates.

5 Enzymic Hydrolysis to Toxic Metabolites

An important feature of glucosinolate chemistry, and one which has been the subject of extensive study, is the enzymic hydrolysis of glucosinolates under the action of the thioglucosidase enzyme, myrosinase. A brief account of the main findings is given here, in order to set in context further consideration of glucosinolate toxicity.

Glucosinolates are always accompanied in plant tissue by the thioglucosidase enzyme, myrosinase, which catalyses the cleavage of the thioglucoside bond of glucosinolates. In the intact plant, enzyme and substrate occur in separate plant compartments, presumably as an adaptive measure to avoid auto-toxicity but, following cell disruption, enzyme and substrate come into contact. The resulting hydrolysis involves cleavage of the thioglucoside bond and yields free glucose and an aglucone intermediate, which undergoes spontaneous degradation to one of a number of toxic metabolites (Scheme 3). The most commonly cited breakdown products are the isothiocyanates and nitriles but, depending on conditions in the hydrolysis medium and the structure of the parent molecule, a number of other products may arise. For example, a β-OH group on the R-chain of progoitrin facilitates cyclization of the corresponding isothiocyanate to yield 5-vinyl-2-oxazolidinethione (5-OZT). Furthermore, isothiocyanates of the indole glucosinolates may undergo spontaneous degradation to yield molar quantities of thiocyanate ion. Protein co-factors may direct rearrangement of the aglucone intermediate to produce cyanoepithioalkanes at the expense of aliphatic nitriles.[65] Similarly, organic thiocyanates may be produced in certain cases instead of the more usual isothiocyanates.[66]

Of particular interest in relation to the toxicology of glucosinolates is the way in which hydrolysis is under the influence of conditions in the hydrolysis medium such as pH, temperature, metallic ion concentrations, and the presence of various protein co-factors. These interact in a complex way to determine the proportions of toxic products arising from hydrolysis and so may influence ultimate toxicity. Consideration of the factors affecting glucosinolate hydrolysis is important therefore to any discussion of glucosinolate toxicity.

Low pH tends to favour nitrile production, while neutral and high pH leads to isothiocyanate production (Figure 1).[67,68] Uda *et al.*[69] included an assessment of pH effects in a study of sinigrin hydrolysis *in vitro*. Increasing pH was found

---

[65] R. A. Cole, *Phytochemistry*, 1978, **17**, 1563.
[66] V. Gil and A. J. Macleod, *Phytochemistry*, 1980, **19**, 1369.
[67] V. Gil and A. J. Macleod, *Phytochemistry*, 1980, **19**, 2547.
[68] C. H. VanEtten, M. E. Daxenbichler, J. E. Peters, and H. L. Tookey, *J. Agric. Food Chem.*, 1966, **14**, 426.
[69] Y. Uda, T. Kurata, and N. Arakawa, *Agric. Biol. Chem.*, 1986, **50**, 2735.

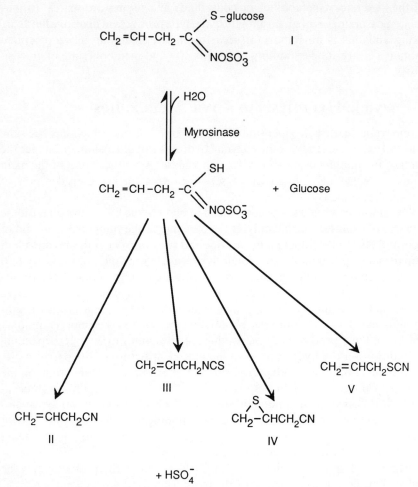

Scheme 3 *Formation of breakdown products following enzymatic hydrolysis of sinigrin* (Reproduced with permission from ref. 76, *J. Agric. Food Agric.*, 1989, **46**, 211)[76]

to increase the rate of glucose release up to a pH of 6.5 as well as favouring formation of allyl isothiocyanate at the expense of allyl cyanide. Thus, pH was shown to influence both enzymatic aglucone formation and the non-enzymatic rearrangement of the aglucone to form the final products. In a recent experiment, in which the effects of pH on the hydrolysis of indole glucosinolates in rapeseed were examined *in vitro*, low pH (< pH 3) resulted in significant production of indole acetonitriles while higher pH values gave rise to quantitative production of thiocyanate ion, presumably by way of isothiocyanate degradation.[70]

Temperature appears to have little direct effect on the proportions of glucosi-

[70] B. A. Slominski and L. D. Campbell, *J. Sci. Food Agric.*, 1989, **47**, 75.

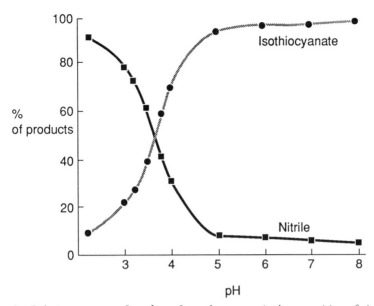

Figure 1 *Relative amounts of products formed on enzymic decomposition of sinigrin at different pH values*
(Reproduced with permission from ref. 67, *Phytochemistry*, 1980, **19**, 2547)[67]

nolate hydrolysis products[66] although indirect temperature effects have been noted, associated with the denaturation of certain heat-labile factors which are important in glucosinolate hydrolysis. The most well known of these is epithiospecifier protein (ESP), a heat-labile factor demonstrated by Tookey[71] to direct hydrolysis toward epithioalkane production in experiments with *Crambe* seed. The existence of ESP in a range of genera was later demonstrated.[72]

The presence of various other compounds in the hydrolysis medium is also known to influence glucosinolate breakdown. Ferrous ions are known to direct hydrolysis towards nitriles[73] and more recently a range of other metallic ions have been shown to have similar properties.[74] Thiol compounds, such as cysteine and glutathione, enhance the action of Fe^{2+} in favouring nitrile production.[75] Interactions also exist between the Fe^{2+} ion and pH in their influence on glucosinolate hydrolysis.[69] Although quantitative data were lacking, addition of mercaptoethanol to the hydrolysis medium was recently shown to favour hydrolysis of sinigrin to epithioalkanes and allyl cyanide at the expense of isothiocyanate production, with activation of epithiospecifier protein by the mercaptoethanol being proposed as the likely mechanism.[76]

Glucosinolate hydrolysis is clearly a complex process with the range of

[71] H. L. Tookey, *Can. J. Biochem.*, 1973, **51**, 1654.
[72] R. J. Petroski and H. L. Tookey, *Phytochemistry*, 1982, **21**, 1903.
[73] H. L. Tookey and I. A. Wolff, *Can. J. Biochem.*, 1970, **48**, 1024.
[74] L. M. Searle, K. Chamberlain, and D. N. Butcher, *J. Sci. Food Agric.*, 1983, **35**, 745.
[75] Y. Uda, T. Kurata, and N. Arakawa, *Agric. Biol. Chem.*, 1986, **50**, 2741.
[76] M. B. Springett and J. B. Adams, *J. Sci. Food Agric.*, 1989, **46**, 211.

hydrolysis products produced being highly dependent on the nature of the medium. This is important in any consideration of the potential toxic effects of glucosinolates since conditions in the digestive tract are likely to greatly influence which toxic products are produced following plant ingestion. An obvious example is the effect of gastric pH on glucosinolate hydrolysis; the acid conditions characteristic of the monogastric stomach provide a very different chemical environment to that of the ruminant fore-stomach. Comparative studies are likely to lead to an improvement in our understanding of glucosinolate toxicology.

6 Glucosinolate Toxicity

The effects of glucosinolates on the productive performance of animals have been studied mainly in the context of the feeding of glucosinolate-containing rapeseed to farm livestock. Rapeseed is grown primarily as a source of vegetable oil for human consumption but, following extraction of the oil, the protein-rich meal is a valuable source of nutrients for farm livestock. The main problem associated with the use of rapeseed meal is the presence of glucosinolates, which reduce the palatability of the meal and elicit toxic symptoms. Considerable progress has been made by plant breeders in reducing levels of glucosinolates in rapeseed and also of erucic acid, which reduces oil quality. The resulting 'double-low' rapeseed varieties show considerably improved nutritional characteristics. The effects of including rapeseed meals in livestock diets on productive characteristics such as growth, egg production, and milk production have been the subject of exhaustive review[77-80] and these aspects will not be re-examined in detail here. Work on rapeseed meal will be considered, however, where it provides understanding of the biochemical mechanisms underlying glucosinolate toxicity. The value and limitations of various approaches to glucosinolate toxicity assessment will also be discussed to highlight areas in which future work would be valuable.

The simplest means of assessing the toxicity of glucosinolates is to correlate toxicity with the glucosinolate content of the diet. This approach has been particularly used in the evaluation of the new 'double-low' rapeseed varieties.[81-83] The studies are characterized by their emphasis on gross productive characteristics with little attention to the elucidation of the toxic action of the glucosinolates, apart from measurement of organ weights.[84] While useful from an agricultural standpoint in helping to determine appropriate dietary inclusion levels for rapeseed meal, these studies provide little insight into the actual mechanisms of toxicity of glucosinolates.

[77] R. Hill, *Br. Vet. J.*, 1979, **135**, 3.
[78] D. R. Clandinin and A. R. Robblee, *J. Am. Oil Chem. Soc.*, 1981, **58**, 682.
[79] G. R. Fenwick, *Proc. Nutr. Soc.*, 1982, **41**, 277.
[80] M. Rundgren, *Anim. Feed Sci. Technol.*, 1983, **9**, 239.
[81] E. Josefsson and L. Munck, *J. Sci. Food Agric.*, 1972, **23**, 861.
[82] D. Thomas, A. R. Robblee, and D. R. Clandinin, *Br. Poult. Sci.*, 1978, **19**, 449.
[83] B. P. Gill and A. G. Taylor, *Anim. Prod.*, 1989, **49**, 317.
[84] A. Papas, J. R. Ingalls, and L. D. Campbell, *J. Nutr.*, 1979, **109**, 1129.

A more fruitful experimental approach, which takes into account the importance of the hydrolysis of glucosinolates in their toxicity, has been to process the meal in various ways in order to favour production of particular groups of toxic metabolites and to assess resulting toxicity. For example, Srivastava, Philbrick, and Hill[85] controlled hydrolysis of glucosinolates in rapeseed meal to produce meals rich in either nitrile hydrolysis products or 5-OZT. The resulting meals fed to rats or chicks both caused depression in live-weight gain but the toxic mechanisms appeared to be different; the 5-OZT-rich meal caused increased thyroid weight, whilst the nitrile-rich meal resulted in kidney enlargement. More recently, Wight, McCorquordale, and Scougall,[86] in attempting to determine the agent responsible for liver haemorrhage of laying hens, manipulated rapeseed diets in order to produce a nitrile-rich meal. However, the incidence of liver haemorrhage was not increased among animals consuming the nitrile-rich meal and the role of nitriles in producing liver haemorrhage remains unclear. Campbell[87] fed variously treated rapeseed meals to laying hens and related the concentrations of intact glucosinolates and their various breakdown products to the incidence of liver haemorrhage. Liver haemorrhage appeared to be related to the presence of intact glucosinolates in the diet but there were no obvious correlations with specific breakdown products. The importance of hydrolysis conditions in determining glucosinolate toxicity was suggested by the results of a recent experiment in which rapeseed meal diets were supplemented with various mineral salts.[88] The addition of copper and iron salts to the diets may have had simple post-absorptive effects but the presence of such metallic ions in the digestive tract may have affected glucosinolate hydrolysis and consequently altered their toxicity. For example, Fe^{2+} salts added to the diet may have favoured nitrile production in the gut[74] and this may explain the reduced live-weight gains observed on this diet.

Another strategy in the assessment of glucosinolate toxicity has been to subject seed meals to various extraction procedures and to add the extracts to glucosinolate-free diets. Lee, Pittam, and Hill[89] added intact glucosinolate extracts to soyabean meal and demonstrated a depression in the voluntary food intake of growing pigs consuming the diet. This approach has also allowed toxic effects to be attributed to more specific groups of glucosinolate hydrolysis products. Van Etten, Daxenbichler, and Wolff,[90] for example, dosed rats with isolates of *Crambe* seed meal and showed 5-OZT to cause thyroid enlargement while a nitrile fraction had effects on liver and kidney tissues. A similar approach was adopted by Lo and Bell,[91] who prepared a rapeseed isolate composed largely of butenyl isothiocyanate and butenyl cyanide. Rats dosed with this mixture showed depressed live-weight gain and intake and, although

---

[85] V. K. Srivastava, D. J. Philbrick, and D. C. Hill, *Can. J. Anim. Sci.*, 1975, **55**, 331.
[86] P. A. L. Wight, C. C. McCorquordale, and R. K. Scougall, *Res. Vet. Sci.*, 1987, **43**, 351.
[87] L. D. Campbell, *Nutr. Rep. Int.*, 1987, **36**, 491.
[88] M. Vermorel and J. Evrard, *Reprod. Nutr. Dev.*, 1987, **27**, 769.
[89] P. A. Lee, S. Pittam, and R. Hill, *Br. J. Nutr.*, 1984, **52**, 159.
[90] C. H. VanEtten, W. E. Gagne, D. J. Robbins, A. N. Booth, M. E. Daxenbichler, and I. A. Wolff, *Cereal Chem.*, 1969, **46**, 145.
[91] M. T. Lo and J. M. Bell, *Can. J. Anim. Sci.*, 1972, **52**, 295.

thyroid weight was unaffected, tracer studies with ^{125}I suggested that biosynthesis of thyroid hormones was affected by the treatment. These studies have emphasized the fact that glucosinolate toxicity depends to a large extent on the route of enzymic hydrolysis. Thus, hydrolysis of 2-hydroxy-3-butenyl glucosinolate to 5-OZT was found to result in impaired thyroid function as measured by reduced levels of plasma thyroid hormones and increased thyroid weight.[85] Hydrolysis of progoitrin and other glucosinolates to nitriles, however, resulted in more acute toxic symptoms to liver and kidney tissues, although the biochemical basis for this toxicity remains unclear.

The development of improved methods of isolating intact glucosinolates from plant material has facilitated a series of experiments in which individual pure glucosinolates have been fed to animals to determine their antinutritional effects.[92-95] The emphasis of this work was on performance aspects, although organ weights were recorded as crude measures of toxicological effect. In a recent experiment, Vermorel, Heaney, and Fenwick[94] showed that administration of a range of intact, isolated glucosinolates to rats over a period of 29 days at levels representative of those in low glucosinolate rapeseed had little effect on dry matter intake or live-weight gain; indeed there were trends towards increased intakes on the glucosinolate diets. In a further experiment, progoitrin was again added to a basal diet, this time accompanied by exogenous myrosinase.[96] Voluntary food intake and live-weight gain were initially unaffected by the presence of progoitrin and myrosinase but after 16 days intake declined significantly; increased thyroid weight and reduced levels of circulating T4 suggested impaired thyroid function as a result of goitrin release as being the cause of progoitrin toxicity.

The somewhat equivocal and often confusing results of these toxicity studies with individual glucosinolates illustrate the difficulties inherent in assessing the toxicity of compounds which are consumed in benign form and which are toxic by virtue of their digestive hydrolysis to toxic metabolites. Because this hydrolysis can be influenced both by conditions within the digestive tract, such as pH, and by the nature of the diet with respect to mineral composition and the presence or absence of active myrosinase, there is a need to clarify the digestive fate of glucosinolates in relation to their toxicity. In the above experiments, knowledge of the extent of glucosinolate breakdown and the proportions of toxic products produced in the alimentary tract could have aided in their interpretation. Research on the toxicology of individual glucosinolate breakdown products together with studies on factors affecting the digestive fate of the parent compounds would be useful. The current state of knowledge in these areas will now be reviewed.

[92] N. Bille, B. O. Eggum, I. Jacobsen, O. Olsen, and H. Sorensen, *Z. Tierphysiol., Tierernaehr. Futtermittelkd.*, 1983, **49**, 195.
[93] B. O. Eggum, O. Olsen, and H. Sorensen, Proceedings of the 6th International Rapeseed Congress, Paris, 1983.
[94] M. Vermorel, R. K. Heaney, and G. R. Fenwick, *J.Sci. Food Agric.*, 1987, **37**, 1197.
[95] B. O. Eggum, O. Olsen, and H. Sorensen, in 'Advances in the Production and Utilization of Cruciferous Crops', ed. H. Sorensen, Nijhoff/Junk, Rotterdam, 1985, p. 50.
[96] M. Vermorel, R. K. Heaney, and G. R. Fenwick, *J.Sci. Food Agric.*, 1988, **44**, 321.

7 Toxic Properties of Breakdown Products

Nitriles

There is mounting evidence that nitriles are the predominant hydrolysis products when glucosinolates undergo hydrolysis in natural environments. For example, nitriles have been found to predominate in the digestive tract of poultry following rapeseed ingestion[97] and autolysis of plant tissues appears to favour nitrile release.[98] Certainly nitriles appear to be among the most acutely toxic of the glucosinolate breakdown products, although the biochemical basis of this toxicity is, as yet, unclear.

The toxicity of nitriles, arising from glucosinolates, is generally attributed to their effects on liver and kidney tissues and increased organ weights in toxicity studies have been reported.[90,99] The mode of toxicity has seldom been investigated, although a number of histological studies have been conducted.

Rats dosed with a crude nitrile extract of *Crambe* seed meal showed enlarged kidneys and more detailed histological examination revealed hypertrophy and enlarged nuclei among the epithelial cells lining the convoluted tubules.[90] Similar lesions were reported when rats were chronically dosed with the β-OH nitrile resulting from progoitrin hydrolysis, 1-cyano-2-hydroxy-3,4-epithiobutane (CHEB),[100] and acute doses of the related nitrile compound, 1-cyano-epithiobutane (CHB), again produced histological lesions of the kidney tubule epithelial cells of rats.[99] Kidney damage was indicated by the elevated serum urea nitrogen and serum creatinine concentrations which accompanied kidney lesions in rats dosed with CHEB.[101]

Similar disruption to the cellular integrity of liver tissues has been found in a number of studies. Van Etten *et al.*[90] noted disruption to the normal lobular structure of liver tissues along with irregular bile duct proliferation in rats dosed with a crude nitrile extract. Bile duct proliferation along with hepatocyte necrosis was also a feature of the histopathology of rats dosed with CHEB in a later study.[100] Elevated levels of plasma alkaline phosphatase and γ-glutamyl transpeptidase provided further indication of hepatocyte damage and cholestasis in CHEB-dosed rats.[100]

Although rarely considered in the context of glucosinolate toxicity, release of free cyanide in the tissues as a result of nitrile biotransformation may be an important route of toxicity. The acute toxicity of a number of related nitrile compounds has been attributed to release of free cyanide as indicated both by direct detection of the cyanide ion and by depressed activity of tissue cytochrome

[97] T. K. Smith and L. D. Campbell, *Poult. Sci.*, 1976, **55**, 861.
[98] R. A. Cole, *Phytochemistry*, 1976, **15**, 759.
[99] K. Nishie and M. E. Daxenbichler, *Food Cosmetic Toxicol.*, 1980, **18**, 159.
[100] D. H. Gould, R. M. Gumbmann, and M. E. Daxenbichler, *Food Cosmetic Toxicol.*, 1980, **18**, 619.
[101] D. H. Gould, M. F. Fettmann, M. E. Daxenbichler, and B. M. Bartuska, *Toxicol. Appl. Pharmacol.*, 1985, **78**, 190.

Table 1 Cytochrome oxidase activities 1 hour after oral LD_{50} of various nitriles to rats
(Reproduced with permission from ref. 104, *Toxicol. Lett.*, 1982, **12**, 157)[104]

	Liver		Kidney		Brain	
	Activity[a]	% Control	Activity[a]	% Control	Activity[a]	% Control
Control	1.01 ± 0.02	100	1.56 ± 0.16	100	1.78 ± 0.13	100
KCN	0.37 ± 0.03	36	1.04 ± 0.18	66	0.53 ± 0.10	29
Acetonitrile	0.97 ± 0.05	96	1.60 ± 0.22	102	1.64 ± 0.21	92
Propionitrile	0.50 ± 0.01	49	1.03 ± 0.18	66	0.84 ± 0.14	47
Butyronitrile	0.84 ± 0.12	83	1.39 ± 0.16	89	0.97 ± 0.16	54
Malononitrile	0.28 ± 0.03	27	0.58 ± 0.20	37	0.73 ± 0.13	41
Acrylonitrile	0.56 ± 0.06	55	0.85 ± 0.19	54	0.81 ± 0.12	45
Allylcyanide	0.90 ± 0.11	89	1.47 ± 0.20	94	1.32 ± 0.23	74
Fumaronitrile	1.04 ± 0.05	102	1.38 ± 0.20	88	1.80 ± 0.22	101

[a] log[ferrocytochrome c] per minute for a 1:100 tissue dilution
Values are means ± S.D. of 6 animals

oxidase (Table 1).[102,103] Further studies have shown that the structure of nitriles and, in particular, their degree of saturation, are important in influencing their toxicity, presumably by altering their metabolic fate.[104] *In vitro* studies have indicated liver microsomes to be an important site of biotransformation of nitriles to cyanide.[103] The capacity of nitriles to bind the sulphydryl groups of glutathione and so depress tissue glutathione concentrations has been suggested as a route of toxicity by Szabo *et al*.[105] and Ahmed and Farooqui[104] although neither author proposed specific mechanisms and the phenomenon is more likely to aid metabolism and excretion of nitriles than to enhance their toxicity. Other toxic effects of nitriles reported in the literature include haemorrhage and necrosis of the adrenal cortex of rats dosed with acrylonitrile[106] and ulcerogenic effects.[107]

There is considerable evidence for an association between the levels of rapeseed in the diet and the incidence of liver haemorrhage[108,109] but attempts to correlate pathology with specific toxic agents, including nitriles, have been unfruitful. Campbell[87] fed rapeseed meal diets containing varying concentrations of intact glucosinolates and breakdown products to laying hens; there appeared to be no clear association between the concentrations of intact glucosinolate, goitrin, or CHB and the occurrence of liver haemorrhage. Wight, McCorquordale and Scougall[86] attempted to eludicate the causative

[102] H. Ohkawa, R. Ohkawa, I. Yamamoto, and J. E. Casida, *Pestic. Biochem. Physiol.*, 1972, **2**, 95.
[103] C. C. Willhite and R. P. Smith, *Toxicol. Appl. Pharmacol.*, 1981, **59**, 589.
[104] P. Ahmed and M. Y. H. Farooqui, *Toxicol. Lett.*, 1982, **12**, 157.
[105] S. Szabo, K. A. Vailey, P. J. Boor, and R. J. Jaeger, *Biochem. Biophys. Res. Commun.*, 1977, **79**, 32.
[106] S. Szabo, I. Huttner, K. Kovacs, E. Horvath, D. Szabo, and H. C. Horner, *Lab. Investig.*, 1980, **42**, 533.
[107] S. Szabo and E. S. Reynolds, *Environ. Health Perspect.*, 1975, **11**, 135.
[108] S. Yamashiro, T. Umemura, M. K. Bhatnagar, L. David, and M. Sadiq, *Res. Vet. Sci.*, 1974, **23**, 179.
[109] A. Papas, L. D. Campbell, and P. E. Cansfield, *Can. J. Anim. Sci.*, 1979, **59**, 133.

mechanism of liver haemorrhage by supplementing rapeseed meal and soyabean meal diets with various additives with known effects on liver metabolism and determining their influence on the incidence of liver haemorrhage. For example, it was suggested that liver haemorrhage might be caused by weakening of hepatic blood vessels due to inhibited collagen synthesis caused by hypothyroidism, or more directly by lathyrogenic compounds such as nitriles. These possibilities were tested by adding the goitrogen thiouracil and the lathyrogen β-aminopropionitrile to RSM diets and looking histologically at effects on liver haemorrhage. However, none of the agents produced liver haemorrhages characteristic of those produced on rapeseed meal diets and the aetiology of the syndrome remains obscure.

Isothiocyanates

Isothiocyanates are electrophilic in nature and studies of the chemistry of the isothiocyanate group have shown the compounds to be extremely reactive and to undergo a wide range of reactions which may shed light on the nature of their toxicity.[110] Biochemical studies on the specific toxic actions of isothiocyanates in nature are scarce, however, and current knowledge of isothiocyanate toxicity is based on limited experimental evidence.

Early toxicity studies showed allyl isothiocyanate (AITC) to possess goitrogenic properties with increased thyroid weight being recorded in rats dosed with AITC.[111,112] The degree of goitrogenicity was found to be weak, however, compared with more well-known goitrogens, such as thiouracil.[113] The goitrogenic activity of the isothiocyanates may be related to their metabolism to thiocyanate ion, which competes with iodide for uptake by the thyroid by simple competitive inhibition. Reduced uptake of ^{131}I has indeed been noted in rats dosed with AITC.[112,114] As well as gross changes in thyroid weight, dose-related depression in total plasma T4 and free plasma T4 concentrations have recently been measured in rats dosed with phenyl isothiocyanates.[115]

Possibly related to their goitrogenic properties, a number of other biochemical changes have been observed in response to isothiocyanate administration. Idris and Ahmad[116] found AITC dosing to increase plasma phospholipid concentrations in rats while Muztar et al.,[117] showed AITC to depress plasma glucose and uric acid concentrations. Effects on carbohydrate metabolism may have been due to hypothyroidism although more direct effects on key enzymes cannot be discounted. For example, altered activity of liver succinic dehydroge-

[110] L. Drobnica, P. Kristian, and J. Augustin, in 'The Chemistry of Cyanates and Their Thioderivatives' ed. S. Patai, J. Wiley and Sons, Chichester, p. 1003.
[111] P. Langer, *Physiol. Bohemoslov.*, 1964, **3**, 542.
[112] P. Langer and V. Stolc, *Endocrinology*, 1965, **76**, 151.
[113] A. Ahmad and A. J. Muztar, *Pak. J. Biochem.*, 1971, **4**, 72.
[114] P. Langer and M. A. Greer, *Metabolism*, 1968, **17**, 596.
[115] G. J. A. Speijers, L. H. J. C. Danse, F. X. R. VanLeewen, and J. G. Loeber, *Food Chem. Toxicol.*, 1985, **23**, 1015.
[116] R. Idris and K. Ahmad, *Biochem. Pharmacol.*, 1975, **24**, 2003.
[117] A. J. Muztar, T. Huque, P. Ahmad, and S. J. Slinger, *Can. J. Physiol. Pharmacol.*, 1979, **57**, 504.

nase and kidney xanthine oxidase were reported when rats were dosed with allyl isothiocyanate.[118]

As well as their effects on thyroid function, a number of *in vitro* studies have demonstrated more direct potential toxic actions of isothiocyanates based on their affinity for sulphydryl groups. Thus, inactivation of papain was attributed to the cleavage of the disulphide bond by benzyl isothiocyanate in an *in vitro* incubation experiment[119] and similar explanations were proposed for the inhibition of a proton-potassium ATP-ase by AITC.[120] By analogy, a number of other toxic actions can be conceived for the isothiocyanates by virtue of their capacity to cleave the disulphide bonds of important proteins and thus disrupt their tertiary structure. Thus AITC-mediated inactivation of anti-diuretic hormone has been suggested as the explanation for altered fuel metabolism in rats dosed with AITC[117] and addition of AITC onto the sulphydryl group of tyrosine was cited as a possible reason for reduced thyroid hormone synthesis in a parallel experiment.[121]

Clearly, a number of potential reactions of AITC with important biological molecules have been demonstrated *in vitro*, though the extent to which these operate *in vivo* is, as yet, unclear. *In vitro* observations will need to be critically tested *in vivo* before the mode of toxicity of isothiocyanates is more fully understood.

5-Vinyl Oxazolidinethione

The goitrogenic effects of the cyclic isothiocyanate derivative, 5-vinyl oxazolidine thione which is formed following the hydrolysis of 2-hydroxy-3-butenyl glucosinolate, are well documented and appear to be mediated by interference with the organic iodination of thyroxine in the biosynthesis of thyroid hormones. Although this toxic mechanism distinguishes 5-OZT from the other brassica-derived goitrogens, SCN^- and the isothiocyanates, the gross manifestations are the same with increased thyroid weight being a consistent feature even at low dose rates (1–2 μg rat^{-1} day^{-1}).[122–124] Some confusion exists in the literature regarding the effects of 5-OZT on the uptake of iodine by the thyroid with some authors reporting increases,[123] some decreases,[122] and others no change in the rate of thyroidal ^{131}I uptake.[125] Conflicting results may have reflected differences in the iodine content of the experimental diets.[125] As well as effects on thyroid hormone biosynthesis in the thyroid, peripheral changes in

---

[118] K. Ahmad, F. M. M. Rahman, A. Rahman, and R. Begum, Proceedings of the 7th International Congress on Nutrition, 1967, Vol. 5, p. 815.
[119] C. S. Tang, *J. Food Sci.*, 1974, **39**, 94.
[120] N.Takeguchi, Y. Nishimura, T. Watanabe, and M. Morii, *Biochem. Biophys. Res. Commun.*, 1983, **112**, 464.
[121] A. J. Muztar, P. Ahmad, T. Huque, and S. J. Slinger, *Can. J. Physiol. Pharmacol.*, 1979, **57**, 385.
[122] P. Langer and N. Michajlovskij, *Endocrinol. Exp.*, 1969, **62**, 21.
[123] Y. Akiba and T. Matsumoto, *Poult. Sci.*, 1976, **55**, 716.
[124] F. E. Krusius and P. Peltola, *Acta Endocrinol.*, 1966, **53**, 342.
[125] S. Elfving, *Ann. Clin. Res.*, 1980, **12**, 7.

T3 and T4 kinetics were shown in response to 5-OZT administration although relatively high dose rates were involved (100 μg rat^{-1} day^{-1}).[125]

The importance of any of the possible toxic routes described is dependent on the digestive fate of the parent compounds and the rate and extent to which the compounds can be modified or excreted before exerting their toxic effects in the animal.

8 Detoxification

An understanding of the metabolic fate and potential excretory routes of glucosinolates following their ingestion is important for a better understanding of their potential toxicity when present in different diets and when fed to different animal species. Unfortunately, research in this area has been rather limited and knowledge of the digestive fate of glucosinolates is particularly poor. With regard to the systemic fate of glucosinolate hydrolysis products, more is known and studies with chemically similar compounds have helped to provide insight into the likely metabolic routes and detoxification processes.

Metabolism in the Digestive Tract

The fate of glucosinolates in the digestive tract following ingestion from plant sources is an area of research which has been neglected. The lack of research in this area may be attributed, in part, to the practical analytical problems associated with the determination of reactive breakdown products in the complex medium of digestive fluid; improvements in isolation techniques and the use of isotopically labelled compounds may aid progress in this area.

The most significant findings in relation to the digestive fate of glucosinolates have been with poultry fed on rapeseed; Smith and Campbell[126] determined the proportions of 5-OZT and nitrile hydrolysis products arising from progoitrin in various portions of the digestive tract of hens fed on rapeseed and found nitriles to predominate throughout. Since absolute concentrations were not determined, however, the extent of glucosinolate hydrolysis was unknown. More recently, quantitative balance studies have shown glucosinolate recovery to be in the range of 15–50 % in the faeces and urine of intact hens with the unrecovered fraction presumably undergoing hydrolysis in the digestive tract.[127,128] Caecectomy of hens as well as inclusion of antibiotics in the experimental diets dramatically increased glucosinolate recovery to approximately 80 %, indicating the importance of hind-gut micro-organisms in glucosinolate hydrolysis.[128,129] This was further corroborated by *in vitro* experiments which showed a caecal fraction to have significant activity with respect to glucosinolate hydrolysis.[130] (See also ref. 131.)

Work in other species has been fragmentary. Minimal recovery (1–2 %) of

---

[126] T. K. Smith and L. D. Campbell, *Poult. Sci.*, 1976, **55**, 861.
[127] B. A. Slominski, L. D. Campbell, and N. E. Stanger, *Can. J. Anim. Sci.*, 1987, **67**, 1117.
[128] B. A. Slominski, L. D. Campbell, and N. E. Stanger, *J. Sci. Food Agric.*, 1988, **42**, 305.
[129] L. D. Campbell and B. A. Slominski, *J. Sci. Food Agric.*, 1989, **47**, 61.
[130] A. A. H. Freig, L. D. Campbell, and N. E. Stanger, *Nutr. Rep. Int.*, 1987, **36**, 1337.

intact glucosinolates was found in the faeces of rats fed on rapeseed.[131,132] Analysis for hydrolytic products of glucosinolates in the gut contents of rats fed rapeseed showed low concentrations of 5-OZT, while other products were not detected and this may have reflected the difficulties associated with the detection of reactive hydrolysis products in digestive fluids as well as their rapid absorption and transformation to non-toxic metabolites. Bacterial myrosinase activity was demonstrated in human gut contents using isolated progoitrin in an *in vitro* study[133] and sheep rumen fluid was also found to show hydrolytic activity towards progoitrin with production of 5-OZT and unidentified polymeric compounds.[134]

A more rigorous characterization of glucosinolate hydrolysis in the digestive tract would aid our understanding of glucosinolate toxicity. In particular, work with ruminants is virtually non-existent and the fate of glucosinolates as present in vegetative material following ingestion of forage brassicas, has not been studied. In addition, the subsequent fate of glucosinolate breakdown products following their production in the gut is unknown. Evidence from work on other secondary plant compounds indicates that significant detoxification may occur particularly in the rumen[135] and this may also occur with the glucosinolates.

Systemic Metabolism

The metabolic fate of glucosinolate hydrolytic products following their absorption is better known and a number of pharmacological studies exist. Attempts to detect hydrolytic products of glucosinolates in the blood and tissues of animals fed on rapeseed have been largely fruitless and uninformative.[136,137] Failure to detect compounds in their unchanged form is not surprising, however, in the light of their often highly reactive chemical nature and the capacity of hepatic metabolic processes to transform xenobiotics into more readily excretable compounds.

Studies in which laboratory animals have been administered with accurate doses of isolated, pure compounds have proved useful in elucidating excretory routes. Such studies with aliphatic isothiocyanates have shown glutathione conjugation to be a major excretory route[138,139] with consequent excretion of mercapturic acid derivatives of isothiocyanates in the urine. Species differences exist, however, with dogs excreting mainly hippuric acid derivatives,[138] while in guinea pigs and rabbits a cyclic mercapturic acid derivative predominates.[140] In

[131] A. Marangos and R. Hill, *Proc. Nutr. Soc.*, 1974, **33**, 90A.
[132] M. T. Lo and D. C. Hill, *Can. J. Physiol. Pharmacol.*, 1972, **50**, 962.
[133] E. L. Oginsky, A. E. Stein, and M. A. Greer, *Proc. Soc. Exp. Biol. Med.*, 1965, **119**, 360.
[134] A. Lanzani, G. Piana, G. Piva, M. Cardillo, A. Rastelli, and G. Jacini, *J. Am. Oil Chem. Soc.*, 1974, **51**, 517.
[135] J. R. Carlson and R. G. Breeze, *J. Anim. Sci.*, 1984, **58**, 1040.
[136] M. T. Lo and D. C. Hill, *Can. J. Anim. Sci.*, 1971, **51**, 187.
[137] C. H. VanEtten, M. E. Daxenbichler, W. Schroeder, L. H. Princen, and T. W. Perry, *Can. J. Anim. Sci.*, 1977, **57**, 75.
[138] G. Brusewitz, B. D. Cameron, L. F. Chasseaud, K. Gorler, D. R. Hawkins, H. Koch, and W. H. Mennicke, *Biochem. J.*, 1977, **162**, 99.
[139] H. Mennicke, K. Gorler, and G. Krumbiegel, *Xenobiotica*, 1983, **13**, 203.
[140] K. Gorler, G. Krumbiegel, W. H. Mennicke, and H. U. Siehl, *Xenobiotica*, 1982, **12**, 535.

mice a range of unidentified urinary metabolites were detected by TLC as well as the conventional mercapturic acid derivative.[141]

Glutathione conjugation of isothiocyanates was shown to be reversible in an *in vitro* experiment in which an equilibrium was demonstrated between free and conjugated benzyl isothiocyanate.[142] Initial conjugation of isothiocyanates following absorption with subsequent release of the free toxin at chemically favourable sites in the body was suggested as being significant toxicologically. The importance of glutathione conjugation as an excretory route for isothiocyanates was further suggested by the increased activity of glutathione transferase in the intestinal mucosa and liver of mice following benzyl isothiocyanate consumption.[143] Similarly, glutathione transferase activity was enhanced in the liver of rats fed on Brussels sprouts[144] and cabbage[145] and isothiocyanate release may have contributed to this phenomenon.

The metabolic fate of glucosinolate-derived nitriles is less clearly understood although work with chemically similar aliphatic nitriles provides useful indications as to their likely catabolism. For example, acrylonitrile metabolism has received considerable attention because of the importance of this compound in the manufacture of plastics[105,146–148] and metabolism of similar unsaturated nitriles such as allyl cyanide is likely to follow analogous pathways.[104]

A consistent feature of nitrile metabolism experiments is the increase in urinary thiocyanate ion excretion following nitrile administration.[149,150] This has been attributed to the release of free cyanide in the tissues followed by conventional excretion of cyanide as SCN^- under the action of hepatic rhodanese. While *in vivo* transformation of nitriles to free cyanide might appear unlikely in the light of the lability of the $C\equiv N$ bond, a number of workers have demonstrated the phenomenon both by direct detection of CN^- ion and by monitoring cytochrome oxidase activity.[103] *In vitro* experiments have localized the biotransformation to the microsomes[103,151] and shown that cellular integrity is required for the catalysis to occur.[149] The requirement of NADPH, O_2, and $MgCl_2$ for release of free cyanide from acrylonitrile *in vivo* strongly suggested cytochrome P-450 as the enzyme responsible for catalysing the reaction.[151]

A further feature of nitrile excretion is the depletion of tissue glutathione concentrations following nitrile dosing.[104,105] The importance of glutathione conjugation in nitrile excretion has been further supported by the detection of

---

[141] Y. M. Ioannou, L. T. Burka, and H. B. Matthews, *Toxicol. Appl. Pharmacol.*, 1984, **75**, 173.
[142] I. M. Bruggemann, J. H. M. Temmink, and P. J. van Bladeren, *Toxicol. Appl. Pharmacol.*, 1986, **83**, 349.
[143] V. L. Sparnins, P. L. Venegas, and L. W. Wattenberg, *J. Nat. Cancer Inst.*, 1982, **68**, 493.
[144] C. E. Godlewski, J. N. Boyd, W. K. Sharman, J. L. Anderson, and G. S. Stoewsand, *Cancer Lett.*, 1985, **28**, 151.
[145] G. S. Stoewsand, J. L. Anderson, and D. J. Lisk, *Proc. Soc. Exp. Biol. Med.*, 1986, **182**, 95.
[146] P. W. Langvardt, C. L. Putzig, H. Braum, and J. D. Young, *J. Toxicol. Environ. Health*, 1980, **6**, 273.
[147] G. Muller, C. Verkoyen, N. Soton, and K. Norpoth, *Arch. Toxicol.*, 1987, **60**, 464.
[148] R. Tardif, D. Talbot, M. Guerin, and J. Brodeur, *Toxicol. Lett.*, 1987, **39**, 255.
[149] A. R. Contessa and R. Santi, *Biochem. Pharmacol.*, 1973, **22**, 827.
[150] E. H. Silver, S. H. Kuttab, and T. Hasan, *Drug Metab. Dispos.*, 1982, **10**, 495.
[151] M. E. Abreu and A. E. Ahmad, *Drug Metab. Dispos.*, 1980, **8**, 376.

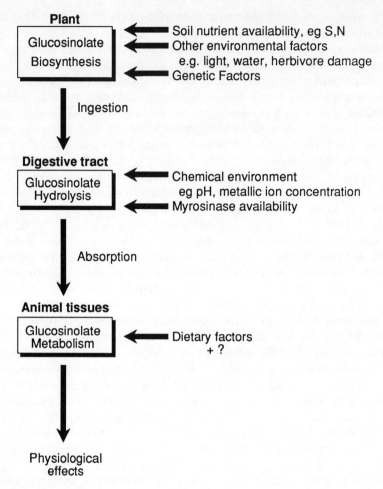

Figure 2 *Factors affecting ultimate physiological effects of glucosinolates in the animal*

urinary mercapturic acid derivatives following administration of acrylonitrile to rats.[146] Although experimental evidence is lacking, a number of catabolic routes for nitriles have been proposed and glutathione conjugation is suggested as an important feature, either by direct conjugation or by conjugation following transformation to an epoxide intermediate.[150] Formation of an epoxide intermediate was also indicated as an important first step in the metabolism of acrylonitrile in more recent experiments in which both cyanoethyl- and hydroxyethyl mercapturic acids were found to be important excretory products. In a radio-isotope study of the metabolism and disposition of the cyanoepithioalkane product of butenyl glucosinolate, 3,4-epithiobutane nitrile (3,4-ETN), excretion as a urinary mercapturic acid was found to be substantially complete with 12 hours. Low levels of residual radioactivity persisted in the tissues after this time which was suggested as being evidence for the alkylating and mutagenic properties of this compound.[152] While knowledge of general nitrile metabolism is useful in understanding the metabolic fate of glucosinolates, only isolated studies have concentrated on the excretion of glucosinolate-derived nitriles.[104,152]

Research on the excretion and metabolism of other glucosinolate metabolites has concentrated on 5-vinyl oxazolidinethione. Elfving[125] extended the earlier work of Peltola and Krusius[153] and Langer and Michajlovskij[122] on the metabolism of 5-OZT, to show that 5-OZT is concentrated in the thyroid following intra-peritoneal injection. 5-Vinyl oxazolidinethione was identified, intact, in the thyroid along with a 4-OH derivative and inorganic sulphate. These compounds were subsequently excreted fairly rapidly in the urine with sulphate showing the most persistence. Excretion of a labelled dose of 5-OZT was substantially complete (85%) within 24 hours.

Finally, the complexity of xenobiotic metabolism is highlighted by studies which followed the finding that consumption of certain cruciferous vegetables was beneficial in inhibition of tumour formation.[154] Examination of the activity of various enzymes of the MFO system indicated the involvement of glucosinolate metabolites, including benzyl isothiocyanate, indole-3-carbinol, and indole acetonitrile, in a non-specific induction of the MFO system and cytoplasmic glutathione-S-transferase with accompanying protection against potential carcinogens.[155] Similar interactions in the metabolism of other xenobiotics doubtless exist and may be significant in relation to the overall toxic effects of the anti-nutritive factors of the Cruciferae.

9 Conclusion

The nature and extent of the physiological effects seen in animals consuming glucosinolates are influenced by many factors beginning with constraints on biosynthesis in the plant through to the efficiency of glucosinolate catabolism within the animal (Figure 2). Research on glucosinolates has, thus far, been centred on aspects of glucosinolate biology, such as plant biosynthesis, and on other aspects of glucosinolate chemistry such as enzymic hydrolysis. Studies with animals have indicated the gross effects of glucosinolates but knowledge of the digestive and systemic fate of glucosinolates is at an early stage. Furthermore, the adaptive significance of the structural diversity seen among the glucosinolates and the relative effects of different glucosinolates are virtually unknown. This review has indicated the potential effects of different glucosinolate breakdown products in the animal but only a limited number of compounds and animal species have been studied. Because of the importance of the physiological effects of glucosinolates in the animal, this is a significant area of research for the future.

[152] J. Luthy and M. H. Benn, in 'Natural Sulfur Compounds', ed. D. Cavallini, G. E. Gaull, and V. Zappia, Plenum, New York, 1980, p. 381.
[153] P. Peltola and F.-E. Krusius, in 'Further Advances in Thyroid Research', ed. K. Fellinger and L. Hofer, Vienna Medical Academy, Vienna, 1971, pp. 149.
[154] L. W. Wattenberg, *J. Nat. Cancer Inst.*, 1977, **58**, 395.
[155] H. G. Shertzer, *Toxicol. Appl. Pharmacol.*, 1982, **64**, 353.

CHAPTER 7

Alkaloids

DAVID S. PETTERSON, DAVID J. HARRIS,
AND DAVID G. ALLEN

1 Introduction

Alkaloids are mostly secondary plant compounds which are found widely in nature. They have been isolated from animals, insects, bacteria, fungi, and mosses.[1] It seems that the alkaloids isolated from animals and insects had their origins in plants; although, in some cases, *de novo* synthesis seems more likely.

Alkaloids are basic nitrogenous compounds that can form salts with acids; they can be classified into three major groups.

(a) True alkaloids, with few exceptions, are basic and contain nitrogen in a heterocyclic ring; they are derived from amino acid precursors. Amongst their basic ring structures are pyridine, isoquinoline, pyrrole, indole, piperidine, and pyrrolidine. These compounds, which are of limited taxonomic distribution, have a wide range of physiological activity and are frequently toxic to animals, *e.g.* nicotine (1) and atropine (2).

(b) Pseudoalkaloids are usually basic, however they are not derived from amino acid precursors, *e.g.* solanidine (3) and caffeine (4).

(c) Protoalkaloids are basic amines that have been derived from amino acids, but their nitrogen is not part of a heterocyclic ring, *e.g.* mescaline (5) and ephedrine (6). For the purpose of this review we shall treat these three groups as being alkaloids.

Not all plant compounds which contain a basic nitrogen atom in a heterocyclic ring system are considered as alkaloids. For example, mimosine (7) is considered as a toxic amino acid rather than an alkaloid (see Chapter 2), and compounds like hordenine are considered to be biogenic amines. Other basic compounds, such as indole, skatole, cadaverine, and histamine, are widespread in both the animal and plant kingdoms, but are not considered as alkaloids as they originate from biological degradation of other nitrogen-containing compounds.

[1] G. A. Cordell, 'Introduction to Alkaloids', John Wiley and Sons, New York, 1981, p. 2.

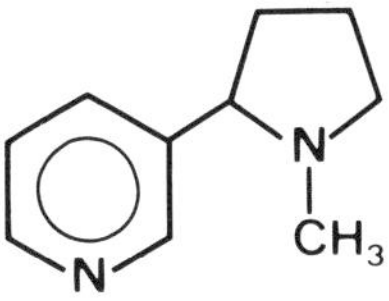

(1) Nicotine

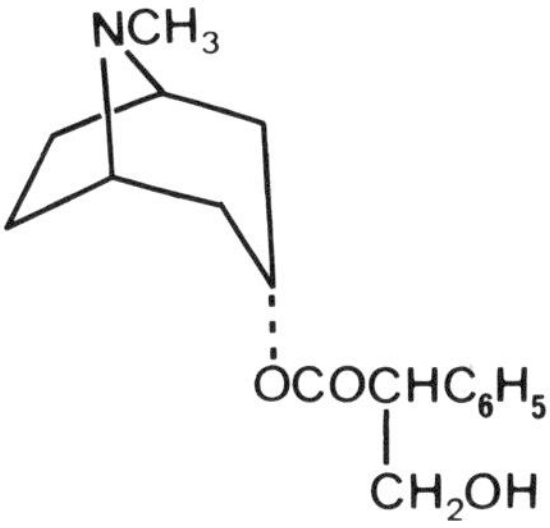

(2) Atropine

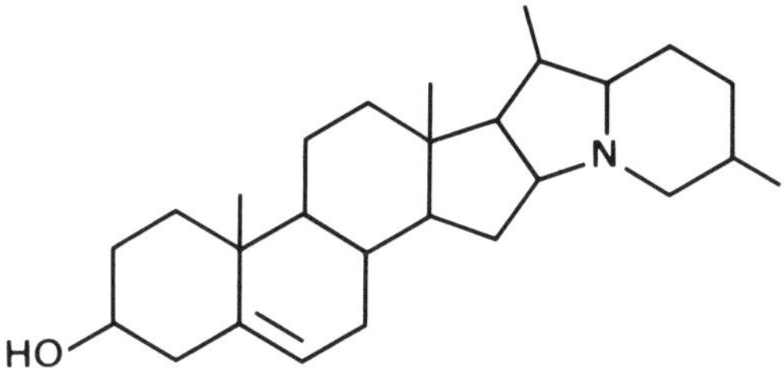

(3) Solanidine

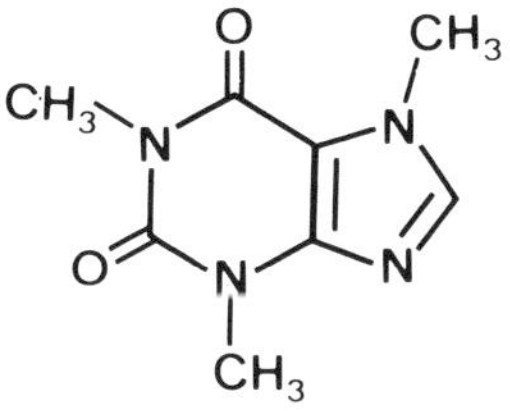

(4) Caffeine

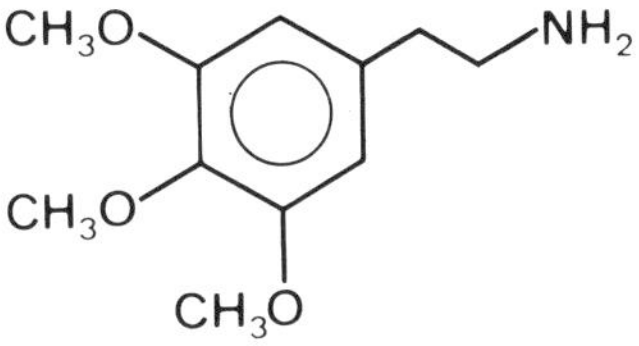

(5) Mescaline

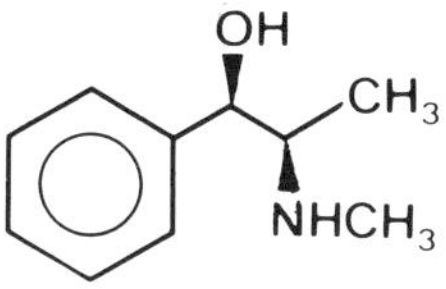

(6) Ephedrine

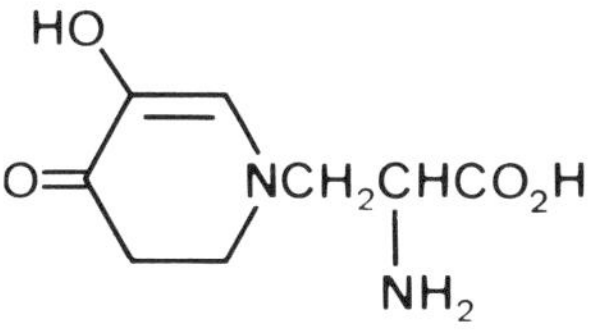

(7) Mimosine

The distribution of alkaloids is uneven within the plant kingdom. Alkaloid-bearing species occur in 34 of the 60 Orders of higher plants, about 40 per cent of all plant families, and yet only about 9 per cent of all genera.[2] Amongst the most important alkaloid-bearing plant families are the Amaryllidaceae, Compositae, Lauraceae, Leguminosae, Liliaceae, Papaveraceae, Rutaceae, and Solanaceae. Of these, the Papaveraceae are unusual in that all the species of all genera studied contain alkaloids.[2]

---

[2] G. A. Cordell, 'Introduction to Alkaloids', John Wiley and Sons, New York, 1981, p. 4.

(**8**) *Noscapine*   (**9**) *Morphine*   (**10**) *Coniine*

Humans have been aware of the effects of alkaloid-bearing plants for thousands of years. Opium was among the first drugs discovered by man. Its behavioural effects are described in Sumerian records dating back to about 4000 BC. The ancient Assyrians, Greeks, and Romans all referred to the use of opium for pain relief and sleep-inducing properties. Lord Nelson was reported to use opium to ease the pain in the stump of his arm.

The first crystalline alkaloid [probably noscapine (8)] was isolated from opium in 1803. Morphine (9), named after Morpheus the Greek god of sleep, was isolated soon after in 1805.[3] Coniine (10) was isolated in 1826 by Pelletier and Caventou, the pioneers of alkaloid chemistry. It was the first alkaloid to be characterized (in 1870) and synthesized (in 1886).[4] Since then, about 5000 alkaloids have been isolated and synthesized. Such is the rate of activity in this area that virtually every issue of *Phytochemistry* for the past few years contained reports on the isolation of new alkaloids. New texts and review series appear regularly.

The toxic nature and bitter taste of alkaloids were deterrents to early man in his search for food to an extent that they are rarely encountered today. The most important exception is the potato (*Solanum tuberosum*) which ranks third in tonnage among food crops grown in the Western world, and is a major source of carbohydrate, vitamins, and minerals. The concentrations of glycoalkaloids in healthy potato tubers are very low and do not normally pose a hazard to consumers. However, the concentration in green tubers and in those that have been physically damaged can be very high, and has been linked with health problems in humans (see Section 5). Glycoalkaloids are found in the tomato plant (*Lycopersicon esculentum*), but the concentrations in the fruit are usually low.

The presence of toxic quinolizidine alkaloids in the many species of lupin (*Lupinus* sp.) hindered the widespread acceptance of this valuable 'grain' legume as a food crop for humans. *L. luteus*, *L. albus*, and *L. angustifolius* have been consumed for centuries in European countries, whereas *L. mutabilis*

---

[3] F. W. Sertürner, *(Trommsdorff's) J. Pharmazie*, 1805, **13**, 234; cited by R. J. Bryant, *Chem. Ind. (London)*, 1988, 146.

[4] A. Ladenburg, *Chem. Ber.*, 1886, **19**, 439.

Table 1 *Commercial uses for some alkaloids*

Alkaloid	Plant source / Common name	Use
Atropine	*Atropa belladonna* / Deadly nightshade	Anti-cholinergic agent
Nicotine	*Nicotiana tabacum* / Tobacco	Insecticide (Black leaf 40)
Tomatine	*Lycopersicon esculentum* / Tomato	Fungicide
Colchicine	*Colchicium autumnale* / Autumn crocus	Gout treatment / Agricultural and medical genetics
Strychnine	*Strychnos nux-vomica*	Rodenticide
Cocaine	*Erythroxylon coca* / Coca bean	Local anaesthetic
Quinine	*Cinchona ledgeriana*	Anti-malarial
Quinidine		Cardiac stimulant
Vincristine	*Catharanthus roseus* / Periwinkle	Leukaemia treatment
Vinblastine		
Morphine	*Papaver somniferum* / Opium poppy	Analgesic
Codeine		Analgesic / Anti-tussive
Emetine	*Cephaelis ipecacuanha*	Emetic
Sparteine	*Lupinus luteus* / Yellow lupin	Oxytocic in obstetrics

(tarwi) is an important component of the diet of South American Indians. Lupin seeds contain 300–450 g kg^{-1} protein, while *L. albus* and *L. mutabilis* also contain 100–150 g kg^{-1} oil. They have a lower content of anti-nutritional factors than soybeans,[5] which dominate the world 'grain' legume market. Traditionally, lupin seeds were debittered to make them palatable. Efforts by plant breeders in Europe, Australia, and South America have resulted in the development of new cultivars that contain very low levels (< 200 mg kg^{-1}) of alkaloids.[6]

The occasional unexpected results of primitive man's sampling of plant species in his search for foods identified other plants that possessed a wide range of useful properties. Experiments that ended in tragedy led to the discovery of potent toxins which greatly increased his weaponry for catching prey. Consumption of some plants produced euphoria while others diminished the sense of pain, or cured ailments. By the end of the Nineteenth century, pure alkaloids and alkaloidal preparations from plant extracts accounted for a considerable proportion of the pharmacopoeia of the day. Since then, the introduction of synthetic drugs has increased dramatically. Nonetheless, alkaloids and their synthetic derivatives are still widely used in modern medicine.

---

[5] D. S. Petterson, D. G. Allen, B. N. Greirson, G. R. Hancock, D. J. Harris, and F. M. Legge, *Pro. Nutr. Soc. Aust.*, 1986, **11**, 118.
[6] J. S. Gladstones, *Field Crop Abstr.*, 1970, **23**, 123.

Indeed, new alkaloids are being continuously investigated as potential medicinal agents; the most topical example being that of testing castanospermine as a treatment for AIDS.[7] Although many alkaloids have been synthesized in the laboratory, most alkaloids of commercial importance are still prepared from extracts of crop plants. Selected examples are listed in Table 1.

Our aim in this chapter is to provide an overview of various aspects of alkaloid chemistry including methods of analysis, biosynthesis, and toxicology. To illustrate these topics, we have selected several examples of plants grown for their food and pharmacological value. The steroidal glycoalkaloids (Solanaceae) and the quinolizidine alkaloids (*Lupinus* spp.) represent the food crops. Space does not permit discussion of the alkaloids found in peppers (pyridine alkaloids), capsicum, buckwheat (fagomine), yams (dioscorine), the betel nut (piperidine alkaloids), herbs (*e.g.* comfrey), and others. The coffee, tea, and tobacco alkaloids, which have neither nutritive nor therapeutic value, but are important for their mood-affecting properties, are discussed. Alkaloids derived from the opium poppy, autumn crocus, coca beans, periwinkle, and some others will be discussed briefly. The wide range of topics discussed necessitated our being selective with literature references.

2 Structure and Biosynthesis

The structures of the majority of alkaloids, despite their complex nature, were determined by classical means. In many cases, structures were solved from empirical formulae, the nature of functional groups, and the identity of products derived from degradation procedures such as oxidation and the Hoffman degradation. Although it was long suspected that amino acids were precursors for most alkaloids, it was not possible to prove the proposed biosynthetic pathways until the advent of radioactive isotopes of carbon and hydrogen. In a typical experiment, an amino acid containing the radioactive isotope of carbon (^{14}C) is fed to a plant or tissue culture, and the alkaloid is extracted after a suitable period. If the selected amino acid is indeed a precursor, the alkaloid will be radioactive. Further information can be obtained by degrading the alkaloid to determine the position of the incorporated radioactive label. The use of precursors containing magnetically active nuclei, such as ^2H, ^{13}C, or ^{15}N, has greatly facilitated these biosynthetic studies as the labelling patterns in the alkaloid can be determined by nuclear magnetic resonance (NMR) spectroscopy without resorting to degradation techniques. Infrared spectroscopy and mass spectrometry have also greatly facilitated identification, frequently as support to the classical methods.

X-Ray crystallography is another powerful tool for the determination of alkaloid structures.[8] It was necessary to use X-ray diffraction methods to determine the structure of colchicine (11)[9] as previous chemical work could not

---

[7] V. A. Johnson, B. D. Walker, M. A. Barlow, T. J. Paradis, T-C. Chow, and M. S. Hirsch, *Antimicrob. Agents Chemother.*, 1989, **33**, 53.
[8] J. Finer-Moore, E. Arnold, and J. Clardy, in 'Alkaloids: Chemical and Biological Perspectives', ed. S. W. Pelletier, Wiley, New York, 1983, Vol. 2, p. 1.
[9] L. Lessinger and T. N. Margulis, *Acta Crystallogr.*, 1978, **34**, 578.

(11) *Colchicine*

distinguish the relative positions of the methoxyl and ketone groups. X-Ray analysis provides information on the conformational stereochemistry of individual alkaloids and can be used to determine the absolute configuration of optically active alkaloids.

Solanaceae Alkaloids

The glycoalkaloids found in plants of the Solanaceae, including the potato (*Solanum tuberosum*) and the tomato (*Lycopersicon esculentum*), consist of a steroidal aglycone to which various sugar moieties are attached. The major glycoalkaloids of the potato, α-solanine (12), and α-chaconine (13) comprise > 95 % of the total; they contain solanidine (3) as the aglycone. The corresponding saturated aglycone, demissidine, is found in several minor glycoalkaloids.

There have been a number of reviews on the biochemistry and possible biogenetic relationships of steroids and steroidal alkaloids of the Solanaceae.[10–12]

Biosynthesis

The first labelling studies with potato sprouts showed that acetate was incorporated into the aglycone portion of the glycoalkaloid.[13] It was further shown that mevalonate was incorporated more efficiently than acetate.[14,15] Mevalonic acid is incorporated into cycloartenol which is metabolized to cholesterol and then to the glycoalkaloids.[16–19]

[10] S. J. Jadhav, R. P. Sharma, and D. K. Salunkhe, *CRC Crit. Rev. Tox.*, 1981, **9**, 21.
[11] E. Heftmann, *Lloydia*, 1968, **31**, 293.
[12] K. Schreiber, in 'The Alkaloids', Vol. 10, ed. R. H. F. Manske, Academic Press, New York, 1978, Chapter 1, p. 1.
[13] A. R. Guseva and V. A. Paseshnickenko, *Biokhimiya*, 1958, **23**, 1958.
[14] A. R. Guseva and V. A. Paseshnickenko, *Biokhimiya*, 1961, **26**, 723.
[15] M. T. Wu and D. K. Salunkhe, *Biol. Plant*, 1978, **20**, 149.
[16] D. F. Johnson, E. Heftmann, and G. V. C. Houghland, *Arch. Biochim. Biophys.*, 1964, **104**, 102.
[17] H. Ripperger, W. Moritz, and K. Schruber, *Phytochemistry*, 1971, **10**, 2699.
[18] M. A. Hartmann and P. Benveniste, *Phytochemistry*, 1974, **13**, 2667.
[19] C. R. Tschesche and H. Hulpke, *Z. Naturforsch., Teil B*, 1967, **22**, 791.

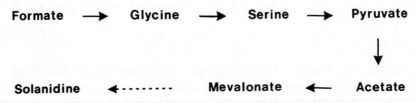

(12) R = D-galactose — L-rhamnose / D-glucose   α-Solanine

(13) R = D-glucose — L-rhamnose / L-rhamnose   α-Chaconine

Ramaswary et al.[20,21] demonstrated the involvement of chlorophyll synthesis with the synthesis of alkaloids, and showed that formate, glycine, and pyruvate were intermediates in the biosynthesis of solanidine (Scheme 1).

Heftmann[22] hypothesized that the nitrogen in glycoalkaloids may be introduced before the formation of 27-hydroxycholesterol, by replacing the hydroxyl group with an amino group. However, to this day (1991) the origin of the nitrogen in the glycoalkaloid remains a mystery.[10]

Formate → **Glycine** → **Serine** → **Pyruvate**
↓
Solanidine ←------ **Mevalonate** ← **Acetate**

Scheme 1 *Intermediates in the biosynthesis of solanidine*

Studies on α-tomatine (14) biosynthesis have shown the incorporation of a label from acetate, mevalonate, cycloartenol, and cholesterol.[10]

Biosynthesis of the potato glycoalkaloids appears to be mostly in the periderm and cortex of the tuber, and in the eye regions. At flowering the concentration of alkaloid is lowest in the tuber and highest in the flowers. Synthesis is influenced by light, and physiological, nutritional, and physical stress.[10,23]

[20] P. M. Nair, A. G. Behere, and N. K. Ramaswary, *J. Sci. Ind. Res.*, 1981, **40**, 529.
[21] N. K. Ramaswary, A. G. Behere, and P. M. Nair, *Eur. J. Biochem.*, 1976, **67**, 275.
[22] E. Heftmann, *Lloydia*, 1967, **30**, 209.
[23] J. A. Maga, *CRC Crit. Rev. Food Sci. Nutr.*, 1980, 371.

(14) α-Tomatine

De novo synthesis of α-tomatine, the only glycoalkaloid of *L. esculentum*, occurs in the young fruit and shoots. The peak concentration in fruit occurs just before ripening commences, then the concentration rapidly falls away. At all stages the highest concentrations are in the leaves.[10]

Lupin Alkaloids

The alkaloids of *Lupinus* species are usually bicyclic, tricyclic, or tetracyclic derivatives of quinolizidine (15).

The bicyclic alkaloids are typified by (−)-lupinine (16), the tricyclics by angustifoline (17), and the tetracyclic series by (−)-sparteine (18). Lupanine (19), the 2-oxo derivative of sparteine, is found in most lupin species and related genera.[24] The hydroxylated derivatives 13-hydroxylupanine (20) and 4-hydroxylupanine (21) occur in most lupin species. Table 2 lists the major alkaloids present in commercial species of lupins.

Esters of hydroxylated alkaloids have been reported. For example, Wink *et al.*[25] identified the tigloyl, angeloyl, and other esters of 13-hydroxylupanine in extracts of *L. mutabilis*.

The structural determination and synthesis of the quinolizidine alkaloids have been comprehensively reviewed.[24,26]

Both enantiomers of lupanine occur in extracts of lupins and related genera; (+)-lupanine is the more common form among the lupin species, whereas the (−) form occurs in *Thermopsis* sp. The racemate has been isolated from *L. albus*. The absolute configuration of (+)-lupanine (19) was determined by X-ray crystallography.[27]

[24] N. J. Leonard, in 'The Alkaloids', ed. R. H. F. Manske and H. L. Holmes, Academic Press, New York, 1953, Vol. 3, Chapter 19, p. 123.
[25] M. Wink, L. Witte, T. Hartmann, C. Theuring, and V. Volz, *J. Med. Plant Res.*, 1983, **48**, 253.
[26] M. F. Grundon, *Nat. Prod. Rep.*, 1989, **6**, 523.
[27] E. Skrzypozak-Jankun, *Acta Crystallogr. Sect B*, 1978, **34**, 2651.

(15) Quinolizidine

(16) Lupinine

(17) Angustifoline

(18) Sparteine

(19) R¹ = H,   R² = H;   Lupanine

(20) R¹ = H,   R² = OH; 13-Hydroxylupanine

(21) R¹ = OH, R² = H;   4-Hydroxylupanine

Table 2 *The major alkaloids in domesticated* Lupinus *species*

Alkaloid	Species			
	L. albus	L. angustifolius	L. luteus	L. mutabilis
Lupanine	**	**		**
Sparteine			**	**
13-Hydroxylupanine	**	**		**
4-Hydroxylupanine				*
Lupinine			**	
Angustifoline	*	*		

** Major alkaloid (> 10% of total alkaloids).
* Minor alkaloids (1–10% of total alkaloids).

Biosynthesis

The basic building block of the quinolizidine alkaloids is lysine. The biosynthesis proceeds via cadaverine (22).[28-30]

A detailed study on the biosynthesis of the quinolizidine alkaloids was done

[28] H. R. Schutte, H. Hudorf, K. Mathis, and G. Habner, *Liebigs Ann. Chem.*, 1964, **680**, 93.
[29] E. K. Nowacki and G. R. Waller, *Rev. Latinoam. Quim.*, 1977, **8**, 49.
[30] M.Wink, T. Hartmann, and H-M. Scheibel, *Z. Naturforsch., Teil C*, 1979, **34**, 704.

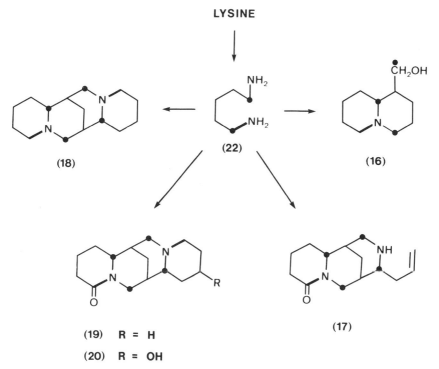

Scheme 2 *Labelling patterns of five quinolizidine alkaloids from* L. polyphyllus *fed with labelled cadaverine*

by Robins' group.[31-36] They used labelled cadaverine (^{13}C, ^{14}C, ^{15}N, and ^{2}H) and modern analytical techniques (*i.e.* ^{13}C NMR and ^{2}H NMR) to study the incorporation of cadaverine units into the quinolizidine alkaloids of *L. luteus* and *L. polyphyllus*.

The pulse feeding of [1-amino ^{15}N, 1-^{13}C] cadaverine and [1,5-^{14}C] cadaverine to *L. polyphyllus* resulted in the isolation of five labelled quinolizidine alkaloids; sparteine (18), lupanine (19), 13-hydroxylupanine (20), angustifoline (17), and lupinine (16)[36] (Scheme 2).

The assignment of the carbon labelling was confirmed by ^{13}C NMR spectroscopy. This showed that lupanine, sparteine, and 13-hydroxylupanine were formed from three cadaverine units. Angustifoline was also formed from three cadaverine units, but one of these was cleaved to form a side chain with loss of a labelled carbon atom from position 15. However, only two cadaverine units were incorporated into lupinine.[37]

[31] J. Rana and D. J. Robins, *J. Chem. Soc., Chem. Commun.*, 1983, 1335.
[32] A. M. Fraser and D. J. Robins, *J. Chem. Soc., Chem. Commun.*, 1984, 1477.
[33] J. Rana and D. J. Robins, *J. Chem. Res.*, 1985, 196.
[34] J. Rana and D. J. Robins, *J. Chem. Soc., Perkin Trans.*, 1986, 113.
[35] A. M. Fraser and D. J. Robins, *J. Chem. Soc., Perkin Trans.*, 1987, 105.
[36] D. J. Robins and G. N. Sheldrake, *J. Chem. Res.*, 1987, 256.
[37] D. J. Robins and G. N. Sheldrake, *J. Chem. Res.*, 1987, 159.

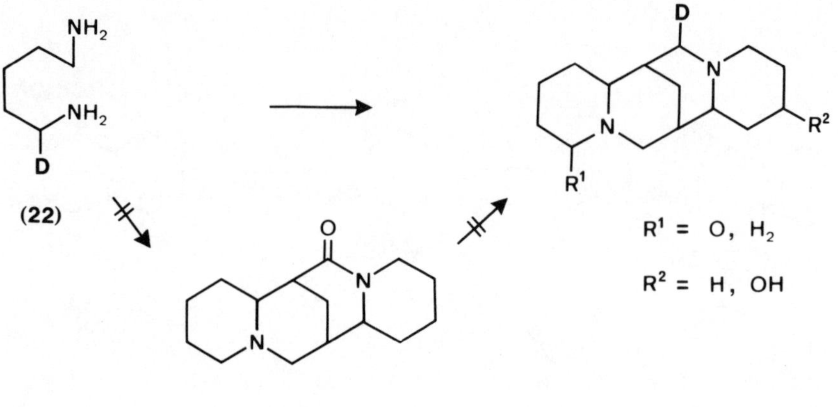

Scheme 3 *Biosynthetic pathway of labelled cadaverine fed to various* Lupinus *species.*

Wink et al.[30] added cadaverine to an enzyme preparation from a cell suspension of *L. polyphyllus* to give 17-oxosparteine (23), which was proposed as an intermediate in the biosynthesis of all quinolizidine alkaloids by the *Lupinus* genus. However, this was disproved by feeding 1R-[1-^{2}H]-cadaverine to several *Lupinus* species and isolating ^{2}H quinolizidine alkaloids.[38] In all of the isolated tetracyclic alkaloids the 17 position contained ^{2}H, so a carbonyl compound could not be an intermediate in the biosynthesis of tetracyclic quinolizidine alkaloids (Scheme 3).[35]

To date (1991) no intermediates beyond cadaverine have been isolated in the biosynthesis of quinolizidine alkaloids.[37–39] These intermediates are believed to be enzyme-bound and can not be isolated in standard quenching experiments.[31]

The stereochemistry of some of the processes involved in the biosynthesis of quinolizidine alkaloids was demonstrated by feeding studies using enantiomeric [1-^{2}H]-cadaverine hydrochloride.[33,35,36,38,40–42]

The biosynthesis of the lupin alkaloids occurs in green parts of the plant, especially in the leaves. The key enzymes are located in chloroplasts and synthesis follows a light-dependent diurnal rhythm.[43] Translocation to all parts of the plant occurs via the phloem and storage occurs in the epidermal and subepidermal tissues of stems and leaves. The seeds are especially rich,[44] with the alkaloids concentrated in the cotyledons. The alkaloid concentration in the

[38] W. M. Golebiewski and I. D. Spenser, *J. Am. Chem. Soc.*, 1984, **106**, 1441.
[39] D. J. Robins, personal communication, 1990.
[40] W. M. Golebiewski and I. D. Spenser, *Can. J. Chem.*, 1988, **66**, 1734.
[41] W. M. Golebiewski and I. D. Spenser, *Can. J. Chem.*, 1985, **63**, 2707.
[42] W. M. Golebiewski and I. D. Spenser, *J. Chem. Soc., Chem. Commun.*, 1983, 1509.
[43] M. Wink and L. Witte, *Planta*, 1984, **161**, 519.
[44] M. Wink, 'Allelochemicals: Role in Agriculture and Forestry', ACS Symposium Series No. 330, 1987, p. 524.

(24) $R^1 = H$, $R^2 = H$, $R^3 = H$; *Xanthine*

(25) $R^1 = H$, $R^2 = CH_3$, $R^3 = CH_3$; *Theobromine*

(26) $R^1 = CH_3$, $R^2 = CH_3$, $R^3 = H$; *Theophylline*

(4) $R^1 = CH_3$, $R^2 = CH_3$, $R^3 = CH_3$; *Caffeine*

leaves increases until flowering and then remains relatively constant until maturity.[45] In the seed the peak concentration occurs at maturity.

Alkaloids in Beverage Crops

The purine alkaloids of crop plants are methylated derivatives of xanthine (24). Caffeine (4) is found in about 60 genera, most notably in the seeds of *Coffea* and *Cola* genera and in the leaves and leaf buds of *Ilex paraguensis* and *Camellia sinensis*. Theobromine (25) is a minor constituent of the above and occurs in significant concentrations in the seed of *Theobroma cacao*. Theophylline (26) is only a minor constituent of these crops.

Biosynthesis

In reviewing the biosynthesis of the methylxanthines, Suzuki and Waller[46] concluded that the pathway led from the nucleotide pool (the 5' monophosphates of adenosine, guanosine, inosine, and xanthosine) to xanthosine and then, in turn, to 7-methylxanthosine, 7-methylxanthine, theobromine, and caffeine. Theophylline is presumed to form via a different mechanism.[47]

In *Coffea arabica*, caffeine is synthesized in the pericarp and is transported into the seed where it accumulates,[48] firstly in the seed coat and then in the cotyledon.[46] The alkaloid levels are 10–30 g kg^{-1} in developing leaflets protected by a resin layer and the stipules increase to about 40 g kg^{-1} when the leaflet is fully open, and then decrease as the leaves grow bigger and become less palatable.[49] Theobromine follows a similar pattern, whereas theophylline is only found in the pericarp of ripe fruits.

In the fruits of *Camellia sinensis*, the alkaloids accumulate in both pericarp

---

[45] W. Williams and J. E. M. Harrison, *Phytochemistry*, 1983, **22**, 85.
[46] T. Suzuki and G. R. Waller, *Ann. Bot. (London)*, 1985, **56**, 537.
[47] G. A. Cordell, 'Introduction to Alkaloids', Wiley, New York, 1981, Chapter 12, p. 958.
[48] T. W. Baumann and H. Wanner, *Planta*, 1972, **108**, 11.
[49] P. M. Frischknecht, J. Ulmer-Dufek, and T. W. Baumann, *Phytochemistry*, 1986, **25**, 613.

Tobacco Alkaloids

Tobacco (*Nicotiana tabacum*) is one of the world's major crop plants, with annual production of approximately 5 million tonnes. The major alkaloid of tobacco, nicotine (1), is probably the most widely studied of all alkaloids. Extracts of tobacco were used for treatment of skin diseases and as an insecticidal spray in Europe in the Seventeenth century, long before the alkaloidal properties were identified (in 1828). Since then, a plethora of papers about nicotine and its effect on human health have appeared in the literature.

Although nicotine is most commonly associated with tobacco, it is one of the more widely distributed alkaloids, having been identified in 24 genera (12 families) of plants. Notable occurrences include *Erythroxylon coca* (coca), *Lycopersicon esculentum* (tomato), *Atropa belladonna* (deadly nightshade), and *Asclepia syriaca* (milkweed). Other alkaloids of *N. tabacum* include anabasine (27), which is the major alkaloid of tree tobacco (*N. glauca*), nornicotine (28), and anatabine (29).

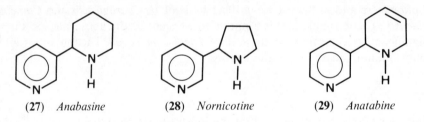

(**27**) *Anabasine*   (**28**) *Nornicotine*   (**29**) *Anatabine*

In comparison with other alkaloids, the structure of nicotine is relatively simple. Nonetheless, it was not established until 1893, some 65 years after it was first isolated in pure form.

The structure of nicotine was established following various oxidative degradations of the molecule. For example, the presence of the pyridine ring was established by oxidation of the free alkaloid with chromic acid to nicotinic acid (pyridine-3-carboxylic acid). The presence of the pyrrolidine ring was demonstrated by identifying hygric acid (30) as one of the products of sequential

(**30**) *Hygric acid*

oxidation of the methiodide salt. The elucidation of the structure of nicotine has been reviewed.[50,51]

Biosynthesis

The biosynthesis of nicotine and related alkaloids has been widely reviewed. The presence of the pyrrolidine ring system in nicotine suggested to earlier workers that proline was a probable precursor. However, subsequent labelling studies revealed that ornithine was involved in the formation of the pyrrolidine ring.[52] Significantly, when [2-^{14}C]-ornithine was used as a labelled precursor, the labelled atoms in the resulting nicotine were equally distributed between C-2' and C-5', suggesting the presence of a symmetrical intermediate. The most likely candidate was the diamine putrescine, which is formed by decarboxylation of ornithine. This hypothesis was supported by the incorporation of [1,4-^{14}C$_2$]-putrescine to produce nicotine containing labelled C-2' and C-5' atoms.[53]

The next step of the sequence is believed to be methylation to form N-methylputrescine, which undergoes oxidative deamination to form 4-methylaminobutanal, which in turn cyclizes spontaneously to the N-methylpyrrolinium salt (31).

The pyridine ring is believed to originate with the formation of nicotinic acid. Although nicotinic acid is formed from tryptophan in animals and microorganisms, this is not the case for *Nicotiana* species in which the amino acid precursor is aspartic acid.

Use of labelled nicotinic acid showed that the only hydrogen atom lost was that from C-6 and that the pyrrolidine ring system was attached to the carbon atom to which the carboxylic acid group was previously attached. These observations were explained by a mechanism in which the labelled nicotinic acid is converted to the dihydropyridine (32) which reacted with the N-methylpyrrolinium salt followed by expulsion of carbon dioxide and the labelled proton to form nicotine (Scheme 4). Detailed elucidations of the biosynthesis were reviewed.[51,54,55]

Nicotine does not appear to be the final product of alkaloid production in the tobacco plant. Nornicotine (28) is produced from nicotine by transmethylation of the N-methyl group. Various post-harvest reactions that occur in the curing of tobacco leaves result in further degradation of nicotine to various alkaloids including its N-oxide, myosmine, and cotinine.

---

[50] L. Marion, in 'The Alkaloids', ed. R. H. F. Manske and H. L. Holmes, Academic Press, New York, 1953, Vol. 1, Chapter 5, p. 236.
[51] C. R. Enzell, I. Wahlberg, and A. J. Aasen, *Fortschr. Chem. Org. Naturst.*, 1977, **34**, 1.
[52] E. Leete, in 'Alkaloids: Chemical and Biological Perspectives', ed. S. W. Pelletier, Wiley, New York, 1983, Vol. 1, Chapter 3, p. 85.
[53] E. Leete, *J. Am. Chem. Soc.*, 1958, **80**, 2162.
[54] R. B. Herbert, in 'Comprehensive Organic Chemistry', ed. D. H. R. Barton and W. D. Ollis, Pergamon Press, Oxford, 1978, Vol. 5, p. 1045.
[55] E. Leete, in 'Alkaloids: Chemical and Biological Perspectives', ed. S. W. Pelletier, Wiley, New York, 1983, Vol. 1, Chapter 3, p. 91.

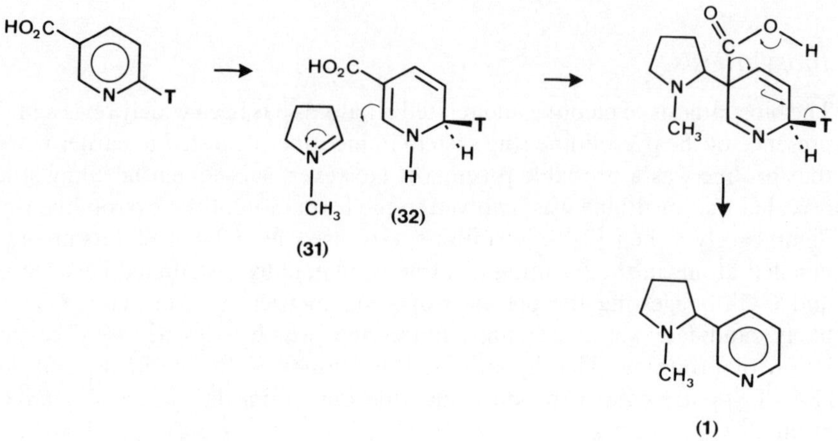

Scheme 4 *Proposed biosynthetic mechanisms for the formation of nicotine from nicotinic acid and the N-methylpyrrolinium ion*

Nicotine is produced in the roots of the tobacco plants and is then translocated to the leaves where it accumulates.

Opium Alkaloids

The opium poppy, *Papaver somniferum*, contains morphinan alkaloids such as morphine (9), codeine (33), thebaine (34), papaverine (35), and noscapine (8). These compounds have few rivals as analgesic compounds, and also serve as a source for synthesis of ethylmorphine, pholocodine (an anti-tussive), etorphine (a widely used veterinary drug), heroin, and the narcotic antagonists naltrexone and naloxone. About 30 alkaloids have been identified: of these codeine is the most important commercially. Codeine has about one quarter the analgesic potency of morphine, but is much less addictive. It is also a valuable anti-tussive (cough suppressant). The amounts of codeine present in opium are not sufficient to meet demand and supplies are supplemented by methylation of morphine. Codeine can also be prepared from thebaine, the major alkaloid of *P. bracteatum*. The seed of the opium poppy is a valuable animal foodstuff as it contains high concentrations of oil and protein. The capsule of the opium poppy, not the seed, is the main site of alkaloid accumulation and the time of peak accumulation can occur during the three weeks before maturity depending upon environmental factors.[56] Opium is derived from incision of the capsules and allowing the exudate to dry.

---

[56] J. C. Laughlin, *J. Agric. Sci., Cambridge*, 1985, **104**, 559.

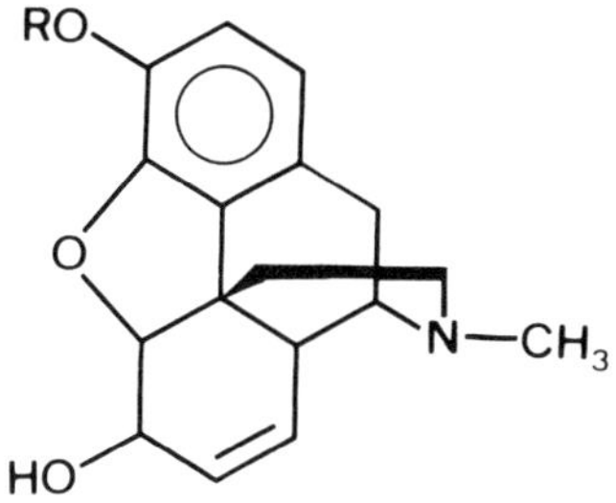

(9) R = H; *Morphine*

(33) R = CH$_3$; *Codeine*

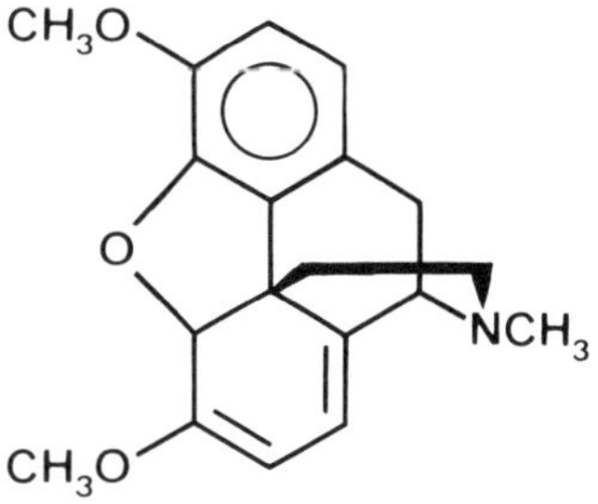

(34) *Thebaine*

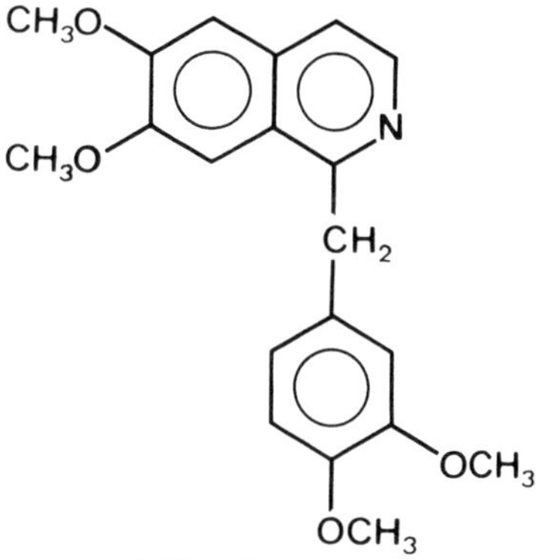

(35) *Papaverine*

The structural elucidation of morphine presented a major challenge to chemists of the first half of the Twentieth century as degradation products were complicated by various skeletal rearrangement reactions. The correct structure was proposed by Gulland and Robinson in 1923[57] but was not confirmed for almost 30 years when an unambiguous structure was determined by X-ray crystallography.[58] The biosynthesis of morphinan alkaloids has been exten-

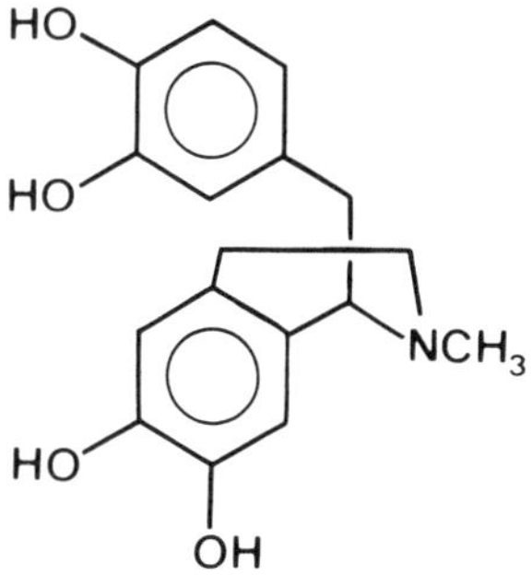

(36) *Norlaudanosoline*

[57] J. M. Gulland and R. Robinson, *Mem. Proc. Manchester Lit. Philos. Soc.*, 1925, **69**, 79.
[58] D. Mackay and D. Hodgkin, *J. Chem. Soc.*, 1955, 3261.

sively reviewed.[54,59] Early labelling studies showed that the morphine skeleton is composed of two tyrosine molecules.[60] (S)-Norlaudanosoline (36) was shown to be a key intermediate in the biosynthesis.[61–63] The degree of incorporation into various alkaloids increased as the precursor was changed from (36) to its di-*O*-methyl and its *N,O,O*-trimethylether [reticuline]. Reticuline is converted to thebaine which is demethylated to codeine and then to morphine.

Cocaine

Cocaine (37) is the major alkaloid present in the leaves of *Erythroxylon coca*, which is grown commercially in South America and South East Asia. For centuries South American Indians have used a mixture of coca leaves and lime as a masticatory to relieve the symptoms of fatigue when travelling. It is used medicinally as a local anaesthetic, however it is best known as a substance of abuse.

(37) Cocaine

Cocaine belongs to the tropane group of alkaloids, which includes the pharmacologically important alkaloids hyoscyamine (from henbane), atropine (racemic hyoscyamine, from deadly nightshade), and scopolamine.

Early attempts to elucidate the biosynthesis of cocaine were frustrated by the low incorporation of labelled precursors. Leete[64] finally achieved success by painting leaves of *E. coca* with a solution of DL[5-^{14}C]-ornithine. Degradation of the resulting alkaloid showed the activity was located at the bridgehead atoms and equally divided between the two positions, and that the biosynthesis of cocaine followed a different pathway to that of hyoscyamine. Similar labelling studies with the latter gave an alkaloid with the label located at only one bridgehead atom. These observations were explained by the presence of a symmetrical intermediate in the biosynthesis of cocaine and an asymmetrical intermediate in that of hyoscyamine. The proposed mechanism requires decarboxylation of ornithine to putrescine, which then undergoes methylation. In the

[59] R. B. Herbert, in 'The Alkaloids', ed. J. E. Saxton, Specialist Periodical Reports, The Chemical Society, London, 1973, Vol. 3, p. 157.
[60] E. Leete, *J. Am. Chem. Soc.*, 1959, **81**, 3948.
[61] A. R. Battersby, R. Binks, and B. J. T. Harper, *J. Chem. Soc.*, 1962, 3534.
[62] A. R. Battersby, R. Binks, R. J. Francis, D. J. McCaldin, and H. Ramuz, *J. Chem. Soc.*, 1964, 3600.
[63] T. Lorenz, R. L. Legge, and M. Moo-Young, *Enzyme Microb. Technol.*, 1988, **10**, 219.
[64] E. Leete, *J. Am. Chem. Soc.*, 1982, **104**, 1403.

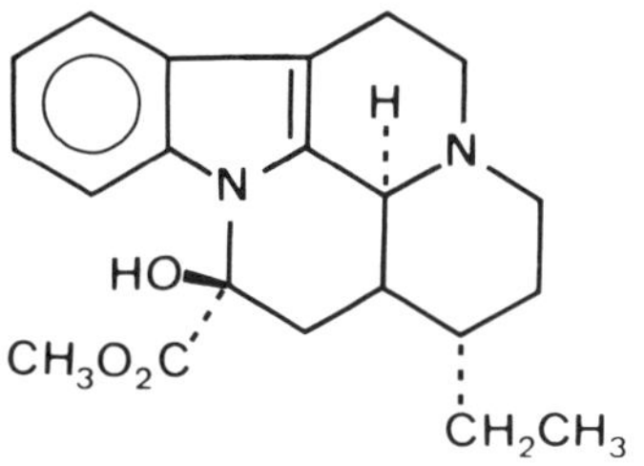

(38) *Vincamine*

biosynthesis of hyoscyamine, N-methylputrescine is formed by methylation of ornithine prior to decarboxylation. N-Methylputrescine then undergoes oxidative cyclization to the N-methylpyrrolinium ion (31) which condenses with acetoacetate.

Alkaloids in Other Crop Plants

The plants commonly known as periwinkles comprise two distinct genera, both of which contain alkaloids of pharmaceutical value. The periwinkles of the temperate zone (Europe, Western and Central Asia, and North America) belong to the genus *Vinca*, the most important species being *Vinca minor*. The alkaloids of *V. minor* have been used in European countries to treat various ailments including hypertension and angina. The most important alkaloid is vincamine (38); minor alkaloids include vincine (11-methoxyvincamine), vincaminine, and vincinine. The chemistry of these compounds was reviewed by Stern.[65]

The alkaloids of the Madagascan periwinkle, *Catharanthus roseus* (sometimes incorrectly referred to as *Vinca rosea*) are more important. More alkaloids (over 90) have been isolated from *C. roseus* than any other plant. Although vindoline is the major alkaloid of *C. roseus*, the minor alkaloids vincaleukoblastine (vinblastine) (39) and leurocristine (vincristine) (40) have attracted most attention through their cytotoxic properties. Cytotoxicity was first recognized in the P1534 leukaemia system.[66] Several other tumours have been reported to respond to these alkaloids. These alkaloids have been used to treat Hodgkin's disease. The chemistry and biosynthesis of the bisindole alkaloids of *C. roseus*, which occurs via tryptophan, have been reviewed.[67]

Colchicine (11), the major alkaloid of the autumn crocus or meadow saffron (*Colchicum autumnale*), has been used in medicine since antiquity. Although it has been used to treat gout and Familial Mediterranean Fever, current interest is focused on its anti-neoplastic properties. Colchicine is effective against chronic myelocytic leukaemia, but only at toxic or near toxic doses. It is used in agricultural and genetic research for its ability to induce polyploidy (doubling

---

[65] 'The Vinca Alkaloids, Botany, Chemistry and Pharmacology', ed. W. I. Taylor and N. R. Farnsworth, Marcel Dekker, New York, 1973.
[66] I. S. Johnston, J. G. Armstrong, M. Gorman, and J. P. Burnett, *Cancer Res.*, 1963, **23**, 1390.
[67] G. Blasko and G. A. Cordell, *Alkaloids (N.Y.)*, 1990, **37**, 1.

(39) R = CH₃; *Vinblastine*   (40) R = CHO; *Vincristine*

of chromosome groups). The plant has also been responsible for the death of grazing animals. The unusual tropolonic structure was originally proposed in 1945[68] and confirmed by subsequent work (see Section 2).

For a review of the biosynthesis of colchicine, which originates from one mole each of tyrosine and phenylalanine, via a phenylisoquinoline, see Herbert.[69] Earlier reviews covered the chemistry and biology.[70,71]

Alkaloids from the bark of the cinchona, *Cinchona officinalis*, and related trees (*C. ledgeriani* and *C. succirubra*) are still widely used in the treatment of malaria and cardiac arrhythmias. The major alkaloid quinine (41) represents 70 % of the total. Its diastereomer, quinidine, is an anti-arrhythmic drug. Other alkaloids of pharmaceutical interest are cinchonine and cinchonidine.

For a general review of the biosynthesis of plant alkaloids see Herbert.[69]

3 Methods of Analysis

Colorimetric Procedures

The earliest diagnostic tests used were based upon the ability of alkaloids to form insoluble complex metal–iodide salts. The most widely used reagent was that proposed by Dragendorff, who used a bismuth iodide solution which produced distinctive red precipitates in the presence of a wide range of alkaloids. Other examples include Mayer's ($HgCl_2/KI$), Marme's (CdI_2/KI), and Wagner's (I_2/KI) reagents.

Lupin plant breeders still use Dragendorff's reagent to identify bitter genotypes. The test involves the application of plant sap to a test paper impregnated with the reagent: the presence of alkaloid is indicated by an orange-brown

[68] M. J. S. Dewar, *Nature (London)*, 1945, **155**, 141.
[69] R. B. Herbert, *Nat. Prod. Rep.*, 1987, **4**, 423.
[70] D. J. Eigsti and P. Dustin, in 'Colchicine in Agriculture, Medicine, Biology and Chemistry', Iowa State College Press, Ames, Iowa, 1955.
[71] H. G. Capraro and A. Brossi, *Alkaloids (N.Y.)*, 1984, **23**, 1.

(41)   Quinine

colour. Although the test is rapid and easy to use, its limit of detection is equivalent to 3 g kg^{-1} alkaloid in the seed which is inadequate for routine purposes with low alkaloid (< 500 mg kg^{-1}) cultivars.

Colorimetric methods for the analysis of glycoalkaloids were widely used before the introduction of gas chromatography. Sanford and Sinden[72] used a solution of antimony trichloride in hydrochloric acid as the chromogenic reagent for analysis of glycoalkaloids in potato. Disadvantages of this method include a lengthy extraction process, the use of toxic and corrosive reagents, and the lack of response to fully saturated alkaloids such as demissine. Clarke's reagent (paraformaldehyde in 85 % phosphoric acid) was used as a colorimetric reagent for glycoalkaloids.[73] Bromocresol purple[74] and bromothymol blue[75] have been used for the colorimetric determination of lupin alkaloids and total glycoalkaloids, respectively.

An acid–base titration with *p*-toluenesulphonic acid and tetrabromophenolphthalein ethyl ester as indicator was used for the analysis of alkaloids in low-alkaloid (or 'sweet') lupins.[76] The limit of detection was 100 mg kg^{-1} alkaloid in the seed, which was considered suitable for a screening test to identify new cultivars (varieties) of lupin that might contain unacceptable levels of alkaloid.

A procedure was developed for glycoalkaloids based on titration of the purified aglycones with bromophenol blue.[77] The method, however, was not reliable because of low and variable recoveries of added solanine, and the use of 'recovery factors' was advocated to compensate for this.[78]

Diethylthiobarbituric acid has been used as a spectrophotometric agent for the tobacco alkaloids.[79]

Chromatographic Procedures

Chromatographic procedures have largely replaced the classical colorimetric

---

[72] L. L. Sanford and S. L. Sinden, *Am. Potato J.*, 1972, **44**, 209.
[73] H. Ross, P. Pasemann, and W. Nitzsche, *Z. Pflanzenzuecht.*, 1978, **80**, 64.
[74] D. von Baer, E. H. Reimeroles, and W. Feldheim, *Z. Lebensm. Unters. Forsch.*, 1979, **169**, 27.
[75] D. T. Coxon, K. R. Price, and P. G. Jones, *J. Sci. Food Agric.*, 1979, **30**, 1043.
[76] L. P. Ruiz, *N.Z. J. Agric. Res.*, 1976, **20**, 51.
[77] T. J. Fitzpatrick and S. F. Osman, *Am. Potato J.*, 1974, **51**, 318.
[78] H. Butcher, *N.Z. J. Exp. Agric.*, 1978, **6**, 127.
[79] C. L. Smith and M. Cook, *Analyst (London)*, 1987, **112**, 1515.

Table 3 *Chromatographic techniques used for the analysis of alkaloids in crop plants*

Alkaloid/plant	Method	Reference
Purine alkaloids	HPLC	1
	GLC	2
Glycoalkaloids/potato	Capillary GC	3
	HPLC	4, 5
	GC/MS	6
Lupin alkaloids	Capillary GC, GC/MS	7
	Capillary GC	8
Catharanthus alkaloids	HPLC	9
	HPLC/MS	10
Opium poppies	SCFC	11
	HPLC	12, 13

[1] L. C. Trugo, R. Macrae, and J. Dick, *J.Sci. Food Agric.*, 1983, **34**, 300.
[2] B. Y. Guo and H. B. Wan, *J. Chromatogr.*, 1990, **505**, 435.
[3] W. M. J. van Gelder, H. H. Jonker, H. J. Huizing, and J. J. C. Scheffer, *J. Chromatogr.*, 1988, **442**, 133.
[4] S. C. Morris and T. H. Lee, *J. Chromatogr.*, 1981, **219**, 403.
[5] K. Kobayashi, A. D. Powell, M. Toyoda, and Y. Saito, *J. Chromatogr.*, 1989, **462**, 357.
[6] W. M. J. van Gelder, L. G. M. T. Tuinstra, J. van der Greef, and J. J. C. Scheffer, *J. Chromatogr.*, 1989, **482**, 13.
[7] D. von Baer, E. H. Reimberdes, and W. Feldheim, *Z. Lebensm. Unters. Forsch.*, 1979, **169**, 27.
[8] C. R. Priddis, *J. Chromatogr.*, 1983, **261**, 95.
[9] T. Naaranlahti, M. Nordstrom, A. Huhtikangas, and M. Lounasmaa, *J. Chromatogr.*, 1987, **410**, 488.
[10] S. Auriola, V. P. Ranta, T. Naaranlahti, and S. P. Lapinjoki, *J. Chromatogr.*, 1989, **474**, 181.
[11] J. L. Janicot, M. Caude, and R. Rosset, *J. Chromatogr.*, 1988, **437**, 351.
[12] N. R. Ayyangar and S. R. Bhide, *J. Chromatogr.*, 1988, **436**, 455.
[13] V. K. Srivastava and M. H. Maheshwari, *J. Assoc. Off. Anal. Chem.*, 1985, **68**, 801.

and titrimetric procedures. Reviews have covered glycoalkaloids[80] and quinolizidine alkaloids.[81]

Capillary gas chromatography coupled with mass spectrometry (GC/MS) provides a powerful tool for identification and quantitation of alkaloids. GC/MS procedures are used for the tobacco alkaloids.[82] Retention times and mass spectra of the major quinolizidine alkaloids have been documented[83,84] and provide a useful database for identification purposes.

Thin-layer chromatography (TLC) is also a useful technique for identifying alkaloids, at a fraction of the cost of GC/MS. The R_f values of a wide range of alkaloids in various TLC solvent systems have been documented[83,85] and can be used for identification and semi-quantitation of alkaloids. One can use TLC to

[80] R. Verpoote and A. Baerheim-Svendsen, in 'Chromatography of Alkaloids', Journal of Chromatography Library, 23B, Elsevier, Amsterdam, 1984, Chapter 19, p. 185.
[81] R. Verpoote and A. Baerheim-Svendsen, in 'Chromatography of Alkaloids', Journal of Chromatography Library, 23B, Elsevier, Amsterdam, 1984, Chapter 7, p. 55.
[82] M. A. Scheijen, B. Brandt-de Boer, J. J. Boon, W. Haas, and V. Heeman, *Beitr. Tabakforsch. Int.*, 1989, **14**, 261.
[83] Y. D. Cho and R. D. Martin, *Anal. Biochem.*, 1971, **44**, 49.
[84] T. Hatzfold, I. Elmadfa, R. Gross, M. Wink, T. Hartmann, and L. Witte, *J. Agric. Food Chem.*, 1983, **31**, 934.
[85] A. B. Svendsen, *J. Planar Chromatogr.*, 1989, **2**, 8.

identify isomeric alkaloids that are not readily distinguished by retention times or mass spectra; e.g. lupinine and epilupinine have similar retention times on a 3% OV-17 column and identical mass spectra, but can be distinguished by their different R_f values on TLC plates.[83]

Dragendorff's reagent and iodoplatinate solution are commonly used to detect the presence of alkaloids on TLC plates. The latter reagent has the added advantage of producing distinctive colours with different alkaloids.

The non-volatile nature of the glycoalkaloids limits the utility of gas chromatography for quantitative analysis. This can be partially countered by hydrolysis to the aglycones which are readily determined, or by derivatization.

Liquid chromatography has been used to analyse crop plant alkaloids (Table 3). A major difficulty associated with analysis of quinolizidine alkaloids is the relatively poor sensitivity associated with the lack of characteristic chromophores. Liquid chromatographic assays are described in the literature for the determination of most of the alkaloids discussed here in biological fluids.

Immunoassays

Immunoassays are gaining in popularity for the analysis of foodstuff contaminants, such as mycotoxins,[86] pesticides,[87] and other small molecules such as glucosinolates[88] and hormones.[89] Immunoassays are generally sensitive, specific for the analyte of interest and require minimal sample preparation. They are well suited to applications involving large numbers of samples.

The first immunoassay to be developed for glycoalkaloids was a radioimmunoassay.[90] Antisera to the aglycone, solanidine (3), was prepared by injecting rabbits with a conjugate between bovine serum albumin (BSA) and the hemisuccinate ester of the aglycone. The assay was of limited value because it was necessary to hydrolyse the glycoalkaloid extracts to the aglycones, the titre of the antisera was low, and radio-labelled solanine was unavailable commercially.

A much improved assay in the form of an enzyme-linked immunosorbent assay (ELISA) was developed by Morgan et al.[91] Antisera were raised in rabbits against a conjugate between BSA and solanine prepared by a periodate cleavage method. Cross-reactions showed that the antibodies recognized α-solanine (12), α-chaconine (13), solanidine, and demissidine with equal affinity. Sample preparation was simple, no hydrolysis was required and recovery of added α-chaconine was quantitative. Results obtained by ELISA were highly correlated with those obtained by colorimetry and HPLC.[92]

[86] M. R. A. Morgan, A. S. Kang, and H. W-S. Chan, J. Sci. Food Agric., 1986, 37, 873.
[87] S. I. Wie and B. D. Hammock, J. Agric. Food Chem., 1984, 32, 1294.
[88] F. Hassan, N. E. Rothnie, S. P. Yound, and M. V. Palmer, J. Agric. Food Chem., 1988, 36, 398.
[89] G. Davis, M. Hein, B. Neely, R. Sharp, and M. Carnes, Anal. Chem., 1985, 57, 638A.
[90] R. P. Vallejo and C. D. Ercogovich, Proceedings of the Ninth Materials Research Symposium, National Bureau of Standards (USA), Special Publication No. 519, 1979, p. 333.
[91] M. R. A. Morgan, R. McNerney, J. A. Matthew, D. T. Coxon, and H. W. S. Chan, J. Sci. Food Agric., 1983, 34, 593.
[92] K. F. Hellenas, J. Sci. Food Agric., 1986, 37, 776.

The assay was improved to enable visual assessment of potato samples containing total glycoalkaloid (TGA) levels greater than 20 mg 100 g^{-1} fresh tuber weight.[93]

An ELISA for lupin alkaloids was developed in one of our (D. J. Harris and D. G. Allen) laboratories. Antibodies were raised in sheep against a conjugate between keyhole limpet haemocyanin (KLH) and the hemisuccinate ester of 13-hydroxylupanine.[94] The antibody recognizes lupanine (19) and 13-hydroxylupanine (20) with equal affinity, but has less affinity for angustifoline (17), lupinine (16), sparteine (18), and α-isolupanine. The assay is used for determining alkaloid concentrations in low-alkaloid cultivars of *L. angustifolius* in which lupanine and 13-hydroxylupanine comprise about 90% of total alkaloids. The assay can also be used for *L. albus*, which has a similar alkaloid profile.

ELISAs are available for the *Catharanthus* alkaloids.[95,96] Radioimmunoassays are available for the opium alkaloids.[97,98]

Several organizations manufacture ELISA kits for determination of the morphinan alkaloids, cocaine (36), and derivatives and the methylxanthines in biological fluids.

The development phase of an immunoassay requires considerable effort, especially in the preparation of conjugates of high specificity and antigenicity. To justify this a large number of routine assays would need to be anticipated. Such assays would not be suited for plant breeding programmes with new species, or with hybrids, until all of the development work has been established by GC/MS.

Other Procedures

Wink *et al.*[99] used laser microprobe mass analysis (LAMMA 1000) to determine the distribution of lupin alkaloids in plant tissue. They found that the alkaloids were concentrated in the epidermis and one or two sub-epidermal cell layers in fresh stems of *L. polyphyllus*.

Near infrared reflectance spectroscopy (NIRS) may prove to be a valuable method for analysis of alkaloids in plant material. Advantages include minimal sample preparation and speed. The sensitivity is a major disadvantage. The method is now used routinely to determine nicotine (1) in tobacco.[100]

Radioisotope dilution methods have been used for the methylxanthines.[101]

[93] C. M. Ward, J. G. Franklin, and M. R. A. Morgan, *Food Addit. Contam.*, 1988, **5**, 621.
[94] N. F. Gare, B. N. Greirson, D. G. Allen, S. C. Baseden, and I. Watson, Proceedings of the 10th Australian Symposium on Analytical Chemistry, Brisbane, 1989, Abstract 249.
[95] S. Lapinjoki, H. Verajankorva, J. Heiskanen, M. Niskanen, A. Huhtikangas, and M. Lounasmaa, *Planta Med.*, 1987, **53**, 565.
[96] S. Lapinjoki, H. Verajankorva, A. Huhtikangas, T. J. Lehltola, and M. Lounasmaa, *J. Immunoassay*, 1986, **7**, 113.
[97] E. Gurkan, *Fitoterapia*, 1984, **6**, 349.
[98] G. T. F. Galasko, K. I. Furman, and E. Alberts, *Food Chem. Tox.*, 1989, **27**, 49.
[99] M. Wink, J. H. Heinen, H. Vogt, and H. M. Schiebel, *Plant Cell Rep.*, 1984, **3**, 230.
[100] B. G. Osborne and T. Fearn, 'Near infrared spectroscopy in food analysis', Longman Scientific, Harlow, 1986, p. 179.
[101] T. W. Baumann and H. Gabriel, *Plant Cell Physiol.*, 1984, **25**, 1431.

The concentration of alkaloids in high-alkaloid (bitter) lupin species has been measured using a gravimetric procedure.[102]

Fast-atom-bombardment mass spectrometry of *Solanum* glycoalkaloids has been used to analyse crude extracts from potatoes.[103]

4 Biological Role

There is strong evidence given in the literature for the primary role of all alkaloids in the plant being one of chemical defence. To fulfil this role, the plant would be expected to accumulate these secondary metabolites in relation to the risk of predation.[104] Tissues with a high dietary value, such as seeds, buds, and young leaves, would be most at risk and therefore expected to have the highest alkaloid concentrations. In addition, the bitter flavour of the alkaloids, even at very low concentrations, would be expected to act as a deterrent to vertebrates. Vertebrate predators that overcome the bitter taste of these compounds then have to cope with the pharmacological effects such as inducing vomiting (ipecacuanha alkaloids), fibrillations (lupin alkaloids), anticholinesterase activity (steroidal alkaloids), and cardiac arrest (strychnine). Toxic effects are not confined to vertebrates. Many alkaloids have fungitoxic and bacteriostatic properties and cause aversive behaviour in molluscs. An allelopathic effect has also been demonstrated, showing a defence against other plant species as well.

Wink[105] reported that the quinolizidine alkaloids deterred the feeding of herbivorous mammals, insects, and molluscs. They were lethal to insects at concentrations of 3–50 mM, and inhibited the growth of fungi and bacteria at 1–50 mM. These levels are higher than those in modern cultivars of *L. angustifolius* grown in Western Australia.[106] Wink[105] also reported that yields were reduced if the alkaloid levels were any lower than about 5 mM kg^{-1}. Western Australian plant breeders aim to keep the level between about 0.5 and 1 mM kg^{-1} because aphid damage becomes too great at any lower concentrations.[107]

There is a case for the alkaloids of tea and coffee having a defence role. Firstly, they are excreted during germination[101] and appear to have an allelopathic effect.[108] Secondly, they have been shown to have insecticidal properties, as inhibitors of phosphodiesterase activity,[109] and anti-fungal properties.[110–112] Furthermore, the concentrations are highest in those plant parts most likely to

---

[102] M. Muzquiz, I. Rodenas, J. Villaverde, and M. Cassinello, Proceedings of the Second International Lupin Conference, Torremolinos, 1982.
[103] K. R. Price, F. A. Mellor, R. Self, G. R. Fenwick, and S. F. Osman, *Biomed. Mass Spectrom.*, 1985, **12**, 79.
[104] D. F. Rhoades, in 'Herbivores', ed. G. A. Rosenthal and D. H. Janzen, Academic Press, New York, 1979, p. 3.
[105] M. Wink, *Plant Syst. Evol.*, 1985, **150**, 65.
[106] D. J. Harris, Proceedings of the Fifth International Lupin Conference, Poznan, 1988, p. 593.
[107] J. S. Gladstones, personal communication, 1990.
[108] C. H. Chou and G. R. Waller, *J. Chem. Ecol.*, 1980, **6**, 643.
[109] J. A. Nathanson, *Science*, 1984, **226**, 184.
[110] S. J. H. Rizvi, V. Jaiswal, D. Mukeiji and S. N. Mathur, *Naturwissenschaften*, 1980, **67**, 459.
[111] R. L. Buchanan, G. Tice and D. Marino, *J. Food Sci.*, 1981, **47**, 319.
[112] S. K. Prabhuji, G. C. Srivastrava, S. J. H. Rizvi, and S. N. Mathur, *Experientia*, 1983, **39**, 177.

be eaten by predators. The use of nicotine as an insecticide and fungicide shows its value to the defence of the plant.

A secondary role in nitrogen metabolism and storage seems probable for there is a diurnal variation in the alkaloid content of the developing lupin plant (D. G. Allen and D. J. Harris, unpublished observations). The alkaloid content of the germinating lupin seed falls quickly during the first few days and then slowly increases again. This may be due to metabolism as a source of nitrogen or to efflux whereby an allelopathic function could be performed. The latter is reputed to be the case for the purine alkaloids.[113]

In the domestication of plants for cropping purposes, two opposing needs emerged. The plant breeders working with potatoes and lupins, have subjected the crops to a selection process aimed at reducing the alkaloid content of the tuber and the seed respectively to increase their palatability. On the other hand, plant breeders working with *Papaver* and *Nicotiana* would not want to reduce alkaloid levels in parallel with any improved agronomic performance, since it is the alkaloid that is of commercial interest. However, a low-alkaloid producing cultivar, *P. somniferum* L. var. album D.C., was developed for the production of oil and baking ingredients.[114] During the 1980s interest was aroused in developing low-alkaloid cultivars of *Coffea* sp.

5 Pharmacology and Toxicology

The *Solanum* alkaloids have strong anticholinesterase activity on the central nervous system[115-117]; they are more inhibitory than quinine (41)[115] although a-tomatine (14) is a weak inhibitor. The lethal effect of glycoalkaloids in mice was reduced by treatment with atropine sulphate, a stimulant of the central nervous system.[116] These glycoalkaloids also have saponin-like properties and can disrupt membrane function in the gastro-intestinal tract leading to haemorrhagic damage.[118] This damage can be severe enough to cause death with the extent of necrosis far outweighing the inhibitory affects on acetylcholinesterase activity.[119] The pathology of this condition was recently described.[120] There are statements in the literature to the effect that the glycoalkaloids can disrupt mitosis.[121] The uptake of glycoalkaloids by the mammalian body could be affected by the presence of saponins, pH, and the amount of glycoalkaloid in the diet.[122] Peak accumulation in the rat occurs about 12 h after ingestion, and excretion of the aglycone occurs rapidly.[123]

Reported cases of glycoalkaloid poisoning in humans have all involved the

[113] G. Attaguile, C. Barbagallo, and F. Savoca, *Pharmacol. Res. Commun.*, 1988, **20**, Suppl. 5, 129.
[114] T. Suzuki and G. R. Waller, *Plant Soil*, 1987, **98**, 131.
[115] W. H. Orgell, *Lloydia*, 1963, **26**, 36.
[116] B. C. Patil, R. P. Sharma, D. K. Salunkhe, and K. Salunkhe, *Food Cosmet. Toxicol.*, 1972, **10**, 395.
[117] M. McMillan and J. C. Thompson, *Quart. J. Med.*, 1979, **48**, 227.
[118] H. König and A. Staffe, *Dtsch. Tieraerztl. Wochenschr.*, 1953, **60**, 150.
[119] D. C. Baker, R. F. Keeler, and W. P. Garfield, *Toxicol. Pathol.*, 1988, **16**, 333.
[120] D. Baker, R. Keeler, and W. Garfield, *Toxicol. Pathol.*, 1988, **16**, 33.
[121] P. Danneberg and D. Schmähl, *Arzneim. Forsch.*, 1953, **3**, 151.
[122] S. C. Morris and T. H. Lee, *Food Technol. Aust.*, 1984, **36**, 118.
[123] K. Nishie, M. R. Gumbmann, and A. C. Keyl, *Toxicol. Appl. Pharmacol.*, 1971, **19**, 81.

Table 4 Acute oral toxicity of major alkaloids in crop plants

Alkaloid	Subjects	Lethal dose mg kg^{-1}	LD$_{50}$	Reference
Glycoalkaloids	Rats		590	1, 2
Glycoalkaloids	Mice		590	3
Glycoalkaloids (solasodine)	Hamsters		>1500	4
Lupin alkaloids	Rats		2279	5
Lupinine			1464	5
Caffeine	Rats		200	6a
Caffeine	Humans	150–200		8
Tomatine	Rats	900–1000		7
Nicotine	Rats	50–60		6b
Theobromine	Rats		950	8
Theophylline	Mice		332	8

[1] H. König and A. Staffe, *Dtsch. Tieraertztl. Wochenschr.*, 1953, **60**, 150.
[2] D. D. Gull, F. M. Isenberg, and H. H. Bryan, *Hort. Sci.*, 1970, **5**, 316.
[3] B. C. Patil, R. P. Sharma, D. K. Salunkhe, and K. Salunkhe, *Food Cosmet. Toxicol.*, 1972, **10**, 395.
[4] R. F. Keeler, D. R. Douglas, and D. F. Stallknecht, *Am. Potato J.*, 1975, **52**, 125.
[5] D. S. Petterson, Z. L. Ellis, D. J. Harris, and Z. L. Spadek, *J. Appl. Tox.*, 1987, **7**, 51.
[6] Merck Index, 11th edn., (a) Monograph 1623; (b) Monograph 6242.
[7] R. H. Wilson, G. W. Poley, and F. De Eds, *Toxicol. Appl. Pharmacol.*, 1961, **3**, 39.
[8] S. M. Tarka, *CRC Crit. Rev. Toxicol.*, 1982, **9**, 275.

potato alkaloids. Typical signs of poisoning include elevated body temperature, drowsiness, apathy, abdominal pain, diarrhoea, vomiting, weakness, and depression.[117,124] Humans appear to be more susceptible than laboratory animals to poisoning by glycoalkaloids. Recorded toxic and lethal doses to humans are about 2–5 mg kg^{-1} body weight, *i.e.* about the same as for strychnine, whereas in rodents it is about 600 mg kg^{-1} (Table 4). The amount of alkaloids consumed in a 100 g serve of potatoes with a glycoalkaloid level of 20 mg kg^{-1} would be about 1/100th of a lethal dose. However, consumption of greater quantities of damaged or green potatoes could see a dramatic reduction in the safety margin, especially since there is considerable variation in individual susceptibility to the glycoalkaloids.[117,125] Although a concentration of 200 mg TGA kg^{-1} has frequently been cited as 'safe', Slanina[126] recommends that the average TGA concentration in new cultivars of potatoes should not exceed 100 mg kg^{-1}. This would help to increase the safety margin over the lowest reported dose to induce toxic symptoms in humans, 2 mg kg^{-1} day^{-1}.

Studies have shown that feeding potatoes with high alkaloid levels to laboratory animals can result in an increased incidence of neurological defects in the young; increased foetal mortality; increase in resorptions; and a decreased conception rate.[127–129] The potential risk for pregnant women consuming green or damaged potatoes would seem to be high.

[124] L. A. Reelah and A. Keem, *Sov. Med.*, 1958, **22**, 129.
[125] S. G. Willimott, *Analyst (London)*, 1933, **58**, 431.
[126] P. Slanina, *Var Föda*, 1990, Supplement 1,3.
[127] D. E. Poswillo, D. Sopher, S. J. Mitchell, D. T. Coxon, R. F. Curtis, and K. R. Price, *Teratology*, 1973, **8**, 339.
[128] B. E. Kline, H. von Elbe, N. A. Dahle, and S. M. Kupchan, *Proc. Soc. Exp. Biol. Med.*, 1961, **107**, 807.
[129] C. A. Swinyard and S. Chaube, *Teratology*, 1973, **8**, 349.

For a detailed review of the toxicity of the potato glycoalkaloids to mammals, insects, nematodes, and fungi see Morris and Lee.[122]

The various common lupin alkaloids and their derivatives have been shown to arrest cardiac action and induce contractions of isolated rabbit intestine and of guinea pig uterus[130]; decrease the amplitude of heart contractions[131]; induce a marked general depression and an arching of tails in rats[132]; induce cyanosis, cramps, and shallow breathing in humans[133]; cause respiratory depression and neuromuscular blockage in cats and dogs[134]; and cause fibrillations and cyanosis in rats.[134,135] Sparteine (18), which was once widely used in obstetrics, has an antifibrillatory effect.[136] All of these phenomena have a threshold level and are reversible.

Caffeine (4) is one of the most widely used drugs in the world[137] and there is a plethora of information, much of which is conflicting, in the literature about its pharmacological and physiological effects.[138-142] In small doses, < 3 mg kg^{-1} day^{-1}, caffeine increases attention and improves mood, allowing the consumer to sustain mental and physical performance.[138] Doses exceeding 500 mg may elicit neuroendocrine effects in humans.[143] The diuretic effects are well documented.[144] The increase in urinary excretion of minerals[145] is of particular concern for women. The methylxanthines are powerful anorexic agents.[146] The reported teratogenic effects[139,140] seem to be at levels unlikely to ever be reached; at the equivalent of more than 40 cups of coffee daily.

Interest in the scientific and popular media about an association between caffeine and heart disease has continued for decades. A careful examination of the literature to the end of 1982 consistently failed to show a direct association between coffee intake and myocardial infarction.[147] Schreiber's group[148]

[130] M. Mazur, P. Polakowski, and A. Szadowska, *Acta Physiol. Polon.*, 1966, **17**, 299.
[131] M. Mazur, P. Polakowski, and A. Szadowska, *Acta Physiol. Polon.*, 1966, **17**, 311.
[132] J. E. Peterson, *Aust. J. Exp. Biol.*, 1963, **451**, 123.
[133] J. Schmidlin-Meszaros, *Mitt. Geb. Lebensmittelunters. Hyg.*, 1973, **64**, 194.
[134] G. G. Lu, *Toxicol. Appl. Pharmacol.*, 1964, **6**, 328.
[135] D. S. Petterson, Z. L. Ellis, D. J. Harris, and Z. E. Spadek, *J. Appl. Tox.*, 1987, **7**, 51.
[136] G. Zetler and O. Strubelt, *Arzneim. Forsch.*, 1980, **30**, 1497.
[137] J. D. Lane and D. C. Manus, *Psychosom. Med.*, 1989, **51**, 373.
[138] S. M. Tarka, *CRC Crit. Rev. Toxicol.*, 1982, **9**, 275.
[139] S. M. Tarka and C. A. Shively, in 'Toxicological Aspects of Food', ed. I. Miller, Elsevier, Barking, 1987, Chapter 11.
[140] G. M. Al-Hackim, *Eur. J. Obstet. Gynecol. Reprod. Biol.*, 1989, **31**, 237.
[141] T. W. Rall, in 'The Pharmacological Basis of Therapeutics', ed. A. R. Gilman, L. S. Goodman, T. W. Rall, and F. Muraals, 7th edn, 1985, Macmillan, New York, Chapter 25, p. 589.
[142] N. W. Russ, E. T. Sturgis, R. J. Malcolm and L. Williams, *J. Clin. Psychiatry*, 1988, **49**, 457.
[143] E. R. Spindel, R. J. Wartman, A. McCall, D. B. Carr, L. Conley, L. Griffith, and M. A. Arnold, *Clin. Pharmacol. Ther.*, 1984, **36**, 402.
[144] L. J. Dorfman and M. E. Jarvik, *Clin. Pharmacol. Ther.*, 1970, **11**, 869.
[145] L. K. Massey and T. Berg, *Fed. Proc.*, 1985, **44**, 1149.
[146] J. H. Gans, R. Korson, M. R. Cater, and C. C. Ackerly, *Toxicol. Appl. Pharmacol.*, 1988/89, **53**, 481.
[147] P. W. Curaltoo and D. Robertson, *Am. Int. Med.*, 1983, **98**, 641.
[148] G. B. Schreiber, M. Robins, C. E. Maffeo, M. Masters, A. Bond, and D. Morganstein, *Prev. Med.*, 1988, **17**, 295.
[149] M. L. Burr, J. E. J. Gallagher, B. Butland, C. H. Bolton, and L. G. Downs, *Eur. J. Clin. Nut.*, 1989, **43**, 477.

checked on two design factors, imprecise measurements and confounding variables, that may account for discrepancies in reported findings on the effects of caffeine on disease. Of 32 risk factors analysed by linear and logistic regression only sex and smoking were found to be important confounders of caffeine intake. However they did point to other factors that should be taken into account when considering (future) study designs. For men, they suggested intake of dietary fat and vitamin C, and body mass index; for women, vitamin use, alcohol intake, stress, and perceived health status.[148] A 1989 study concluded that the effect of coffee on the risk of heart disease is very low in the United Kingdom.[149] This may be because of the type of coffee used by the subjects. The British tend to consume instant coffee with a low alkaloid content per cup whereas most American studies would have been on subjects who drank strong boiled coffee containing possibly twice as much caffeine. In one study on caffeine consumption, in 58 homes studied the range of caffeine intake of individuals varied from 49–1022 mg day^{-1}.[150] The same authors also noted a high variability of caffeine content within and between national brands. It seems that a low caffeine intake poses little risk of heart disease, but that additional factors such as stress may increase the likelihood of heart disease. Maternal caffeine intake during breast feeding could contribute to infants' restlessness.[151,152]

The structure–activity relationships and pharmacological properties of the opium alkaloids were reviewed.[153,154] Other reviews covered the medicinal chemistry of the *Catharanthus* alkaloids[155]; the biological activity of *Erythroxylon coca* alkaloids[156,157] and the biology,[158] pharmacology,[159] and therapeutic uses[160] of the *Cinchona* alkaloids.

The properties and biological activity of colchicine were reviewed by Brossi[161]; the suppression of microtubular function[162] and inhibitory role of fibroblast migration (hence a possible role in glaucoma surgery)[163] are some of the topics recently covered in the literature.

6 Detoxification

The mammalian body uses three main processes to deal with foreign compounds (xenobiotics).

---

[150] B. Stavric, R. Klassen, B. Watkinson, A. Karpinski, R. Stapley, and P. Fried, *Food Chem. Toxicol.*, 1988, **26**, 111.
[151] M. I. Clement, *Br. Med. J.*, 1989, **298**, 1461.
[152] J. Rustin, *Br. Med. J.*, 1989, **299**, 12.
[153] E. Lindner, in 'Chemistry and Biology of the Isoquinoline Alkaloids', ed. J. D. Phillipson, M. F. Roberts, and M. H. Zenk, Springer, Berlin, 1985, p. 38.
[154] G. W. Pasternak, *Ann. N.Y. Acad. Sci.*, 1986, **467**, 130.
[155] H. L. Pearse, *Alkaloids (N.Y.)*, 1990, **37**, 145.
[156] G. Fodor and R. Dharanipragada, *Nat. Prod. Rep.*, 1986, **3**, 181.
[157] M. Novak, C. A. Salemink, and I. Khan, *J. Ethnopharmacol.*, 1984, **10**, 261.
[158] R. Verpoorte, J. Schripsema, and T. van der Leer, *Alkaloids (N.Y.)*, 1987, **34**, 331.
[159] J. J. McCormack, *Alkaloids (N.Y.)*, 1990, **37**, 205.
[160] R. Neuss and M. N. Neuss, *Alkaloids (N.Y.)*, 1990, **37**, 229.
[161] A. Brossi, *J. Med. Chem.*, 1990, **33**, 2311.
[162] Y. Ouyang, W. Wang, S. Huta, and Y. H. Chang, *Clin. Exp. Rheumatol.*, 1989, **7**, 397.
[163] J. P. Joseph, I. Grierson, and R. A. Hitchings, *Curr. Eye Res.*, 1989, **8**, 203.

(a) They may be excreted unchanged in the urine and faeces.
(b) They may undergo a biotransformation, most commonly a hydrolysis or oxidation, to give a more polar product.
(c) They may be conjugated to form products that are more readily excreted, most commonly a glucuronide or a sulphate.

The principal site of detoxification is the liver; the kidney and lungs are lesser sites.

The enzymes involved in the biotransformation of the alkaloids belong to two of the six main groups that deal with xenobiotics.[164] These are the hydrolases, for the hydrolysis of ester, amide, and glycoside functional groups and the oxidoreductases, or mixed function oxidases, which catalyse hydroxylation at saturated and aromatic carbon atoms as well as oxidation of groups containing nitrogen and sulphur.

Studies on morphine (9) showed that up to 80 % of the drug is excreted in the urine within 24 h, mostly as the free drug.[165,166] There is a negligible N-demethylation *in vivo*: most of the remainder is glucoronide. Morphine is converted to the 3-glucuronide by rat liver homogenates.[167,168] There is some evidence in literature that the glucoronide can cross the blood–brain barrier and that there could be an enhancement of biological activity. The direct synthesis of a morphinan alkaloid in an animal tissue has been demonstrated,[169] so morphine and codeine in brain and adrenal gland could be of endogenous origin. Codeine (33) first undergoes O-demethylation to morphine before it is metabolized. Thebaine (34) is eliminated unchanged and as the glucuronide in the rat.[170]

Nicotine (1) is oxidized to various metabolites, the main ones being cotinine and nicotine-N-oxide. Nicotine detoxification has been reviewed.[171,172] There is an involvement of prostaglandin hydrogen synthase and the mixed function oxidases.

There is little information on the detoxification of quinolizidine alkaloids in the literature. Early studies showed that about 80 % of the total alkaloid ingested was excreted in the urine within 48 h of ingestion.[173] Studies on sparteine (18) showed that it is metabolized by an N-oxidation and a rearrangement; in 5 % of the population studied the alkaloid was excreted unchanged.[174] In cattle, the quinolizidine alkaloids are also excreted unchanged in the milk.[175]

[164] H. L. Holland, *Alkaloids (N.Y.)*, 1981, **18**, 324.
[165] L. M. Mellet and L. A. Woods, *Proc. Soc. Exp. Biol. Med.*, 1961, **106**, 221.
[166] A. L. Misra, S. J. Mulé, and L. A. Woods, *J. Pharmacol.*, 1961, **132**, 317.
[167] Q. Yue, C. von Bahr, I. Odar-Cederlöf, and J. Saive, *Pharmacol. Toxicol.*, 1990, **66**, 221.
[168] L. S. Abrams and H. W. Elliot, *J. Pharmacol. Exp. Ther.*, 1974, **189**, 285.
[169] C. J. Weitz, K. F. Faull, and A. Goldstein, *Nature (London)*, 1987, **330**, 674.
[170] A. L. Misra, R. B. Pontani, and S. J. Mule, *Xenobiotica*, 1974, **4**, 17.
[171] J. W. Gorrod and P. Jennur, *Essays Toxicol.*, 1975, **6**, 35.
[172] H. Nakayama, *Drug Metab. Drug Interact.*, 1988, **6**, 95.
[173] H. Wittenberg and K. Nehring, *Pharmazie*, 1965, **20**, 156.
[174] M. Eichelbaum, N. Spannbrucker, B. Steincke, and H. J. Dengler, *Eur. J. Clin. Pharmacol.*, 1979, **16**, 183.
[175] K. E. Panter and L. F. James, *J. Anim. Sci.*, 1990, **68**, 892.

The major excretory product of quinine (41) is the 3-hydroxy product. The major metabolism of caffeine and theophylline (26) is via N-demethylation; C-8 hydroxylation is a minor pathway.[176] The acetylation of caffeine has been widely reported and reviewed.[177] Theobromine (25) also undergoes N-demethylation and hydroxylation.[178] Variable amounts of these alkaloids are excreted unchanged.

7 Alternative Methods of Production

Considerable attention has been directed towards alternative methods to the extraction and purification of alkaloids of commercial significance from plant matter.

(a) Direct synthesis is rarely feasible because of the complex structures of these compounds. Several steps would be needed and overall yields are likely to be very small.
(b) Enzyme preparations from plants would seem to be impractical with present-day knowledge and technology. The biosynthetic pathway for these compounds usually requires several stages (see Section 2) and it is possible that transient, as yet unrecognized, intermediates are necessary.
(c) The cloning of genes for alkaloid (or any other secondary product) enzymes into bacterial or yeast cells would require the cloning of many genes into the same organism. To achieve this, and then to co-ordinate the regulation of product gene expression in the host cell may be impossible.
(d) The production of alkaloids, and many other secondary plant products, in cell culture systems has been demonstrated.

The advantages of using cell cultures are many, however, there has been very little commercial success because of the low yields achieved.[179] At present only one plant metabolite, shikonin, a non-alkaloid pigment and antibiotic, is produced commercially.[180,181]

The requirements for growth of the cells in culture and for the production of secondary metabolites appear to be quite different. The accumulation of secondary metabolites often starts when the cultured cells are put under nutritional stress.[182-184] Other stresses can influence productivity. For caffeine (4) production by *C. arabica* cultures, high light intensity led to an increased alkaloid production. It seems that if alkaloid production is mainly in the leaf chloroplasts, as with the methylxanthines and many of the quinolizidine alkaloids, then the production will be proportional to chlorophyll content and

[176] F. Berthou, D. Ratanasavanh, D. Alix, D. Carlhart, C. Riche, and A. Guillouzo, *Biochem. Pharmacol.*, 1988, **37**, 3691.
[177] W. Kalow, *Prog. Clin. Biol. Res.*, 1986, **214**, 331.
[178] C. A. Shively and S. M. Tarka, *Toxicol. Appl. Pharmacol.*, 1983, **67**, 376.
[179] E. J. Staba, *J. Nat. Prod.*, 1985, **48**, 203.
[180] R. Ganapathi and F. Kargi, *J. Exp. Bot.*, 1990, **41**, 259.
[181] A. J. Parr, *J. Biotechnol.*, 1989, **10**, 1.
[182] A. J. Parr, personal communication, 1990.
[183] K-H. Knoblock and J. Berlin, *Z. Naturforsch., Teil C.*, 1980, **335**, 551.
[184] G. B. Lockwood, *Z. Pflanzenphysiol.*, 1984, **114**, 361.

the amount of illumination.[185,186] For codeine (33) and morphine (9) production various stresses increased the yields from *P. somniferum* cultures from around 60 μg L^{-1} to about 4 mg L^{-1}.[187] For other morphinans from *Papaver* spp. cultures the highest yields were from cultures subjected to temperature stress.[184] Similar observations were made on cultivars of *Cinchona*[188] and *Catharanthus*.[189]

Crude preparations of ginseng and 'rodozin', an extract of *Rodiola rosea*, are made commercially in the Soviet Union.[182]

Ganapathi and Kargi[180] reviewed the effect of cultivation techniques and environmental factors on the production of alkaloids by *C. roseus* cells. Some of their more important conclusions, which may also apply to the production of other secondary metabolites by their host cell cultures, are mentioned below. Hormones are important: they recommend the exclusion of 2,4-D from culture systems and suggest that abscisic acid will give increased yields. Ammonium ion is preferred to nitrate. Various nutritional, osmotic, or salt stresses might increase production. The optimal pH for maintenance of *C. roseus* cultures is 5.8, raising or lowering the pH can be used to release intra-cellular alkaloids into the culture medium. Product accumulation is stimulated by stress factors such as ultra-violet light; the presence of fatty acids, inorganic salts, heavy metal ions, and fungal wall components. Production in response to those factors, or biotic elicitors, is akin to the plant cell defence mechanism acting in response to an offending foreign substance. Cultures from high yielding plants give the highest alkaloid yields. The age of the culture is also an important factor.

These authors[180] also reviewed cultivation methods. The main methods used in the past were suspension and immobilized cell cultures. Biofilm culture techniques were developed more recently, they are essentially an advance on immobilized cultures. In analysing the merits of the various systems it was considered that immobilized cells, entrapped on membrane surfaces such as hollow fibre reactors or on a biofilm of inert support particles such as alginate beads, or encapsulated in inert particles, in a low shear bioreactor operated in the perfusion mode would be most likely to give the greatest yields. However, for a process to be economically feasible the rate of product formation would need to increase about 40-fold and the final product concentration would need to be at least 10 times greater than presently achievable. More research and technological breakthroughs will be needed before the use of cultured cells for alkaloid production is to become a commercial reality.

8 Conclusions

The presence of alkaloids in plants confers benefits to the plant through protection against predators. The survival of some plant species may well have

---

[185] P. M. Frischknecht and T. W. Baumann, *Phytochemistry*, 1985, **10**, 2255.
[186] M. Wink and T. Hartmann, *Planta Med.*, 1980, **40**, 149.
[187] P. F. Heinstein, *J. Nat. Prod.*, 1985, **48**, 1.
[188] E. J. Staba, *Basic Life Sci.*, 1988, **44**.
[189] M. Lounasmaa and J. Galambos, *Prog. Chem. Org. Nat. Prod.*, 1989, **55**, 89.

depended upon the presence of these compounds. Humanity has benefited greatly from their presence in the potato and, to a lesser extent, from their presence in tomatoes, lupins, and a number of other crops. Additional, non-nutritive benefits were derived from plants containing alkaloids with therapeutic properties, such as tea, coffee, coca, tobacco, opium poppy, periwinkle, and cinchona. Unfortunately the addictive nature of some of these alkaloids and their deleterious side effects have led to major health and social problems.

The alkaloids of crop plants, and of some species not yet domesticated, could provide an interesting challenge to researchers in agriculture and agrochemicals in the next few decades. There is an increasing disenchantment on some parts of society with the synthetic chemicals that are used to protect crops. Whilst these are mostly consumers who are unaware of the problems that face modern farmers, they do have an impact on policy. Now may be the time to look to a new 'green' revolution; one in which these naturally occurring, readily biodegradable compounds are used as a first line of defence in the battle against pests and disease.

For edible seed crops, such as lupins, there is a challenge to the plant breeder and biotechnologist to keep the alkaloids in the leaves and stems and out of the seeds.

The chemist will need to meet the challenge of extracting the alkaloids from other species to be used in sprays for protecting a range of crops. Only those alkaloids that will not themselves leave untoward residues in the crop could be used for this purpose. A desirable extraction procedure would leave no solvent residues and allow further preparation of the leaf, or seed, protein to be used in foods or animal feedstuffs. Research into the extraction of leaf protein from *Nicotiana* sp. has already shown promising results.

Another challenge to agriculture could be to devise farming systems in which cash crops are surrounded by a 'barrier crop' of a species which is a high alkaloid producer. Only in the event of this 'barrier' being breached would the farmer need to resort to the use of pesticides to protect the crop against insect and mollusc predators.

This form of agriculture might allow a similar scale of agriculture to the present one; and could be a high yielding alternative to organic or biodynamic farming. The benefit from this could be a greater amount of land available for wildlife habitat. One argument against organic farming is that yields are about one third lower than in agrochemical systems so more land has to be taken for food production and less is available for wildlife.[190] The final balance will depend upon population pressures on the earth's limited resources.

Further challenges await the chemist. New derivatives of greater, or more specific biological activity may be needed to ensure the long-term success and economic viability of these systems.

---

[190] N. Adams, *New Scientist*, 1990, **127**(1734), 52.

CHAPTER 8

Condensed Tannins

D. WYNNE GRIFFITHS

1 Introduction

Phenolic compounds constitute a wide and diverse group of secondary plant compounds, which can vary in molecular complexity from the comparatively simple phenolic acids such as p-hydroxybenzoic acid to the polymeric structures associated with the lignins and tannins.

The term 'tannin' was first introduced by Seguin in around 1796[1] being derived from the latin form of a Celtic word for oak[2] and was used to describe any substance with the ability to convert animal hides and skins to leather. However, in the early to mid 1960s a more rigorous definition was proposed[3] and consequently the term vegetable tannin is now more generally used to describe any naturally occurring compound of plant origin, which is of a high molecular weight (500–3000) and also contains a sufficiently large number of phenolic hydroxyl groups (1–2 per 100 units of molecular weight) to enable it to form effective cross-linkages between proteins and other macromolecules. Simple phenolic compounds of low molecular weight are normally too small in size to form such effective cross-linkages and although they may be absorbed onto proteins and other polymers, the stability constants of such complexes are usually low.

In the early decades of this century investigations into the molecular structures of vegetable tannins led to a number of proposed schemes for the classification of tannins into various sub-groups on the basis of their chemical composition.[4] In 1920 Freudenberg[5] proposed the classification, which is still

---

[1] E. Haslam, in 'Biochemistry of Plant Phenolics', ed. T. Swain, J. B. Harborne, and C. F. Van Sumere, Plenum Press, New York, 1979, p. 475.
[2] T. W. Goodwin and E. I. Mercer, 'Introduction to Plant Biochemistry', 2nd edition, Pergamon Press, Oxford, 1983, Chapter 14, p. 561.
[3] T. Swain, in 'Plant Biochemistry', ed. J. Bonner and J. E. Varner, Academic Press, London, 1965, p. 552.
[4] M. Nierenstein, 'The Natural Organic Tannins: History, Chemistry, Distribution', J. and A. Churchill Ltd., London, 1934, Chapter 1, p. 15.
[5] K. Freudenberg, 'Die Chemie der Naturlichen Gerbstaffe', Springer-Verlag, Berlin, 1920, Chapter 1, p. 4.

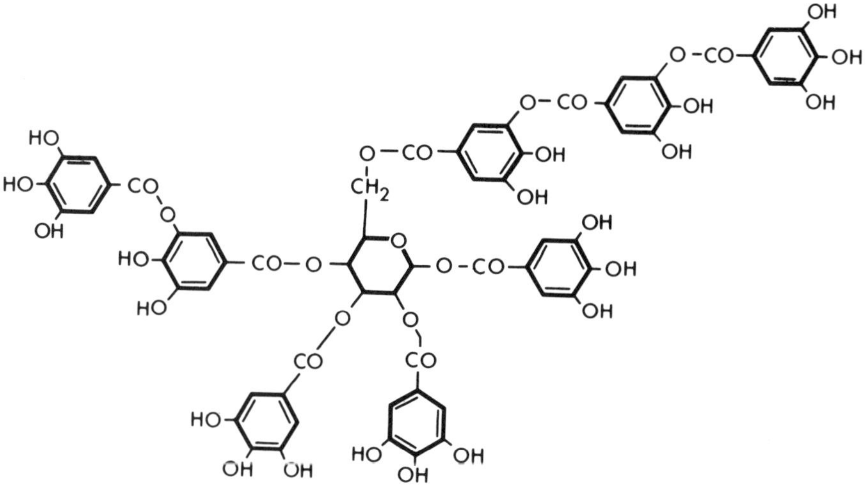

Figure 1 *Typical structure of a hydrolysable tannin*

generally accepted today, whereby the tannins were sub-divided into two major groups, namely the hydrolysable tannins and the non-hydrolysable or condensed tannins. The former are characterized by having a polyhydric alcohol, usually glucose, as a central core, the hydroxy groups of which are partially or wholly esterified with either gallic acid or with hexahydroxydiphenic acid. The attached phenolic acids may also be dimers or higher oligomers of gallic acid linked by depside bonds between the hydroxyl group of one gallic acid molecule and the carboxyl group of another and thus can result in long chains of gallic acid streaming out from a central glucose core. The structure of a comparatively simple hydrolysable tannin, Chinese gallotannin isolated from plant galls on the leaves of *Rhus semiclata*, is shown in Figure 1. On hydrolysis by either acids or enzymes the hydrolysable tannins yield a mixture of the constituent carbohydrate and phenolic acids. Those which produce only a carbohydrate and gallic acid are often defined as gallotannins whilst those which yield hexahydroxydiphenic acid, normally as the dilactone ellagic acid, amongst their hydrolysis products are normally defined as ellagitannins.

The second main goup of tannins, *i.e.* the condensed or non-hydrolysable tannins can be considered as dimers or higher oligomers of variously substituted flavan-3-ols. The monomeric units are usually linked by carbon–carbon bonds, normally between carbon-4 on one flavan-3-ol molecule and carbon-8 on the adjacent molecule (Figure 2). Consequently the stability of the intra-flavan-3-ol bond is considerably greater than either the depside or ester bonds associated with the hydrolysable tannins and, as their name suggests the non-hydrolysable tannins are not readily broken down to their constituent flavan-3-ols by either mild acid or enzymic treatments. On heating with strong

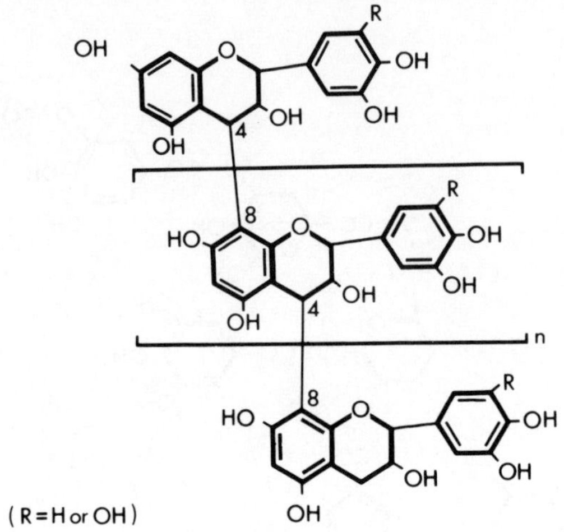

(R = H or OH)

Figure 2 *Structure of a non-hydrolysable or condensed tannin*

acids the condensed tannins polymerize further to produce red amorphous compounds known as phlobaphenes and also yield small quantities of anthocyanidins. This has resulted in a third generic name for the condensed tannins namely proanthocyanidins.

Two additional tannin sub-groups, the oxy-tannins and the β-tannins were defined by Swain[6] in the late 1970s. The oxy-tannins are not normally found in undamaged plants and appear to arise from the auto-oxidation of condensed tannin precursors, such as catechin. The resulting products differ significantly from the condensed tannins particularly with respect to molecular structures, which frequently involve both carbon–carbon and carbon–oxygen bonding as well as cleavage of the central oxygen-containing pyran ring. The β-tannins represent a heterogeneous group of low molecular weight compounds (300–500) whose structures, compared with the other major tannin sub-groups are comparatively simple often containing, as in piceatannol isolated from spruce bark,[7] only two phenyl rings. The tannin-like activity of these compounds, as compared with fairly similar structures with no tannin-like activities, has been associated[6] with their low solubility in water and subsequent ability to form less easily broken hydrogen bonds with proteins and other macromolecules.

The most naturally abundant tannins in the Plant Kingdom are condensed tannins and these polyphenolic compounds have been reported[6] as being present in 54% of all studied angiosperm genera, 74% of all studied gymnosperm genera, and 92% of all studied ferns. Tannins may accumulate in the seed, fruit, stems, leaves, or roots[1] and as shown in Table 1 are frequently found in many commonly utilized constituents of both human and animal diets. It

[6] T. Swain, in 'Herbivores—Their Interaction with Secondary Plant Metabolites', ed. G. A. Rosenthal and D. H. Janzen, Academic Press, New York, 1979, p. 657.
[7] E. Haslam, 'Chemistry of Vegetable Tannins', Academic Press, London, 1966, Chapter 2, p. 17.

Table 1 *Examples of feeds containing condensed tannins consumed by monogastric and/or ruminant animals*

Type	Common Name (species)	Reference†
Non-leguminous seed	Barley (*Hordeum vulgare*)	a
	Oilseed Rape (*Brassica Napus*)	a
	Sorghum (*Sorghum bicolor*)	b
Leguminous seed	Common bean (*Phaseolus vulgaris*)	c
	Faba bean (*Vicia faba*)	d
	Field pea (*Pisum* spp.)	d
Fruits	Apple (*Malus pumila*)	e
	Banana (*Musa* spp.)	e
	Pear (*Pyrus* spp.)	f
Forages	Birdsfoot trefoil (*Lotus corniculatus*)	g
	Sainfoin (*Onobrychis viciaefolia*)	g
	Sericea (*Lespedeza cuneata*)	h

† does not imply first literature citation

[a] B. O. Eggum and K. D. Christensen, in 'Breeding for Seed Protein Improvement using Nuclear Techniques', IAEA, Vienna, 1975, p. 135.
[b] S. I. Chang and H. L. Fuller, *Poult. Sci.*, 1964, **43**, 30.
[c] Y. Ma and F. A. Bliss, *Crop. Sci.*, 1978, **18**, 201.
[d] D. W. Griffiths, *J. Sci. Food Agric.*, 1981, **32**, 797.
[e] L. Y. Foo and L. J. Porter, *J. Sci. Food Agric.*, 1981, **32**, 711.
[f] J. L. Goldstein and T. Swain, *Phytochemistry*, 1963, **2**, 371.
[g] W. T. Jones, R. B. Broadhurst, and J. W. Lyttleton, *Phytochemistry*, 1976, **15**, 140.
[h] R. E. Burns, 'Tannins in *Sericea lespedeza*', Georgia Agricultural Experiment Station, Bulletin N.S. 164, 1966.

should, however, be stressed that within a given tannin-containing species, cultivars may exist that are free of such compounds and indeed seed from totally white-flowered cultivars of both field peas[8] and faba beans[9] contain no tannins and mutant lines of tannin-free barley have been isolated and studied.[10] Tannin content and/or the degree of tannin polymerization may change with plant maturity and the loss of astringency on ripening of many fruits including bananas, persimmons, and peaches has been linked to increased polymerization of the constituent tannins.[11] More recently, Foo and Porter[12] detected no differences in the ^{13}C NMR of tannins extracted from unripe and ripe strawberries and suggested that the apparent loss of astringency on ripening was due to the masking effects of the increased sugar content of ripe fruit. In some fruit, ripening is accompanied by the production of significant quantities of water-soluble polysaccharides, and it has also been suggested[13] that these compounds may reduce astringency by acting as alternative binding sites for the constituent tannins, thus reducing their effect on mucosal proteins.

[8] D. W. Griffiths, *J. Sci. Food. Agric.*, 1981, **32**, 797.
[9] D. G. Rowlands and J. J. Corner, *Eucarpia*, 1962, **3**, 229.
[10] D. von Wettstein, B. Jende-Strid, B. Ahresnt-Larsen, and K. Erdal, *MBAA Tech. Quarterly*, 1980, **17**, 16.
[11] J. L. Goldstein and T. Swain, *Phytochemistry*, 1963, **2**, 371.
[12] L. Y. Foo and L. J. Porter, *J. Sci. Food. Agric.*, 1981, **32**, 711.
[13] T. Ozawa, T. H. Lilley, and E. Haslam, *Phytochemistry*, 1987, **26**, 2937.

2 Condensed Tannin Analysis

Numerous methods have, over the years, been proposed and variously modified for the quantitative estimation of condensed tannin concentration. Many of these originated from the leather industry and consequently were originally designed for the determination of fairly high concentrations of tannins.[6] Such methods were often based on gravimetric, volumetric, or colorimetric estimations of the total phenolic contents of plant extracts before and after precipitation of the constituent tannins by hide powder or other proteinacious substances such as gelatin. The specificity of these tannin-binding agents was often low and many including gelatin have been shown to precipitate non-tannin compounds, such as gallic acid, hydroxyhydroquinone, and phloroglucinol.[14] Greater confusion has perhaps resulted from the use of both quantitative and qualitative methods based only on phenolic type reactions,[3] for although all tannins are phenolic, clearly not all phenols are tannins. Phenolic compounds are indeed generally present in all plants and consequently great care is required in interpreting data where the evidence for the presence of tannin is based solely on characteristic phenolic reaction assays.

Several gravimetric procedures have been described for the quantitative determination of both condensed tannins and total phenols and it has been suggested[15] that such procedures have an advantage over colorimetric methods in as much as no reference compounds are required. The precipitation of condensed tannins by formaldehyde–hydrochloric acid has been developed into a quantitative method for the analysis of tannins in forage[16] and both lead[17] and ytterbium[18] acetates have been used as precipitating agents in the gravimetric determination of total phenols in leaf extracts. Tannin content has also been directly determined by weight in aqueous acetone extracts of oak leaves, following ethanol and ether precipitation, subsequent dialysis and freeze drying.[19]

Volumetric titrations have also been used in the quantitative estimation of tannin content. In particular the so-called Lowenthals permanganate oxidation method has been widely employed and although developed in the last century[20] it was still recommended in 1980 as the 'official method' by the Association of Official Chemists[21] for the determination of the tannin content of cloves and all spice. This method assumes the absence of any other potentially oxidizable compounds in the extract under investigation and consequently is at best to be

---

[14] A. E. Jones, *Analyst (London)*, 1927, 275.
[15] I. Mueller-Harvey, in 'Physio-chemical Characterisation of Plant Residues for Industrial and Feed Use', ed. A. Chesson and E. R. Orskov, Elsevier Science Publishers Ltd., London and New York, 1989, p. 88.
[16] R. E. Burns, 'Methods of Tannin Analysis for Forage Crop Evaluation', Georgia Agricultural Experiment Stations Technical Bulletin N.S. 32, 1963.
[17] A. E. Bradfield and M. Penney, *J. Soc. Chem. Ind.*, 1944, **63**, 306.
[18] J. D. Reed, P. J. Horvath, M. S. Allen, and P. J. Van Soest, *J. Sci. Food. Agric.*, 1985, **36**, 255.
[19] P. P. Feeny and H. Bostock, *Phytochemistry*, 1968, **7**, 871.
[20] *U.S. Dept. Agric. Div. Chem. Bull.*, 1892, **13**, 890.
[21] Association of Official Agricultural Chemists, 'Methods of Analysis', 13th edition, George Banta Co. Inc., Wisconsin, 1980.

regarded as a method for the determination of total phenols.[22] However, it has been used as an indirect method for tannin determination in black tea,[23] where the total phenolic content is determined before and after precipitation of the constituent tannins with gelatin–salt solutions and the tannin concentration thus determined by subtraction. It has also been successfully applied to the quantitative estimation of fern tannins, which were first precipitated from solution using ammoniacal zinc acetate.[24]

Colorimetric assays have been developed for the quantitative determination of tannin content, but as with many such classical techniques, the specificity of such methods is often low. The Folin–Dennis[25] method has been frequently used for the analysis of forage[16] and in numerous ecologically orientated studies[26] but, since the underlying chemistry associated with this technique is dependent on the susceptibility of the constituent phenolic groups to oxidation, the results obtained reflect the total phenolic content of the plant extract rather than true tannin content. The presence of any other reducing agents[27] such as ascorbic acid in the plant extract will also interfere with the assay and can lead to artificially high erroneous absorbance values. However, increased specificity for phenolic groups has been reported[15] for the Folin–Ciocalteau[28] reagent, which supposedly gives a higher colour yield with phenols and is less sensitive to non-phenolic reducing agents.

The susceptibility of phenolic compounds to oxidation by ferric ions has been utilized and developed into the 'Prussian Blue' test for condensed tannins in sorghum.[29] In common with other redox-based reactions the prussian blue method is sensitive to non-tannin phenols and since the number of ferric ions reduced differs from one phenolic compound to another the resulting increase in absorbance is dependent not only on total phenol content but also on the types and relative proportions of the individual phenolic compounds present in the extract analysed. Direct comparisons of reported tannin values based on redox-type methodology is difficult and possibly meaningless unless the phenolic constituents of the species to be compared are similar and the same standards used in the calibration.

Another commonly utilized reaction for the quantitative colorimetric determination of condensed tannins is that with aromatic aldehydes and in particular, vanillin. The specificity of this reaction with various phenolic compounds[11] including flavanoids[11,30] has been extensively studied. The essential structural requirements for the reaction of vanillin with flavanoids include a single bond between carbon 2 and carbon 3 as well as free *meta* orientated hydroxyl groups in the B ring,[30] consequently, vanillin reacts not only with

[22] A. S. Tempel, *J. Chem. Ecol.*, 1982, **8**, 1289.
[23] Association of Official Agricultural Chemists, 'Methods of Analysis', 10th edition, George Banta Co. Inc., Wisconsin, 1965.
[24] S. Laurent, *Arch. Int. Physiol. Biochem.*, 1975, **83**, 735.
[25] O. Folin and W. Dennis, *J. Biol. Chem.*, 1912, **12**, 239.
[26] J. S. Martin and M. M. Martin, *Oecologia*, 1982, **54**, 205.
[27] V. L. Singleton and J. A. Rossi, Jnr., *Am. J. Enol. Vitic.*, 1965, **16**, 144.
[28] O. Folin and V. Ciocalteau, *J. Biol. Chem.*, 1927, **73**, 627.
[29] M. L. Price and L. G. Butler, *J. Agric. Food Chem.*, 1977, **25**, 1268.
[30] S. K. Sarkar and R. E. Howarth, *J. Agric. Food Chem.*, 1976, **24**, 317.

condensed tannins and their flavanol precursors but also with dihydrochalcones and anthocyanins.[30] The sensitivity of the vanillin reagent towards both monomeric and polymeric flavanols varies with the type of solvent used.[31] In the presence of methanol the monomeric flavanols are significantly less sensitive to vanillin than the polymeric condensed tannins and consequently when a monomeric flavanol such as catechin is used as the standard reference compound, over-estimation of the condensed tannin concentration may result.[32] However, in the presence of strongly acidic solvents such as concentrated sulphuric[33] or glacial acetic acids,[31] the extinction coefficient of the vanillin reactions products are broadly similar for both monomers and polymers and are significantly higher than those found in methanol. Although under acidic conditions the sensitivity is therefore greater, inaccuracies may result in plant extracts which contain high proportions of monomeric flavanols. The specificity of the vanillin method is however, considerably greater than that of redox based methods and has been widely used both as a rapid qualitative test for condensed tannins in forage leaves[34] and as a quantitative method for tannin determination in sorghum grain[32] and in fruit.[33]

The release of anthocyanidins from condensed tannins by treatment with hot mineral acids has been developed into the leuco-anthocyanidin method for the quantitative determination of tannin content.[33,35] Traditionally, milled plant tissue or appropriately prepared plant extracts were hydrolysed at 95 °C in a 19:1 v/v solution of butan-1-ol concentrated hydrochloric acid and the concentration of the resulting anthocyanidins determined colorimetrically.[33,35] The colour yields from flavan-3,4-ol monomers have been found to be considerably greater than that of the corresponding trimers, although little change was then observed in high polymeric tannins.[11] Problems of lack of reproducibility have been reported[36,37] but more recently a systematic study[37] of the underlying chemistry of proanthocyanidin hydrolysis has resulted in the modification of the method to include the addition of ferric ions to the hydrolysing reagent and a significant improvement in repeatability achieved.

It has been suggested,[38] that since the physiological significance of plant tannins is largely attributable to their ability to interact with proteins, analytical methods based on protein–tannin interactions should give a better measure of the biological activity of tannins as compared with results obtained from assays dependent on other characteristic tannin reactions. Most of the early methods,[4] which utilize protein–tannin complexation, stem from the leather industry and therefore tend to use animal skin proteins as the precipitating reagent. However, the chemical structure and properties[39] of collagen, the

[31] L. G. Butler, M. L. Price, and J. E. Brotherton, *J. Agric. Food Chem.*, 1982, **30**, 1087.
[32] M. L. Price, S. Van Scoyoc, and L. G. Butler, *J. Agric. Food Chem.*, 1978, **26**, 1214.
[33] T. Swain and W. E. Hillis, *J. Sci. Food. Agric.*, 1959, **10**, 63.
[34] W. T. Jones, L. B. Anderson, and M. D. Ross, *N.Z. J. Agric. Res.*, 1973, **16**, 441.
[35] E. C. Bate-Smith, *Biochem. J.*, 1954, **58**, 122.
[36] P. Ribereau-Gayon, 'Plant Phenolics', Oliver and Boyd, Edinburgh, 1972, p. 192.
[37] L. J. Porter, L. N. Hrstich, and B. G. Chan, *Phytochemistry*, 1986, **25**, 223.
[38] J. McManus, T. H. Lilley, and E. Haslam, in 'Plant Resistance to Insects', ed. P. A. Hedin, American Chemistry Society, Washington D.C., 1983, p. 123.
[39] E. C. Bate-Smith, *Phytochemistry*, 1973, **12**, 907.

major animal skin protein is considerably different from that of proteins present in endogenous animal or insect secretions. Consequently, methods have been developed which utilize more appropriate protein substrates such as casein,[40,41] haemaglobin,[39] and bovine serum albumin.[42] Tannin content can be indirectly determined from the amount of protein precipitated, methods for determining which have included direct colorimetric estimations,[41] the use of various protein dyes such as amido black[26] and Coomassie blue,[43] and hydrolysis of the precipitate followed by the estimation of free amino nitrogen by the ninhydrin reaction.[44] Alternatively, tannin content may be directly estimated by re-dissolving the protein precipitated tannin and subsequently determining tannin content by its reactions with ferric chloride.[42] More recent adaptations which can reduce the necessity for multiple steps involved in many of the above mentioned methods and also increase the sensitivity of the procedures, have used dye-labelled[45] or radioactively labelled proteins[46] as the precipitating reagent.

The effect of tannins on various *in vitro* enzyme systems has been utilized for the measurement and comparison of the biological activities of various tannin extracts. The inhibition of enzymes such as α-amylase,[47] β-glucosidase,[48] and taka-diastase[49] have been developed into quantitative tests for tannin content. Although the results obtained by such methods often parallel those from other tannin methods,[50] enzymatic methods have been criticized particularly with respect to repeatability[44] and the associated need to strictly control the conditions, such as temperature and pH, under which the assays are performed.

Chromatographic techniques employing various column packings such as polyamides,[51] amberlite,[52] and sephadex LH-20[53] have long been utilized in the partial purification of condensed tannins and for the estimation of approximate molecular weights.[54] Separation and purification of individual oligomers has been attempted by high performance liquid chromatography, where a range of packing materials including LiChrosorb 5:60,[55] μ Bondapak C_{18},[56] Spherisorb S-5-ODS,[57] and Fractogel TSK-HW[58] have been reported as successfully

[40] F. Gstirner and G. Korf, *Arch. Pharm.*, 1966, **299**, 763.
[41] W. R. C. Handley, *Plant Soil*, 1961, **15**, 37.
[42] A. E. Hagerman and L. G. Butler, *J. Agric. Food Chem.*, 1978, **26**, 809.
[43] M. M. Martin, D. C. Rockholm, and J. S. Martin, *J. Chem. Ecol.*, 1985, **11**, 485.
[44] D. Marks, J. Glyphis, and M. Leighton, *J. Sci. Food Agric.*, 1987, **38**, 255.
[45] T. N. Asquith and L. G. Butler, *J. Chem. Ecol.*, 1985, **11**, 1535.
[46] A. E. Hagerman and L. G. Butler, *J. Agric. Food Chem.*, 1980, **28**, 944.
[47] K. H. Daiber, *J. Sci. Food Agric.*, 1975, **26**, 1399.
[48] J. L. Goldstein and T. Swain, *Phytochemistry*, 1965, **4**, 185.
[49] H. R. Barnell and E. Barnell, *Ann. Bot. (London)*, 1945, **9**, 77.
[50] R. W. Bullard, J. O. York, and S. Kilburn, *J. Agric. Food Chem.*, 1981, **29**, 973.
[51] L. Horhammer and H. Wagner, *Naturwissenschaften*, 1957, **44**, 513.
[52] R. Vancraenenbroeck, A. Rogirst, H. Lemaire, and R. Lontie, *Bull. Soc. Chim. Belg.*, 1963, **72**, 619.
[53] A. E. Hagerman and L. G. Butler, *J. Agric. Food Chem.*, 1980, **28**, 947.
[54] I. Oh and J. E. Hoff, *J. Food Sci.*, 1979, **44**, 87.
[55] C. W. Glennie, W. Z. Kaluza, and P. J. van Niekereck, *J. Agric. Food Chem.*, 1981, **29**, 965.
[56] I. McMurrough, *J. Chromat.*, 1981, **218**, 613.
[57] W. Kirby and R. E. Wheeler, *J. Inst. Brew. (London)*, 1980, **86**, 15.
[58] G. Derdelinckx and J. Jerumanis, *J. Chromat.*, 1984, **285**, 231.

resolving both dimers and trimers from partially purified condensed tannin mixtures. However, higher oligomers are generally less well resolved and tend to give broad, complex bands.[59] Modification of these separation techniques to achieve quantitative determinations of the concentrations of the various dimers and trimers have been reported for application to both barley and malt grains[60] and it has been suggested[15] that since the extinction values for monomers, dimers, trimers and tetramers were the same on a weight basis, quantification of the broad peaks obtained in HPLC chromatograms could be achieved provided the homogenicity of the proanthocyanidins was known. Generally, quantification is hampered by the lack of suitable reference compounds[15] and possible problems with irreversible adsorptions.[59]

Near infra-red reflectance spectroscopy (NIRS) has become widely accepted as a rapid screening method for the determination of major quality factors in a range of feeds and foods,[61] but although less frequently used as a method for the determination of the concentration of secondary metabolites, screening methods for the determination of the tannin content of both faba bean seeds[62] and sericea forages[63] have been reported. Since NIRS is an indirect method of analysis which requires frequent calibration with samples of known chemical composition, the accuracy and applicability of such an indirect method is directly linked to the suitability and accuracy of the original conventional method used for the calibration.

3 Ecological Significance of Condensed Tannins

It has been suggested[64] that in order to justify the energy expended by plants in synthesizing condensed tannins, the presence of these polyphenolic compounds should infer an ecological advantage to those plants that produce such compounds. In a study of the neo-tropical tree *Cecropia pelata* L.,[65] seedlings grown under similar environmental conditions differed significantly with respect to leaf tannin content. The rate of leaf production was negatively correlated with leaf tannin content and indeed the rate of leaf production was almost 30% less in high-tannin-leaved seedlings as compared with low-tannin-leaved seedlings, possibly reflecting the diversion of both energy and nutrients to tannin biosynthesis. The susceptibility of the seedlings to herbivore damage was also negatively correlated with leaf tannin content, and the area of leaf consumed by the generalist armyworm (*Spodoptera latifascia*) larvae under laboratory conditions was almost five times greater on the low-tannin-leaved seedlings as compared with the high-tannin-leaved seedlings.

[59] L. J. Putman and L. G. Butler, *J. Chromat.*, 1985, **318**, 85.
[60] I. McMurrough, M. J. Loughrey and G. P. Hennigan, *J. Sci. Food Agric.*, 1983, **34**, 62.
[61] I. A. Cowe, J. W. McNicol, and D. C. Cuthbertson, *Analyst (London)*, 1985, **110**, 1227.
[62] A. de Haro, J. Lopez-Medijna, A. Carbrera, and A. Martin, in 'Recent Advances of Research in Anti-nutritional Factors in Legume Seeds', ed. J. Huisman, T. F. B. van der Poel, and I. E. Liener, Pudoc, Wageningen, 1989, p. 172.
[63] W. R. Windham, S. L. Fales, and C. S. Hoveland, *Crop Sci.*, 1988, **28**, 705.
[64] D. F. Rhoades, in 'Herbivores—Their Interactions with Secondary Plant Metabolites', ed. G. A. Rosenthal and D. H. Hanzen, Academic Press, London and New York, 1979, p. 3.
[65] P. D. Coley, *Oecologia*, 1986, **70**, 238.

For both wild and domesticated vertebrate herbivores the effectiveness of condensed tannins in reducing plant damage has been found to be dose dependent, with for example,[66] the voluntary intake by cattle of *Lespedeza cuneata* decreasing by 70% as the tannin content of the forage increased from 4.8% to 12% by dry weight. Both cattle and deer have been reported[67] as avoiding eating bracken fronds during the months of August and September, when their tannin content often exceeds 5% by dry weight and in the Northern Transvaal goats and other ungulates also appear to select their feed on the basis of tannin content, rejecting browse containing over 5% tannin particularly during the wet season.[68]

Similar selectivity for low or tannin-free plant species has been demonstrated by the mountain gorillas of the African Congo,[69] which appear to select plants, simply on the basis of the levels of tannins present.[70] Their diet is consequently restricted to a few dozen plant species and they generally avoid those which contain significant concentrations of tannins in their leaves. The black and white Colobus monkeys of Western Uganda also avoid plants containing over 0.2% tannin and in the Cameroons the black Colobus salonas monkeys ignore readily available high tannin species of plants and systematically hunt out the much scarcer low or tannin-free species.[71] The availability of such plant species is insufficient to meet their nutritional requirements and unlike other Colobus monkeys, the black Colobus salonas monkeys of the Cameroons have diversified their diet to include plant seeds.[72]

The reduced palatability of tannin-containing plant species, particularly with respect to vertebrate browsing herbivores, has primarily been linked to the ability of polyphenolic compounds to act as multidentate ligands.[73] These interact with the glycoproteins present in the mucous secretions of the salivary gland and the resulting precipitate produces a generally unpleasant 'puckering' sensation in the mouth. Consequently high concentrations of protein-precipitating agents, such as condensed tannins, are likely to reduce the palatability and reduce browsing damage to plants and thereby gain an ecological advantage. The protein-binding capacity of various polyphenolic compounds may vary considerably and in those studied,[74] no clear relationship has been found between the predicted energy cost to the plant in their biosynthesis and their ability to precipitate proteins.

The ability of tannins to precipitate proteins, which consequently reduces

---

[66] G. W. Arnold and J. L. Hill, in 'Phytochemical Ecology', ed. J. B. Harborne, Academic Press, London and New York, 1972, p. 72.
[67] G. Cooper-Driver, S. Finch, T. Swain, and E. Bernays, *Biochem. Syst. Ecol.*, 1977, **5**, 177.
[68] S. M. Cooper and N. Owen-Smith, *Oecologia*, 1985, **67**, 142.
[69] J. B. Harborne, 'Introduction to Ecological Biochemistry', 3rd edition, Academic Press, London and New York, 1988, Chapter 6, p. 195.
[70] E. C. Bate-Smith, in 'Phytochemical Ecology', ed. J. B. Harborne, Academic Press, London and New York, 1972, p. 45.
[71] J. F. Oates, T. Swain, and J. Zantovska, *Biochem. Syst. Ecol.*, 1977, **5**, 317.
[72] D. McKey, P. G. Waterman, C. N. Mbi, J. S. Gartlan, and T. T. Struhsaker, *Science*, 1978, **202**, 61.
[73] E. C. Bate-Smith, *Food*, 1954, **23**, 124.
[74] J. E. Beart, T. H. Lilley, and E. Haslam, *Phytochemistry*, 1985, **24**, 33.

palatability and adversely affects the availability of digested nutrients, has led to their being considered as potent anti-feedants to various unadapted insect herbivores. The experimental evidence for such inferences is largely based on reported studies of the seasonal behaviour of insect populations, in particular the winter moth (*Operophtera brumata*) on oak (*Quercus robur*) trees.[75,19] In early spring the insects appear to feed contentedly on oak leaves but in mid-June the number of feeding insects declines abruptly. This sudden reduction in insect numbers did not appear to coincide with any obvious change in environmental conditions, such as an increase in the predatory activities of birds, nor with any major alteration in the nutrient content of the leaves. Detailed chemical analysis of the oak leaves revealed that in mid-June total tannin concentration increased significantly and in particular the decrease in insect population coincided with the commencement of condensed tannin synthesis.[19]

Although there would appear to be a general consensus amongst ecologists that the presence of tannins confer some degree of resistance to damage by herbivorous insects, there is currently considerable debate as to the mechanism of such an effect. Theories based on non-specific tannin-protein interactions[76,77] have been seriously questioned,[78,79] particularly in view of the results obtained *in vitro* using enzymatically active gut fluid from *Manduca sexta* larvae, where no evidence for either dietary protein precipitation or digestive enzyme inhibition was observed.[79] It has also been suggested that the decrease in insect numbers on oak leaves in mid-June may be explained in terms of a rapid alteration in the relative toughness of leaves resulting from changes in cell structure.[80]

In addition to their role as possible chemical defensive agents against browsing mammals and herbivorous insects, tannins have been associated with a decreased susceptibility to microbial and fungal attack.[6,81] Such theories are again based on the ability of tannins to interact with proteins and other macro-molecules and can not only render plant synthesized macro-molecules unavailable to the invading organism but may also inhibit bacterial and fungal extracellular enzymes.[82] Additionally the condensed tannins can bind with macro-molecules present in the cell walls of invading pathogen and interfere with normal cell division.[6]

[75] P. P. Feeny, *Ecology*, 1970, **51**, 565.
[76] P. P. Feeny, in 'Biochemical Interactions between Plants and Insects', ed. J. W. Wallace and R. Mansell, Plenum Press, New York, 1976, p. 1.
[77] D. F. Rhoades and R. G. Cates, in 'Biochemical Interactions between Plants and Insects', ed. J. W. Wallace and R. Mansell, Plenum Press, New York, 1976, p. 168.
[78] W. V. Zucker, *Am. Nat.*, 1983, **121**, 335.
[79] J. S. Martin, M. M. Martin, and E. A. Bernays, *J. Chem. Ecol.*, 1987, **13**, 605.
[80] E. Haslam and A. Scalbert, *Phytochemistry*, 1987, **26**, 3191.
[81] D. Levin, *Ann. Rev. Ecol. Syst.*, 1976, **7**, 121.
[82] A. L. Shigo and W. E. Hills, *Ann. Rev. Phytopathol.*, 1973, **11**, 197.

4 Nutritional Effects of Condensed Tannins

Monogastric Animals

The anti-nutritional effects of condensed tannins in the diets of monogastric animals have been studied in a wide range of animal species including pigs,[83] poultry,[84] hamsters,[85] rats,[86] and mice.[87] Particular emphasis has been given to grain sorghum tannins and to a lesser extent to the nutritional effects of the condensed tannins present in leguminous grain crops such as faba beans (*Vicia faba*). In the case of grain sorghum, nutritional studies undertaken using small laboratory animals have consistently shown reduced weight gains and feed efficiencies with observed deleterious effects increasing as the grain sorghum tannin content of the diets increased.[88] Similar results have been reported for poultry,[89] but at high levels of supplementary protein in the diets no statistically significant difference in animal growth was found between high and low-tannin grain sorghum diets.[90] Egg production in laying hens has been reported[91] to be negatively affected by high-tannin grain sorghum. Contrasting results have been reported from feeding trials undertaken with pigs, in some studies[92,93] the inclusion of high tannin sorghum grain in the diets reduced both weight gain and feed efficiency whilst in others no significant differences in animal performance was observed.[94]

Faba bean condensed tannins fed to rats at a concentration of approximately 0.4 % of the total diet have been shown to reduce significantly live weight gain, feed conversion ratios, and protein efficiency ratios, but had only a comparatively small negative effect on feed intake.[95] The apparent digestibilities of the various constituent dietary components were significantly reduced with the apparent digestibility of crude protein falling from an average value of over 82 % in rats consuming tannin free diets to 72% in rats eating the high tannin diets. Small but statistically significant reductions were found in the apparent digestibilities of both the soluble carbohydrate and lipid fractions. In contrast, biological value was unaffected by the presence of condensed tannins suggest-

---

[83] B. W. Cousins, T. D. Tanksley, D. A. Knabe, and T. Zebrowska, *J. Anim. Sci.*, 1981, **53**, 1524.
[84] R. R. Marquardt and A. T. Ward, *Can. J. Anim. Sci.*, 1979, **59**, 781.
[85] H. Mehansho, J. C. Rogler, L. G. Butler, and D. M. Carlson, *Fed. Proc.*, 1985, **44**, 1960.
[86] W. R. Featherstone and J. C. Rogler, *Nutr. Rep. Int.*, 1975, **11**, 491.
[87] T. N. Asquith, H. Mehansho, J. C. Rogler, L. G. Butler, and D. M. Carlson, *Fed. Proc.*, 1985, **44**, 1097.
[88] R. R. Marquardt, in 'Recent Advances of Research in Anti-nutritional Factors in Legume Seeds', ed. J. Huisman, T. F. B. van der Poel, and I. E. Liener, Pudoc, Wageningen, 1989, p. 141.
[89] W. D. Armstrong, W. R. Featherstone, and J. C. Rogler, *Poult. Sci.*, 1973, **52**, 1592.
[90] E. L. Stephenson, J. O. York, and D. B. Bray, *Feedstuffs*, 1968, **40**, 119.
[91] C. W. Weber, Proc. 13th Ann. Poultry Industry Day., University of Arizona, Tucson, 1969, p. 53.
[92] D. M. Thrasher, E. A. Icaza, W. Ladd, C. P. Bagley, and K. W. Topton, *L. Agric.*, 1975, **19**, 10.
[93] J. C. Hillier, J. J. Martin, and G. R. Walker, Oklahoma Agriculture Experiment Station, Misc. Publication, MP-55, 1959.
[94] C. M. Campabadal, H. D. Wallace, G. E. Combs and D. L. Hammell, Florida Agriculture Experiment Station, Research Rep. No. AL-1976-9, 1976.
[95] G. Moseley and D. W. Griffiths, *J. Sci. Food Agric.*, 1979, **30**, 772.

ing that at the levels fed, condensed tannins from faba beans had little effect on the utilization of absorbed protein but significantly reduced its availability for absorption *in vivo*.

Similar results have been reported for poultry, with statistically significant negative correlations ($r = -0.94$) being found between the weight gain over 20 days of muscovy ducklings and the tannin content of their diets.[96] Additionally, diets containing purified faba bean tannins significantly decreased the digestibility of nitrogenous compounds in growing chickens and the inclusion of high tannin faba beans in poultry diets have been shown to depress growth rates, feed conversion ratios, and the retention of amino acids.[97] In other experiments with hens, significant negative correlations have been reported between the tannin content of grain–legume-based diets and *in vivo* digestibilities of crude protein.[98]

Tannin-nutrient Interactions in Monogastrics

The observed reductions in the nutritional value of compounded diets containing condensed tannins would appear to stem primarily from the ability of these polyphenolic compounds to act as multidentate ligands interacting with the wide range of potential substrates present in the animals intestinal cavity. One of the most abundant potential substrates found *in vivo* are the proteins, which originate not only from dietary sources but also from endogenous salivary and intestinal secretions. The interactions between condensed tannins and proteins have been shown to depend on two predominant mechanisms, hydrogen bonding between the phenolic groups on the tannin molecule, and the keto-imide groups of the protein and hydrophobic bonding resulting in a co-alignment of the condensed tannin phenolic groups and the aromatic side chains of the protein.[99] The stoichiometry of tannin-protein precipitates studied *in vitro* appear dependent on the initial protein concentration.[100] At low protein concentrations, the associating tannin molecules form a surrounding monolayer about the protein molecule thus producing a complex with a relatively reduced hydrophilic characteristic resulting in their subsequent aggregation and precipitation. At higher protein concentrations aggregation and subsequent precipitation result from the formation of effective cross-linkages formed by tannin molecules and two adjacent protein molecules, consequently at elevated protein concentrations, the number of tannin molecules per molecule of protein in the tannin-protein precipitate may be less than in the precipitate formed at lower protein concentrations.

The affinity of a given polyphenol for different proteins has been shown to be dependent on both the amino acid composition and structure of the pro-

---

[96] J. Martin-Tanguy, J. Guillaume, and A. Kossa, *J. Sci. Food Agric.*, 1977, **28**, 757.
[97] A. T. Ward, R. R. Marquardt, and L. D. Campbell, *J. Nutr.*, 1977, **107**, 1325.
[98] E. Lindgren, *Swed. J. Agric. Res.*, 1975, **5**, 159.
[99] L. G. Butler, D. J. Riedl, D. G. Lebryk, and H. J. Blyth, *J. Assoc. Oil Chem.*, 1984, **61**, 916.
[100] E. Haslam, 'Plant Polyphenols', Cambridge University Press, Cambridge, 1989, Chapter 4, p. 154.

Table 2 *The* in vitro *inhibition of enzymes by condensed tannin extracts from various plant sources*

Enzyme	Tannin source	Reference
Cellulase	*Myrica pensylvanica*	a
	Vicia faba	b
β-Amylase	*Quercus petraea*	c
α-Amylase	*Quercus robur*	d
	Vicia faba	e
	Ceratonia siliqua	f
Trypsin	*Vicia faba*	e
	Pisum spp.	e
Lipase	*Vicia faba*	e
	Ceratonia siliqua	f

[a] M. Mandels and E. T. Reese, in 'Advances in Enzymic Hydrolysis of Cellulose and Related Materials', ed. E. T. Reese, Pergamon Press, New York, 1963, p. 115.
[b] D. W. Griffiths and D. I. H. Jones, *J. Sci. Food Agric.*, 1977, **28**, 983.
[c] A. Boudet and P. Gadal, *C. R. Hebd. Seances Acad. Sci.*, Paris, 1965, **260**, 4057.
[d] P. P. Feeney, *Phytochemistry*, 1969, **8**, 2119.
[e] D. W. Griffiths, *J. Sci. Food Agric.*, 1981, **32**, 797.
[f] M. Tamir and E. Alumot, *J. Sci. Food Agric.*, 1969, **20**, 199.

tein.[46,101,102] In general, flexible open-structured proteins and in particular those rich in proline have a higher affinity for tannins than the more compact globular proteins. This has been attributed largely to the increased accessibility of the protein peptide bonds in the open-structured proteins and thus, in an enhanced ability to form strong hydrogen bonds with the interacting condensed tannin.

The affinity of polyphenols for proteins is also affected by the size and structure of the interacting polyphenol. A comparison of the affinities of dimers and trimers of procyanidin for bovine serum albumin revealed that the latter were more strongly bound to the protein and higher oligomers of procyanidin have been shown to be more effective inhibitors of trypsin than either the dimer or trimer.[103] Similar results have been obtained in studies of hydrolysable tannins,[100] where with the addition of an extra gallic acid residue to the central glucose core, the efficiency of association with protein increased reaching a maximum for β-penta-*O*-galloyl-D-glucose. At very high molecular weights the effectiveness of condensed tannins as protein binding agents *in vivo* may, however, be reduced due to problems of molecular flexibility arising from conformational rigidity and decreasing solubility.

Although the observed reductions in the availability of crude protein in monogastric animals consuming condensed tannins may in part be due to the formation of insoluble dietary protein–tannin complexes, it is also possible that

[101] A. E. Hagerman and L. G. Butler, *J. Biol. Chem.*, 1981, **256**, 4494.
[102] T. N. Asquith and L. G. Butler, *Phytochemistry*, 1986, **25**, 1591.
[103] W. E. Artz, B. G. Swanston, B. J. Sendzicki, A. Rasyid, and R. E. W. Birch, in 'Plant Proteins. Applications, Biological Effects and Chemistry', ed. R. L. Ory, ACS Symposium Series No. 312, 1986, p. 126.

reduced nutrient availability may be linked to the ability of tannins to inhibit the activities of the digestive enzymes secreted into the intestinal tract. *In vitro* studies with condensed tannins extracted from various vegetative sources indicate that a wide range of purified enzymes may be effectively inhibited by naturally occurring tannins (Table 2), these include a number of nutritionally important digestive enzymes including amylases, lipases, and trypsin. The presence of condensed tannins in bird-resistant cultivars of grain sorghum has been held responsible for the poor quality of the beer produced by such cultivars. This has been attributed to the ability of these polyphenolic compounds to inhibit the endogenous amylases present in the grain and thereby prevent effective breakdown of the starch during the brewing process.[47] Sorghum tannins have been shown to inhibit two enzymes, alkaline phosphatase and a nucleotide phosphodiesterase isolated and purified from bovine intestinal mucosa.[99] However, when the assays were repeated in the presence of a non-ionic detergent, which apparently simulates an environment more similar to that of the native membrane-bound enzymes *in vivo*, no significant inhibition of the enzymes was observed. Although this might well signify that *in vivo* the opportunities for enzyme inhibition are less than that *in vitro*, it is of interest that the structures of many non-ionic detergents are very similar to that of polyethylene glycol, which is known to have a high affinity for many polyphenolic compounds.

Detailed investigations have been undertaken to determine the mechanism of a-glucosidase inhibition by various polyphenolic compounds.[104] When substrate, enzyme, and inhibitor were simultaneously mixed the observed kinetics corresponded closely with that expected from classical non-competitive inhibition suggesting that both the substrate and inhibitor simultaneously bind with the enzyme to form an enzyme–inhibitor–substrate complex. The binding of the polyphenol to the surface of the enzyme then prevents the latter undergoing the necessary conformational changes to allow the catalysed breakdown of the substrate. However, if the enzyme is pre-equilibrated with the polyphenols mixed competitive and non-competitive kinetics appear to operate. The proposed explanation for such observations assume that the binding of the polyphenol at, or at a point close to the enzyme's active site is comparatively slow compared with attachments to other points on the enzyme surface. Thus, when enzyme and polyphenolic inhibitors are pre-equilibrated the opportunities for the polyphenol to penetrate the active site are greater than when no pre-equilibration is allowed and consequently are in a better position to compete directly with the substrate for the active site.[100]

The results from some animal-based experiments have suggested that enzyme inhibition may be of importance *in vivo*. The intestinal contents of rats fed either high or very low faba bean tannin diets differed significantly with respect to both trypsin and amylase activities.[105] In the case of the rats fed high-tannin-based diets the activities of both enzymes were significantly reduced but when polyvinyl pyrrolidone, a potent tannin-binding agent, was added to the samples

---

[104] T. Ozawa, T. H. Lilley, and E. Haslam, *Phytochemistry*, 1987, **26**, 2937.
[105] D. W. Griffiths and G. Moseley, *J. Sci. Food Agric.*, 1980, **31**, 255.

of intestinal fluid prior to the trypsin assay, the levels of trypsin activity were almost identical in both the high and very low tannin fed rats. This would suggest that the observed reduction in enzymatic activity originally found in intestinal fluid taken from the high tannin diet fed rats was largely due to the presence of enzyme–tannin complexes and did not reflect any differences in the concentrations of proteinacious protease inhibitors in the rat diets.

Further evidence for the possibility that reduced digestibility of proteins *in vivo* in condensed tannin fed animals was not simply due to the formation of dietary protein–tannin complexes has been obtained from other rat feeding experiments using ^{14}C-labelled casein.[106] As might be expected the addition of condensed tannins, from an unspecified origin significantly decreased nitrogen retention. However, an estimation of the ^{14}C content of the faeces revealed that the addition of condensed tannins to the diet did not significantly increase the amount of ^{14}C in the faeces suggesting that the extra protein excreted by the tannin-fed rats could not be attributed to the formation *in vivo* of casein–tannin complexes. The amino acid composition of the faecal proteins was unaffected by the addition of condensed tannins and consequently the authors concluded that the increased protein found in the faeces of tannin-fed rats largely originated from endogenous sources including intestinally secreted digestive enzymes.

In addition to acting as potential polydentate ligands for proteins, condensed tannins may also interact with other macro-molecules such as carbohydrate-based polymers.[100,107] Sorghum tannins isolated from various cultivars of bird resistant sorghum have been shown to be selectively absorbed by starch isolated from both corn and wheat.[108] The amount of tannin absorbed by the starch was dependent on both the source of the starch and the original cultivar from which the tannin was extracted. The inhibition of β-glucosidase by polyphenols has been shown[8] to be significantly reduced by the addition of either cyclodextrins or sodium polygalacturonate. The relative importance of such interactions *in vivo* do not appear to have been extensively studied.

In vitro experiments carried out using tannic acid (a hydrolysable tannin) have shown that some polyphenolic compounds may under certain conditions precipitate a wide range of nutritionally essential minerals, from solution.[109] In trials undertaken with healthy human volunteers no difference in the intestinal absorption of iron was found between those consuming either high- or low-tannin cultivars of grain sorghum.[110] In the case of anaemic patients, initial results did appear to link reduced iron absorption with high tannin content but, when the diets were adjusted to allow for differences in phytate content, any differences previously found in the availability of iron were removed. Consequently the authors concluded that the effect of condensed tannins on iron bioavailability were almost negligible.

[106] Z. Glick and M. A. Joslyn, *J. Nutr.*, 1970, **100**, 516.
[107] H. P. S. Makkar, B. Singh, and R. K. Dawra, *Int. J. Anim. Sci.*, 1987, **2**, 127.
[108] A. B. Davis and R. C. Hoseney, *Cereal Chem.*, 1979, **56**, 310.
[109] N. T. Faithfull, *J. Sci. Food Agric.*, 1984, **35**, 819.
[110] M. R. Radhakrishnan and J. Sivaprasad, *J. Agric. Food Chem.*, 1980, **28**, 55.

Ruminant Animals

In contrast to the investigations carried out with monogastric animals, much of the research undertaken with ruminant animals has been concerned with the potential benefit of the presence of condensed tannins with respect to both the utilization of tannins as a means of protecting dietary protein from rumen microbial breakdown and as a potential adjunct in preventing bloat in susceptible animals. The evidence for the role of condensed tannins in preventing bloat stems in part from the fact that susceptible cattle fed on leguminous forage seldom if ever develop bloat when feeding on pastures containing forage species rich in condensed tannins.[111] The onset of bloat has been linked with the presence of high concentrations of water-soluble proteins in the diet. These on ingestion may result in the development of semi-solid foams in the rumen and thus prevent the escape of the normal gaseous products of fermentation. It has, therefore, been suggested that the presence of condensed tannins in the diet lowers the concentration of soluble proteins by forming insoluble protein–tannin complexes and thereby prevents the build up of stable foams in the rumen. Evidence for such a theory has been reported in an *in vitro* study of the stability of foams prepared from extracts of various leguminous forages.[112] Stable foams were readily prepared from tannin-free species but could only be obtained from species containing high concentrations of condensed tannins when compounds with the ability to displace tannins from the protein–tannin complexes were added to the extraction medium.

The other possible beneficial role of tannin is associated with that of protecting dietary protein from rumen microbial degradation and consequently increasing the quantity of rumen undegradable protein reaching the duodenum. The *in vitro* digestibilities of several animal feeds including faba beans,[113] sorghum grain,[114] and crown vetch forage[115] have been shown to be reduced due to the presence of condensed tannins. In the case of the reduced *in vitro* digestibility of high tannin content faba beans[116] the major reduction in the solubility of dry matter appeared to be associated with the inhibition of the cellulolytic enzymes present in rumen liquor rather than due to any significant reduction in either protein solubility or inhibition of protein digestion.[116] There appears little *in vivo* evidence for the selective binding of condensed tannins to cellulases and under normal feeding conditions the abundant availability of alternative substrates for condensed tannins may render such a mechanism to be of little importance. However, significant reductions in the digestion of organic matter within the rumen of sheep fed varieties of *Lotus pendunculatus* containing high concentrations of condensed tannins have been reported.[117]

[111] W. T. Jones and J. W. Lyttleton, *N.Z. J. Agric. Res.*, 1971, **14**, 101.
[112] W. A. Kendall, *Crop Sci.*, 1966, **6**, 487.
[113] D. A. Bond, *J. Agric. Sci.*, 1976, **86**, 561.
[114] H. B. Haris, D. G. Cummins, and R. E. Burns, *Agron. J.*, 1970, **62**, 633.
[115] J. C. Burns, W. A. Cope, and K. J. Wildonger, *Crop Sci.*, 1976, **16**, 225.
[116] D. W. Griffiths and D. I. H. Jones, *J. Sci. Food Agric.*, 1977, **28**, 983.
[117] T. N. Barry and T. R. Manley, *J. Sci. Food Agric.*, 1986, **37**, 248.

In other sheep-feeding experiments comparing sainfoin with different tannin-free leguminous forages, the presence of condensed tannins have been shown to increase the quantity of nitrogen reaching the duodenum.[118] However, in one experiment where dehydrated sainfoin was fed,[119] the increased nitrogen appeared to be from non-bacterial sources suggesting that the presence of the sainfoin condensed tannin had protected the protein from microbial breakdown whilst in another experiment utilizing fresh sainfoin,[120] the increased nitrogen was of microbial origin, indicating that the increased flow of nitrogen on the sainfoin-based diet was due to increased microbial synthesis. More detailed chemical investigations of the interactions between sainfoin tannins and fraction 1 protein (one of the major leaf proteins) revealed that between pH 3.5 to pH 7.0 the protein–tannin complexes were stable but between pH 1.0 and pH 3.0 up to 95% of the tannin was released from the preformed complexes.[121] Samples taken from sheep fed sainfoin-based diet showed that at normal rumen pH the tannin–fraction 1 protein complexes were stable but at the duodenal end of the abomasum, the pH fell to a value of pH 2.5 and tannin could be readily extracted from the samples taken from this point. This suggested that whilst in the rumen sainfoin tannins protected fraction 1 protein from microbial breakdown but once in the abomasum the associated fall in pH rendered the protein–tannin complex unstable thus releasing the protein for further digestion and subsequent absorption. Although these results would suggest that sainfoin tannins have the potential to protect dietary protein from microbial breakdown their ultimate fate *in vivo* do not appear to have been studied. Consequently there remains a strong possibility that as they pass down the intestinal tract the opportunity for the tannin molecules to interact with either undigested dietary proteins or endogenous protein secretions will increase and some of the benefits gained by increasing the flow of rumen undegradable protein may be lost.

5 Detoxification of Condensed Tannins

Numerous physical and chemical treatments have, over the years, been evaluated as potential methods for removing or partially reducing the anti-nutritional effects associated with the presence of condensed tannins in animal feeds (Table 3). However, more recently it has been shown that some animals have an adaptive response to being fed high concentrations of condensed tannins. In particular, it has been reported that young weanling rats fed high concentrations of sorghum tannins undergo an initial weight loss but after a period of approximately four days begin to gain weight but at a lower rate than

---

[118] I. Mueller-Harvey, A. B. McAllan, M. K. Theodorou, and D. E. Beever, in 'Plant Breeding and the Nutritive Value of Crop Residues', ed. J. D. Reed, B. S. Copper, and P. J. H. Neate, ILCA, Addis Ababa, 1988, p. 97.
[119] D. G. Harrison, D. E. Beever, D. J. Thompson, and D. F. Osbourn, *J. Agric. Sci.*, 1973, **81**, 391.
[120] D. E. Beever and S. C. Siddons, in 'Control of Digestion and Metabolism in Ruminants', ed. L. P. Milligan, W. L. Grovum, and A. Dobson, Prentice Hall, New Jersey, 1985, p. 479.
[121] W. T. Jones and J. L. Mangan, *J. Sci. Food Agric.*, 1977, **28**, 126.

Table 3 *Potential methods for reducing the anti-nutritional effects associated with condensed tannins*

Principle	Method
1 Physical removal of tannin rich tissue	Dehulling
2 Chemical modification of tannins *in situ*	Autoclaving
	Steam
	Alkali
3 Elimination by genetic manipulation	Plant breeding
	Genetic engineering

rats continually fed on tannin-free diets.[122] Closer investigations of the young rats during the initial four day adaptive phase revealed that the inclusion of the sorghum tannins in their diets resulted in the development of hypertrophic parotid glands and a subsequent twelve-fold increase in the concentration of proline-rich proteins. These proteins secreted by the parotid glands had, due to their abnormally high proline content, an open type of structure and consequently a high affinity for condensed tannin. It has therefore been suggested that these proteins function specifically as tannin-binding agents thereby protecting the rat from any toxic effects associated with the consumption of high-tannin-based diets.[123] Similar responses have been obtained with mice fed high-tannin diets[87,124] but no changes in salivary excretions were found in either hamsters[85] or chickens.[88]

Although some animals such as the rat have the ability to detoxify condensed tannins, the necessity to synthesize proline rich proteins in significant quantities must result in reduced animal performance with respect to live weight gain per given quantity of food ingested. Consequently in order to maximize animal performance with respect to dietary intake it would still appear desirable to remove or at least reduce the quantity of tannins in animal feeds prior to their inclusion in formulated diets.

In the case of grain crops such as sorghum and faba beans, where the condensed tannins are concentrated in the testa or seed coat,[113,125] the most effective method of increasing the nutritive value of the crop would appear to be the mechanical removal of the outer seed layer.[126,127] However, even if such a process could be efficiently mechanized with little concomitant loss of the tannin-free fractions, a significant reduction in the economic yield of the crop would result, unless some commercial use could be found for the highly fibrous

---

[122] L. G. Butler, J. C. Rogler, H. Mehansho, and D. M. Carlson, in 'Plant Flavanoids in Biology and Medicine: Biochemical, Pharmacological and Structure—Activity Relationships', ed. V. Cody, E. Middleton, and J. B. Harborne, Alan R. Liss Inc., New York, 1986, p. 141.
[123] H. Mehansho, A. E. Hagerman, S. Clements, L. G. Butler, J. C. Rogler, and D. M. Carlson, *Proc. Nat. Acad. Sci.*, 1983, **80**, 3948.
[124] H. Mehansho, L. G. Butler, and D. M. Carlson, *Ann. Rev. Nutr.*, 1987, 423.
[125] P. Morrall, N. W. Libenberg, and C. W. Glennie, *Scanning Electron Microsc.* 1981, **111**, 571.
[126] B. O. Eggum, L. Monwar, K. E. Bach-Knudsen, L. Munck, and J. Axtell, *J. Cereal Sci.*, 1983, **1**, 127.
[127] A. T. Ward, R. R. Marquardt, and L. D. Campbell, *J. Nutr.*, 1977, **107**, 1325.

tannin-rich residue, which for faba beans could account for well over 14% of the total grain yield.

The use of heat, normally applied by autoclaving or steam heating, has also been studied as potential methods for reducing the anti-nutritional effects associated with condensed tannins. Improved animal performance has been reported for diets containing autoclaved as opposed to untreated ground faba beans,[128,129] and heat treatment has also been shown to inactivate purified faba bean tannins resulting in an increase as compared with unheated purified faba bean tannins in both the digestibility and absorption of protein *in vivo*.[84,97] However, more recent results indicated that autoclaving removed only 57 % of the active tannin fraction present in faba beans.[130] Although the condensed tannins present in sorghum are structurally and chemically similar to those found in faba beans,[96] heat treatment does not appear to increase the nutritional value of high tannin sorghum grain.[88]

Moist alkaline treatments using dilute ammonia, potassium carbonate, or calcium oxide have been shown to reduce the quantity of chemically detectable tannins in sorghum grain by between 80 to 90 % and resulted in significant improvements in the nutritive value of high tannin sorghum.[131,132] Alkaline treatment did not adversely affect the nutritive value of tannin-free sorghums and the observed improvement in the nutritive value of tannin containing cultivars has been attributed to the alkaline conditions promoting oxidative polymerization of the condensed tannins resulting in the formation of highly polymeric nutritionally inactive compounds. The most effective treatment utilized the application of dilute ammonia directly onto whole sorghum grains, which presumably allowed oxidative polymerization to occur prior to the exposure of the condensed tannins to alternative substrates such as seed proteins and other macromolecules.[132] Similar reductions in chemically detectable tannin have been reported for faba beans where treatment with aqueous 4 % sodium hydroxide produced a 97 % reduction in condensed tannin concentration.[130]

The physical and chemical treatment of high tannin grain can under certain circumstances significantly improve the nutritional properties of both faba beans and sorghum but in the long term the use of conventional breeding and genetic engineering methodologies may represent the most promising techniques for reducing and possibly eliminating condensed tannins from agriculturally important crops. In faba beans zero tannin lines are easily identified by the associated white flower characteristic.[9] Some problems with regard to poor establishment and increased susceptibility to disease have also been linked

---

[128] B. J. Wilson and J. M. McNab, *Br. Poult. Sci.*, 1972, **13**, 67.
[129] L. D. Campbell and R. R. Marquardt, *Poult. Sci.*, 1977, **56**, 442.
[130] A. Garrido, A. Gomez, and J. E. Guerrero, in 'Recent Advances of Research in Antinutritional Factors in Legume Seeds', ed. J. Huisman, T. F. B. van der Poel, and I. E. Liener, Pudoc, Wageningen, 1989, p. 235.
[131] M. L. Price, L. G. Butler, J. C. Rogler, and W. R. Featherstone, *J. Agric. Food Chem.*, 1979, **27**, 441.
[132] M. L. Price and L. G. Butler, Purdue University Agricultural Experimental Station, Bulletin No. 272, 1980.

with the zero tannin character but more recent studies indicate no genetic reason why zero tannin cultivars with good agronomic properties could not be produced.[133,134] Additionally low tannin content has been shown to be genetically linked with normal flower colour and red seed coats.[135] The availability of such genetic markers would suggest that rapid progress towards producing faba bean cultivars with lower than average condensed tannin content should also be possible thus minimizing the anti-nutritional effects whilst retaining some of the beneficial agronomic qualities apparently linked with tannin content. Similarly, in the case of sorghum biotechnological techniques have been evaluated in an attempt to optimize the polyphenolic content of cultivars with respect to both agronomic and nutritional characteristics.[136]

6 Conclusion

The anti-nutritional properties associated with the presence of condensed tannins is largely due to the ability of these oligomeric and polymeric flavanoid compounds to interact with nutritionally essential macro-molecules. Although the nature of such interactions, particularly with proteins have been extensively studied and elucidated, further investigations into the significance of tannin structure and degree of polymerization on tannin activities *in vivo* and *in vitro* are required. Over the years many wild animals and insects have learnt to avoid tannin containing plant species, whilst others have developed specific detoxifying mechanisms, which allow them to feed on tannin-rich plants. However, the importance of condensed tannins as plant defense chemicals and their exact mode of action is still a controversial subject. In the case of domesticated farm animals, the presence of condensed tannins in their diets normally results, at least in the case of monogastric animals, in reduced animal performance. This reduction can largely be attributed to reduced protein digestibility but the relative importance of dietary protein–tannin as opposed to endogenous animal protein–tannin interactions has not been fully resolved. Condensed tannins may have a positive effect on the nutrition of ruminants due to their ability to reduce protein breakdown in the rumen and also by possibly preventing the development of bloat in susceptible animals. The ultimate fate of tannins in the lower digestive tract of ruminants does not appear to have been completely resolved. Various chemical and physical treatments have been evaluated as potential methods for removing tannins from animal diets but none would appear to be completely satisfactory. Thus it may well be that the best long-

[133] D. A. Bond and D. B. Smith, in 'Recent Advances of Research in Anti-nutritional Factors in Legume Seeds', ed. J. Huisman, T. F. B. van der Poel, and I. E. Liener, Pudoc, Wageningen, 1989, p. 285.
[134] J. J. A. van Loon, A. van Norel, and L. M. W. Dellaert, in 'Recent Advances in Anti-nutritional Factors in Legume Seeds', ed. J. Huisman, T. F. B. van der Poel, and I. E. Liener, Pudoc, Wageningen, 1989, p. 301.
[135] A. Cabrera, J. Lopez-Medina, and A. Martin, in 'Recent Advances in Anti-nutritional Factors in Legume Seeds', ed. J. Huisman, T. F. B. van der Poel, and I. E. Liener, Pudoc, Wageningen, 1989, p. 297.
[136] L. G. Butler, in 'Biotechnology in Tropical Crop Improvements', ICRISAT, Patancheru, India, 1988, p. 147.

term solution may come from a combination of genetic engineering and conventional breeding techniques, which could be used to optimize both the nutritional and agronomic qualities of various agriculturally important crops.

CHAPTER 9

Cyanogens

RAYMOND HUGH DAVIS

1 Introduction

The poisonous property of cyanogenic plants was known long before any understanding of the chemical nature of hydrogen cyanide (HCN) or of its biochemical and physiological action was possible. Even so the history of the chemistry of cyanogens is a long one, it being more than 150 years since HCN was first obtained from a plant extract and since the first cyanoglycoside was isolated. By the end of the Nineteenth century it was already recognized that cyanogens were very widely distributed among plants. This interest has been maintained, due to a considerable extent to the occurrence of cyanogens in many important food plants.

Cyanogenesis, the formation of free HCN, is particularly associated with cyanohydrins (a-hydroxynitriles) that have been stabilized by glycosylation, the cyanoglycosides. These are found in several plant families. Of more restricted distribution are cyanolipids, which are only cyanogenic if they result from stabilization of a cyanohydrin by ester linkage to a fatty acid. These are found in fruits of the Sapindaceae. There are also so-called pseudocyanoglycosides which are glycoside derivatives of methylazoxymethanol. These appear to be restricted to plants of the Cycadaceae and are not cyanogenic in the same sense although they do release HCN on treatment with alkali. It is also known that cyanide can be released from some nitriles within the animal body but the extent to which this occurs is dependent on structure and it seems unlikely that naturally occurring nitriles are significant sources of free cyanide. Most attention will be given to cyanoglycosides because of their wider distribution, greater diversity, and importance in food plants.

With such a long history of scientific investigation, a vast literature on the natural occurrence of cyanide and cyanide containing compounds now exists. This is exemplified by the publication of two books within the last decade devoted entirely to cyanide and cyanide compounds in biology.[1,2] Other books devoted to poisonous plants in general are useful sources on the regional occurrence of cyanogenic species, some of which may be locally important for food.[3-6] One cyanogenic food of particular economic importance, cassava, has been the subject of intensive, international, interdisciplinary research, and this has given rise to numerous publications relating to the crop.[7-10] Finally, certain aspects of cyanogens have been subjects of reviews.[11-20] Together these provide many citations to supplement those included here, which have of necessity been highly selective.

2 Structures

Although it is frequently stated that more than 2000 plant species are known to be cyanogenic the glycoside or lipid sources of the cyanide have only been identified in a comparatively small proportion of these. A recently quoted number for cyanoglycosides is about 300,[18] yet within these more than 50 distinct structures have been identified. Approximately half of these structures have been confirmed within the last decade, the majority being isomers or modifications of previously known compounds rather than completely

[1] B. Vennesland, E. E. Conn, C. J. Knowles, J. Westley, and F. Wissing, 'Cyanide in Biology', Academic Press, London, 1981.
[2] Ciba Foundation Symposium 140, 'Cyanide Compounds in Biology', Wiley, Chichester, 1988.
[3] M. R. Cooper and A. W. Johnson, 'Poisonous Plants in Britain and Their Effects on Animals and Men', H.M.S.O., London, 1984.
[4] S. L. Everist, 'Poisonous Plants of Australia', Angus and Robertson, Sydney, 1974.
[5] J. M. Kingsbury, 'Poisonous Plants of the United States and Canada', Prentice Hall, Englewood Cliffs, New Jersey, 1964.
[6] J. M. Watt and M. G. Breyer-Brandwijk, 'The Medicinal and Poisonous Plants of Southern and Eastern Africa', 2nd edition, E. and S. Livingstone, Edinburgh, 1962.
[7] B. Nestel and R. MacIntyre, 'Chronic Cassava Toxicity', International Development Research Centre, Ottawa, 1973.
[8] B. Nestel and M. Graham, 'Cassava as Animal Feed', International Development Research Centre, Ottawa, 1977.
[9] A. M. Ermans, N. M. Mbulamoko, F. Delange, and R. Ahluwalia, 'Role of Cassava in the Etiology of Endemic Goitre and Cretinism', International Development Research Centre, Ottawa, 1980.
[10] F. Delange, F. B. Iteke, and A. M. Ermans, 'Nutritional Factors Involved in the Goitrogenic Action of Cassava', International Development Research Centre, Ottawa, 1982.
[11] M. E. Robinson, *Biol. Rev.*, 1930, **5**, 126.
[12] P. Kakes, *Euphytica*, 1990, **48**, 25.
[13] E. E. Conn, in 'International Review of Biochemistry', ed. A. Neuberger and T. H. Jukes, University Park Press, Baltimore, 1979, Vol. 27, p. 21.
[14] E. E. Conn, *Ann. Rev. Plant Physiol.*, 1980, **31**, 433.
[15] E. E. Conn, in 'The Biochemistry of Plants', ed. P. K. Stumpf and E. E. Conn, Academic Press, New York, 1981, Vol. 7, p. 479.
[16] R. D. Montgomery, in 'Toxic Constituents of Plant Foodstuffs', 2nd edition, ed. I. E. Liener, Academic Press, New York, 1980, p. 143.
[17] J. E. Poulton, in 'Handbook of Natural Toxins 1', ed. R. F. Keeler and A. T. Tu, Marcel Decker, New York, 1983, p. 118.
[18] J. E. Poulton, in 'Food Proteins', Proc. Protein Co-prod. Symp., ed. J. E. Kinsella and W. G. Soucie, American Oil Chemical Society, Champaign, Ill., 1989, p. 381.
[19] A. Nahrstedt, in 'Biologically Active Natural Products', ed. K. Hostettmann and P. J. Lea, Clarendon Press, Oxford, 1987, p. 213.
[20] K. L. Mikolajczak, *Prog. Chem. Fats Other Lipids*, 1977, **15**, 97.

novel.[19,21,22] With only a few exceptions cyanoglycosides are formed by glycosylation of the α-hydroxygroup of cyanohydrins, which confers a stability not found in cyanohydrins themselves except in acidic conditions. Another feature is that the monosaccharide unit immediately attached to the aglycone is β-D-glucose in every case, giving the basic structure (1).

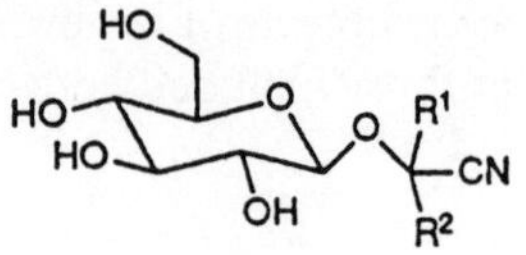

(1)

C_2 of the aglycone is usually asymmetrically substituted making epimeric pairs possible, most of which have been found to occur naturally. Epimers are not usually found together in the same plant except for glycosides derived from cyclopentenyl glycine where several instances have been confirmed. The presence of functional groups in a basic aglycone leads to further variation, as in ester formation with an available hydroxyl group. Alternatively, the basic β-D-glucopyranosyl structure may be modified, either by linkage to a second monosaccharide to give a disaccharide derivative of the cyanohydrin, or by formation of a sugar ester.

With such variation a form of classification is necessary. The currently accepted system is based upon the known or presumed biogenetic precursor of the aglycone moiety, which includes the protein amino acids L-valine, L-isoleucine, L-leucine, L-phenylalanine, and L-tyrosine, the non-protein amino acid L-2-cyclopentenyl-1-glycine, and a single example presumed to originate from nicotinic acid metabolism. Structures are illustrated in Table 1.

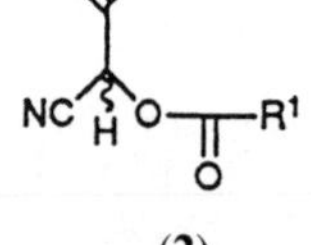

(2) 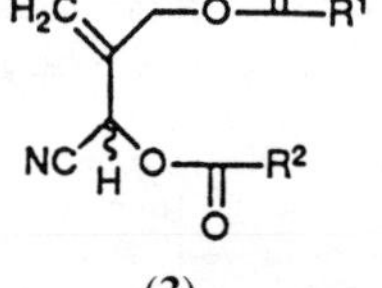   (3)

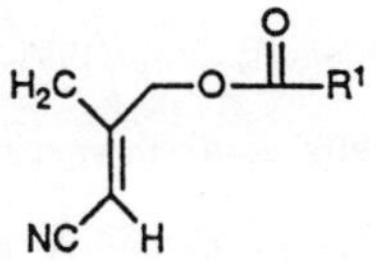

(4) 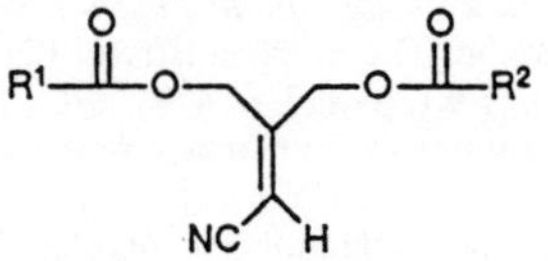   (5)

[21] D. S. Seigler, *Phytochemistry*, 1975, **14**, 9.
[22] D. S. Seigler, in 'Cyanide in Biology', ed. B. Vennesland, E. E. Conn, C. J. Knowles, J. Westley, and F. Wissing, Academic Press, London, 1981, p. 133.

Table 1 *Naturally occurring Cyanoglycosides, grouped according to their known or presumed biogenetic precursor. In each group a typical structure is presented and relationships of others to it are stated. All sugars are in pyranose form except for apiose in 51*

	Valine		
1	linamarin[a]		
2	linustatin[a]	6'-O-β-D-glucosyl	of 1
	Isoleucine		
3	(R)-lotaustralin[a]		
4	(S)-epilotaustralin[b]		epimer of 3
5	(R)-neolinustatin[a]	6'-O-β-D-glucosyl	of 3
6	sarmentosin epoxide[b]		irregular
	Leucine		
7	(S)-heterodendrin[a]		
8	(R)-epiheterodendrin[a]		epimer of 7
9	(S)-3-hydroxyheterodendrin[b]	3-hydroxyl	of 7
10	(S)-proacacipetalin[a]	3,4-dehydro	of 7
11	(R)-epiproacacipetalin[a]		epimer of 10
12	(S)-cardiospermin[a]	5-hydroxyl	of 10
13	(S)-cardiospermin-5-sulphate[a]		
14	(S)-cardiospermin-5-(4-hydroxy benzoate)[a]		
15	(S)-cardiospermin-5-(4-hydroxy-E-cinnamate)[b]		
16	sutherlandin[b]		irregular
17	sutherlandin + proacacipetalin[b]		irregular
18	(S)-proacaciberin[b]	6'-O-α-L-arabinosyl	of 10
	Phenylalanine		
19	(R)-prunasin[a]		
20	(R)-prunasin-6'-malonate[c]		
21	(R)-grayanin[d]	(= (R)-prunasin-6'-caffeate)	
22	(S)-sambunigrin[a]		epimer of 19
23	(R)-amygdalin[a]	6'-O-β-D-glucosyl	of 19
24	(R)-amygdalin-6"-(4-hydroxy benzoate)[c]		

Table 1 (*cont.*)

#	Name	Substituent	Relation
25	(*R*)-amygdalin-6″-(4-hydroxy-*E*-cinnamate)[c]		
26	(*R*)-vicianin[a]	6′-*O*-α-L-arabinosyl	of 19
27	(*R*)-lucumin[a]	6′-*O*-β-D-xylosyl	of 19
28	(*S*)-epilucumin[b]		epimer of 27
29	unnamed diglycoside[b]	2′-*O*-β-D-glucosyl	isomer of 23
30	anthemis glycoside B[b]	4″-*O*-(4-*O*-β-D-glucosylcoumarate)	of 27 or 28
31	anthemis glycoside A[b]	4″″-*O*-β-D-xylosyl	of 30

Tyrosine
32 (*R*)-taxiphyllin[a]

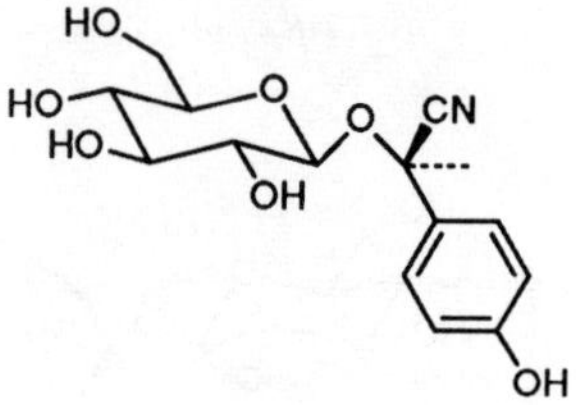

33 (*S*)-dhurrin[a] epimer of 32
34 unnamed[f] 6,7-dihydro of 32 or 33
35 (*S*)-proteacin[a] 6-*O*-β-D-glucosyl of 33
36 *p*-glucosyloxymandelonitrile[a]   irregular
37 nandinin[b] 4′-caffeate of 36

Tyrosine or Phenylalanine
38 (*R*)-holocalin[a]

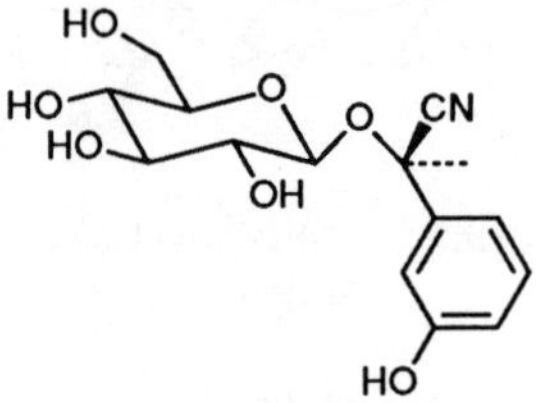

39 (*S*)-zierin[a]   epimer of 38
40 (*S*)-zierinxyloside[b] 6′-*O*-β-D-xylosyl of 39
41 (*S*)-xantherin[g]
     4″-*O*-apiofuranosyl-[5-*O*-(4-β-D-glucosyl)caffeate] of 40

42 triglochin(s)[a]

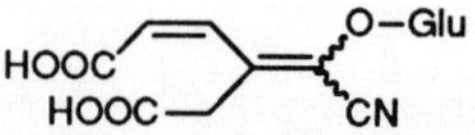

43 isotriglochin(s)[a]

Cyclopentenyl glycine
44 (*R*)-deidaclin[a]

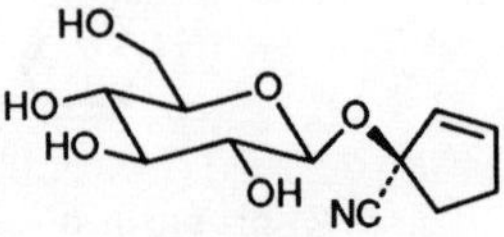

45 (*S*)-tetraphyllin A[a]   epimer of 44
46 (1*R*,4*R*)-volkenin[b] 4-hydroxy of 44
47 (1*R*,4*S*)-taraktophyllin[b] 4-hydroxy of 44

48	(1S,4R)-epivolkenin[h]	4-hydroxy	epimer of 46
49	(1S,4S)-tetraphyllin B[h]	4-hydroxy	epimer of 47
50	(1S,4R)-6'-O-rhamnosylepivolkenin[i]		
51	(1R,4S)-6'-O-rhamnosyltaraktophyllin[i]		
52	(1S,4R)-passicapsin[j]	4-(2,6-dideoxy-β-D-xylosyl)	of 48
53	(1S,4R)-passibiflorin[j]	4-(6-deoxy-β-D-gulosyl)	of 48
55/55	(epi)-passisuberosin[j]	4-hydroxy,2,3 epoxide	
56	one or more sulphate esters[h]		
57	(1S,4R,5S)-gynocardin[a]	4,5-dihydroxy	of 45

Nicotinic acid?
58 acalyphin[b]

[a] D. S. Seigler, in 'Cyanide in Biology', ed. B. Vennesland, E. E. Conn, C. J. Knowles, J. Westley, and F. Wissing, Academic Press, London, 1981, p. 133.
[b] A. Nahrstedt, in 'Biologically Active Natural Products', ed. K. Hostettmann and P. J. Lea, Clarendon Press, Oxford, 1987, p. 213.
[c] A. Nahrstedt, P. S. Jensen, and V. Wray, Phytochemistry, 1989, **28**, 623.
[d] H. Shimomura, Y. Sashida, and T. Adachi, Phytochemistry, 1987, **26**, 2363.
[e] A. Nahrstedt, E. A. Sattar, and S. M. H. El-Zalabani, Phytochemistry, 1990, **29**, 1179.
[f] M. Willems, Phytochemistry, 1988, **27**, 1852.
[g] P. Schwind, V. Wray, and A. Nahrstedt, Phytochemistry, 1990, **29**, 1903.
[h] E. S. Olafsdottir, J. V. Andersen, and J. W. Jaroszewski, Phytochemistry, 1989, **28**, 127.
[i] J. W. Jaroszewski, D. Bruun, V. Clausen, and C. Cornett, Planta Med., 1988, **54**, 333.
[j] E. S. Olafsdottir, C. Cornett, and J. W. Jaroszewski, Acta Chem. Scand., 1989, **43**, 51.

The fatty acids within cyanolipids may vary, although 20 carbon atom acids predominate, but in other respects they correspond closely to cyanoglycosides that also occur in the same plant family, the Sapindaceae, with ester linkages instead of glucosidic links.[20] Four basic structures are known (2–5), all with the same 5 carbon atom skeleton derived from L-leucine but differing either in the position of the double bond or in the number of hydroxyl groups. Only (2) and (3) are cyanogenic because only these give rise to a cyanohydrin after hydrolysis of the ester linkages.

3 Sources and Origins

For some years the presence of cyanoglycosides has been valued as a taxonomic marker in plants. A requirement for this is that close relatives among plant species should contain the same biogenetic type of cyanoglycoside(s) allowing familial relationships to be traced. This holds well in several instances, e.g. ferns contain only glycosides derived from phenylalanine, such as prunasin and vicianin, while only the tyrosine derived taxiphyllin has been detected in gymnosperms.[19] However, relationships are not always so clear and some plant families show greater complexity. It used to be considered that Rosaceae characteristically contained only phenylalanine derived glycosides such as prunasin and amygdalin but it is now known that this is true only for two sub-families. In other sub-families heterodendrin, cardiospermin-p-hydroxybenzoate, cardiospermin-p-hydroxycinnamate, or dhurrin may be found.

The greatest diversity is found among more advanced groups of plants. Different types of glycoside have now been identified not only within families but also within genera and within a single species. This last applies to *Carica papaya* which is unusual in another respect; the leaves not only contain small amounts of tetraphyllin B and prunasin but also benzyl glucosinolate.[23] This is the first recorded instance of cyanoglycosides and a glucosinolate occurring in the same species. More advanced grasses show considerable diversity and cyanoglycosides are now known to be widespread among cereals though not necessarily with toxicological significance.[18,24]

It is usual to find linamarin and lotaustralin co-existing in plants because the biosynthetic enzyme systems are able to use both of the structurally related amino acids valine and isoleucine as precursors.[25] Similarly the corresponding diglucosides, linustatin and neolinustatin, also occur together. The proportions of these can vary widely, according to substrate availability or specificity but in some plants one glucoside may predominate to the extent that the amount of the other pair member is negligible. Thus linamarin is predominant in cassava, *Manihot esculenta*, also known by several other common names including manioc, yuca, and tapioca. Cassava is by far the most important cyanogenic food crop for humans, a major source of dietary energy in many tropical regions. Linamarin is present in leaves and tubers, both of which may be eaten. This glucoside also predominates in beans of the lima or butter type, *Phaseolus lunatus*, an important edible legume source of protein. In the case of rubber, *Hevea brasiliensis*, the seeds of which may be fed to animals, only linamarin has been found. On the other hand, in the important forage crops white clover, *Trifolium repens*, and birds-foot trefoil, *Lotus corniculatus*, either linamarin or lotaustralin may predominate. Both occur also in linen flax plants, *Linum usitatissimum*, but in linseed meal it is the corresponding diglucosides that are found.

Cyanogenesis in grasses is particularly associated with *Sorghum* species in which the cyanogen is dhurrin. This is not found in the seed but can account for up to 30% of the dry weight of young seedlings.[26] The seed is widely used as food for humans in particular but forage species are similarly important for grazing livestock as are arrowgrass species, *Triglochin maritima* and *T. palustris*, in which the cyanogens are triglochin and taxiphyllin. The latter is unusual in being comparatively thermolabile, a significant point for humans used to consuming bamboo shoots, which also contain this glucoside and are highly cyanogenic in the raw state.

The best known of all cyanogens is amygdalin. This is responsible for the toxicity of the seeds of many species of Rosaceae, such as bitter almonds,

[23] K. C. Spencer and D. S. Seigler, *Am. J. Bot.*, 1984, **71**, 1444.
[24] C. Pitsch, M. Keller, H. D. Zinsmeister, and A. Nahrstedt, *Planta Med.*, 1984, **50**, 388.
[25] K. Hahlbrook and E. E. Conn, *Phytochemistry*, 1971, **10**, 1019.
[26] B. A. Halkier and B. L. Møller, *Plant Physiol.*, 1989, **90**, 1552.

peaches, and apricots. It has a bitter taste but considerable reduction by breeding has given rise to a much lower concentration of amygdalin in sweet almonds. Their use for preparation of marzipan is well known and the required processing should eliminate most of the cyanide. Small amounts of cyanide can be detected in products of Rosaceous fruits preserved by canning or as jams and preserves but these are insufficient to have toxicological significance. Instances of human poisoning by Rosaceae have usually, therefore, arisen through misuse, for example through preparing peach leaf tea.[5] In the case of leaves the cyanogen is prunasin since there is the general distinction in this family that the monoglucoside occurs in leaves and the diglucoside in seeds.

In addition to the important cultivated food plants many cyanogenic species are browsed by range animals, including serviceberry, *Amelanchier alnifolia*, in North America[5] and *Acacia* species in Africa and elsewhere.[27] The list of cyanogenic food plants could be extended further with other browse species or plants with more restricted, though locally important, use by humans. The majority of such additions would relate to tropical regions and instances of poisoning by such plants are recorded.[4-6] For some, the cyanoglycoside(s) responsible may not yet have been identified.

Concentrations of cyanogens in food plants are extremely variable. Factors contributing to this include plant variety, age, growing conditions, and post-harvest changes. Consequently quantitative information, although available,[16-18] can only be a rough guide for other situations. For food plants that are processed before consumption, changes in cyanogen content can be expected but change may be inconsistent as has been noted for products from cassava.[27-30] Variation in raw materials together with that arising during processing inevitably leads to unpredictability. Furthermore, some quoted values were obtained before the analytical problems of cyanide and cyanogen estimation were fully recognized. What is abundantly clear, however, is that a fatal amount can be ingested from a wide range of plant foods.

Although it is well established that the aglycones of cyanoglycosides are derived from amino acids, details of biosynthesis are only available for a few cases. These show several common features and relationships with the biosynthesis of glucosinolates and nitro compounds in the early stages.[31] The precursor amino acid is first converted to the *N*-hydroxyamino acid and subsequently to the nitrile via an aldoxime but there are several uncertainties regarding intermediate steps. Decarboxylation occurs at an early stage but a recent proposal is that the hydroxyamino acid is oxidatively converted to the nitroso derivative first in dhurrin biosynthesis.[32] Further questions arise concerning stereochemistry of the aldoxime. For dhurrin biosynthesis, an isomeri-

[27] F. Nartey, in 'Cyanide in Biology', ed. B. Vennesland, E. E. Conn, C. J. Knowles, J. Westley, and F. Wissing, Academic Press, London, 1981, p. 115.
[28] O. C. Olarewaju and Z. Roszormenyi, *West Afr. J. Biol. Appl. Chem.*, 1975, **18**, 7.
[29] V. Ravindran, E. T. Kornegay, and A. S. B. Rajaguru, *Anim. Feed Sci. Tech.*, 1987, **17**, 227.
[30] B. Nambisan and S. Sundaresan, *J. Sci. Food Agric.*, 1985, **36**, 1197.
[31] E. E. Conn, 'Biologically Active Natural Products: Potential Use in Agriculture', ASC Symposium Series No. 380, 1988, p. 143.
[32] B. A. Halkier, C. E. Olsen, and B. L. Møller, *J. Biol. Chem.*, 1989, **264**, 19487.

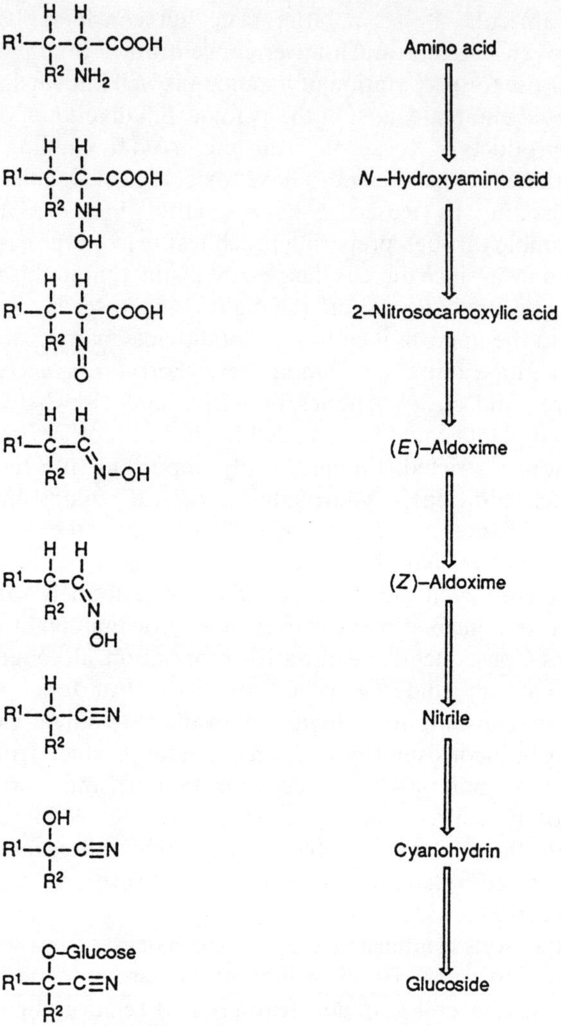

Scheme 1

Amino acid → N-Hydroxyamino acid → 2-Nitrosocarboxylic acid → (E)-Aldoxime → (Z)-Aldoxime → Nitrile → Cyanohydrin → Glucoside

zation has been shown with microsomal preparations from sorghum and it has been suggested that stereochemical considerations have significance for the branching to glucosinolate biosynthesis at this stage (Scheme 1).[32]

A number of factors have contributed to the difficulty in elucidating intermediate stages. Firstly, some intermediates such as the nitroso derivatives of *N*-hydroxyamino acids are extremely unstable. Secondly, most systems investigated have shown channelling of intermediates to some degree so that exogenous substrates may be poorly incorporated.[33,34] The final steps are formation of the cyanohydrin from the nitrile and glycosylation, the former introducing a

[33] E. E. Conn, in 'Cyanide in Biology', ed. B. Vennesland, E. E. Conn, C. J. Knowles, J. Westley, and F. Wissing, Academic Press, London, 1981, p. 183.
[34] B. A. Halkier, H. V. Scheller, and B. L. Møller, in 'Cyanide Compounds in Biology', Ciba Foundation Symposium 140, Wiley, Chichester, 1988, p. 49.

chiral centre with rigid stereospecificity in most instances other than for cyclopentenyl cyanoglycosides. Glycosylation is accomplished by soluble rather than microsomal enzymes.

4 Analytical

Because all plants produce cyanide during formation of ethylene from 1-aminocyclopropane-1-carboxylic acid,[35] cyanide might be detected by very sensitive methods in material from any plant if sampled at an appropriate time. This is not cyanogenesis in the sense used here and so it would be helpful to have some threshold level of cyanide release which could be taken as indicating the presence of cyanogens. It has been argued that plant material should release at least 10 mg HCN kg^{-1} fresh weight to be classed as cyanogenic.[36] Such a level is detectable by the Guignard–Mirande and Feigl–Anger tests, both of which are simple to conduct.[16,37] Cyanide gives a brick-red colour with alkaline picrate in the former and a blue colour with 4,4'-tetramethyldiaminophenylmethane and copper ethylacetoacetate in the latter. The latter is more definite and sensitive and has the additional advantage that dry test papers are used.[38] False positives are possible because neither test is absolutely specific and false negatives can also be obtained if the cyanide that is present in the plant material is not released. Plant materials usually contain enzymes that degrade cyanoglycosides as well as the cyanogens (Scheme 2). However, the enzymes may be lacking or insufficiently active and so exogenous enzymes may be required and would be essential to confirm a negative response. A number of plant glycosidases are available for this purpose but they vary in substrate specificity.[39] A wider spectrum of activity is found with a β-glucuronidase from a snail, *Helix pomatia*, but even so, some glycosides are hydrolysed only slowly by this enzyme and ample time must be allowed for detection.[40] Of course, similar considerations apply to the detection of cyanogens in chromatographic fractions.

Many methods exist for quantitation of cyanide and choice may depend upon source and type of material. Another consideration is convenience and availability of facilities and this has resulted in spectrophotometric methods being widely adopted. Several, though not all, procedures have been based on the König reaction first described for dye synthesis. Cyanide is oxidatively converted to a cyanogen halide which then cleaves pyridine to give glutaconic aldehyde (Scheme 3). This is then coupled with either an amine or a compound possessing an active methylene group to yield a coloured product. Early

[35] G. D. Peiser, T.-T. Wang, N. E. Hoffman, S. F. Yang, H.-W. Liu, and C. T. Walsh, *Proc. Natl. Acad. Sci. USA*, 1984, **81**, 3059.
[36] R. Hegnauer, *Plant Syst. Evol.*, 1977, Suppl. 1, 191.
[37] B. Tantisewie, H. W. L. Ruijgrok, and R. Hegnauer, *Pharm. Weekbl.*, 1969, **104**, 145.
[38] A. Nahrstedt, in 'Cyanide in Biology', ed. B. Vennesland, E. E. Conn, C. J. Knowles, J. Westley, and F. Wissing, Academic Press, London, 1981, p. 145.
[39] L. Brimer, in 'Cyanide Compounds in Biology', Ciba Foundation Symposium 140, Wiley, Chichester, 1988, p. 177.
[40] L. Brimer, S. B. Christensen, P. Molgaard, and F. Nartey, *J. Agric. Food Chem.*, 1983, **31**, 789.

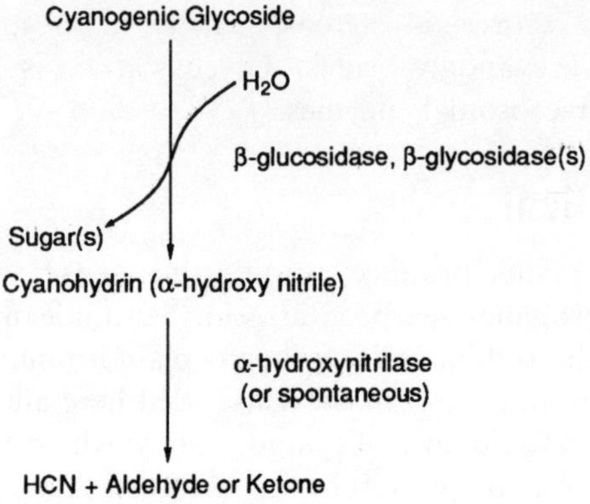

Scheme 2

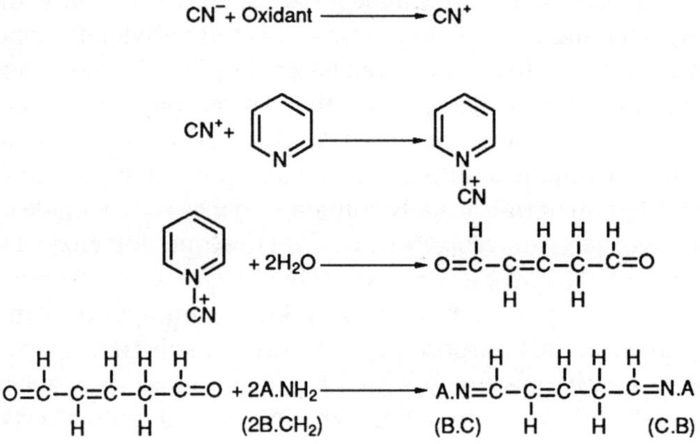

Cyanogenic Glycoside
→ +H₂O, β-glucosidase, β-glycosidase(s)
→ Sugar(s) + Cyanohydrin (α-hydroxy nitrile)
→ α-hydroxynitrilase (or spontaneous)
→ HCN + Aldehyde or Ketone

Scheme 3

$CN^- + \text{Oxidant} \longrightarrow CN^+$

$CN^+ + \text{pyridine} \longrightarrow \text{pyridinium-CN}$

$\text{pyridinium-CN} + 2H_2O \longrightarrow O=CH-CH=CH-CH_2-CHO$

$O=CH-CH=CH-CH_2-CHO + 2A.NH_2 \longrightarrow A.N=CH-CH=CH-CH_2-CH=N.A$
(2B.CH₂) (B.C) (C.B)

examples of such procedures involved benzidine[41] as the amine or pyrazolone[42] as the source of a reactive methylene group. There have been many subsequent developments designed to improve the stability of reagents or their safety, *e.g.* avoidance of carcinogenic amines.[43,44] With these procedures detection down to 0.01 μg cyanide ml⁻¹ is possible. They are well suited to analysis of cyanide that has been separated from biological materials by aeration or distillation but further adaptations can also allow their use directly in reaction mixtures. Although these sensitive and widely used methods are available many others have been developed that match or even exceed in sensitivity. These employ gas chromatography, voltammetry, fluorimetry, cyanide selective electrodes, polarography, or atomic absorption spectrometry.[39,44] An alternative need to sensi-

---

[41] W. N. Aldridge, *Analyst (London)*, 1944, **69**, 262.
[42] J. Epstein, *Anal. Chem.*, 1947, **19**, 272.
[43] J. L. Lambert, J. Ramasamy, and J. V. Paukstells, *Anal. Chem.*, 1975, **47**, 916.
[44] A. Nahrstedt, N. Erb, and H.-D. Zinsmeister, in 'Cyanide in Biology', ed. B. Vennesland, E. E. Conn, C. J. Knowles, J. Westley, and F. Wissing, Academic Press, London, 1981, p. 461.

tivity may be that an assay should be adaptable to field conditions. Developments in this direction have reverted to alkaline picrate for quantitation.[45]

With such a choice of procedures available, the method of estimating cyanide is unlikely to be the limiting factor in analysis of cyanogens. The primary problem will probably arise in handling and storage of the plant materials. This applies particularly if it is required to analyse the separate cyanogenic components individually but applies also to estimation of total cyanide. Autolysis by endogenous enzymes can lead to rapid changes and these are best prevented by low temperature storage.[38] Attempts to inactivate these enzymes by heating and drying may not avoid changes in constituents and considerable loss of cyanide.[46] The need to distinguish between cyanide in a parent compound such as a glycoside and so-called 'free' cyanide is well recognized in relation to food plants because of their differential toxicity. Hence, procedures that allow this distinction have been developed,[47] even to the extent of automation.[48] However, this does not allow the distinction of intermediate products such as cyanohydrins which requires more specialized treatment.[39] Unfortunately there is no single set of conditions suitable for treating plant tissues in which all forms of cyanide are stable. In general, cyanoglycosides are stable in boiling 80 % v/v ethanol but cyanohydrins are stabilized in cold, slightly acidic conditions. Thus, separation of components followed by separate estimations can lead to summation of all components. A wide range of chromatographic procedures have been developed in the course of structure elucidation that can provide not only separation but also quantitation of individual components.[38,39] Many cyanoglycosides can be estimated by either gas–liquid or high performance liquid chromatography. Greater problems are encountered in distinguishing between cyanohydrins and free cyanide but these problems are now soluble.

The existence of so many different cyanogens does not allow recommendations for analysis with universal applicability. Each instance must be considered individually. Fortunately a number of extensive reviews are available and can be strongly recommended.[38,39,44,49]

5 Role in Plants

Although all cyanogens have the property of releasing free cyanide it does not follow that they have a common role. Cyanide gained early notoriety for its toxicity and has been known as a respiratory poison for many years. It is understandable that the assumption should be made that a role of plant cyanogens is in defence against herbivores and disease vectors. Such a suggestion appears to be traceable back to at least 1888[12,50] but it is only comparatively

---

[45] L. Brimer and P. Mølgaard, *Biochem. Syst. Ecol.*, 1986, **14**, 97.
[46] E. Torres, J. F. Pereira, A. M. Brinker, and D. S. Seigler, *J. Sci. Food Agric.*, 1988, **42**, 149.
[47] R. D. Cooke, *J. Sci. Food Agric.*, 1978, **29**, 345.
[48] P. V. Rao and S. K. Hahn, *J. Sci. Food Agric.*, 1984, **35**, 426.
[49] A. Zitnak, in 'Chronic Cassava Toxicity', ed. B. Nestel and R. MacIntyre, International Development Research Centre, Ottawa, 1973, p. 89.
[50] D. A. Jones, in 'Cyanide Compounds in Biology', Ciba Foundation Symposium 140, Wiley, Chichester, 1988, p. 151.

recently that evidence has been obtained to support this and even now its extent and significance is questioned.

From a functional viewpoint it can be argued that cyanogens are not effective in defence because of widespread herbivory among cyanogenic plants. It can further be asked why several plant species are polymorphic for cyanogenesis if defence is the primary role? Yet it is accepted in ecological biochemistry that a defensive function exists.[51] Laboratory evidence indicates that some animal species show a preference for acyanogenic plants and there is some support that preferences also exist in natural habitats.[50,52] It has, though, been argued that too little attention may have been given to negative results.[53] Few plant species have been studied in relation to a defensive role and several of these have been important food sources for men and domestic animals, which may not be the best subjects. The conclusion that can presently be drawn is that cyanogens can confer a defensive advantage in some situations where choice exists and this advantage seems to stem from deterrence to feeding rather than the toxicity of the cyanogens.

Cyanoglycosides generally have a bitter taste, although the unnamed 1,2 diglucoside isomer of amygdalin from a plant used in Japanese cooking[54] has been reported to taste sweet.[19] The bitterness of cassava has been shown to correlate with cyanoglycoside content, although there are exceptions and other factors can contribute to the bitterness but bitterness is reduced by processing that eliminates the cyanoglycosides.[55] Thus, feeding deterrence is explicable in any species that responds adversely to bitter taste. Another possibility is that deterrence is brought about by the HCN or by the carbonyl compounds derived from the aglycones, that are released when plant tissues are damaged. Again, uniformity of response is not found; locusts are deterred by the HCN released from sorghum or cassava[56] whilst slugs[50] and ants[57] find carbonyls more repellent. A third possibility that has been considered is that it is an effect of breakdown products in the digestive tract that serves to deter.[58] There is comparatively little information on a defensive role for cyanolipids but both susceptibility and resistance to them have been shown in tests with certain insects.[59-61]

At the other extreme there are animals that are specialist feeders on cyanogenic plants or polyphagous species that show a preference for cyanogenic plants. The southern armyworm, *Spodoptera eridania*, shows a preference for lima

[51] J. B. Harborne, 'Introduction to Ecological Biochemistry', 3rd edition, Academic Press, London, 1988, p. 98.
[52] D. A. Jones, in 'Cyanide in Biology', ed. B. Vennesland, E. E. Conn, C. J. Knowles, J. Westley, and F. Wissing, Academic Press, London, 1981, p. 509.
[53] A. J. Hruska, *J. Chem. Ecol.*, 1988, **14**, 2213.
[54] M. Aritomi, T. Kumori, and T. Kawasaki, *Phytochemistry*, 1985, **24**, 2438.
[55] S. Sundaresan, B. Nambisan, and C. S. E. Amma, *Indian J. Agric. Sci.*, 1987, **57**, 37.
[56] S. Woodhead and E. Bernays, *Nature (London)*, 1977, **270**, 235.
[57] S. C. Petersen, *Naturwissenschaften*, 1986, **73**, 627.
[58] F. Kaethler, D. J. Pree, and A. W. Bown, *Ann. Entomol. Soc. Am.*, 1982, **75**, 568.
[59] D. H. Janzen, H. B. Juster, and E. A. Bell, *Phytochemistry*, 1977, **16**, 223.
[60] J. C. Braeckman, D. Daloze, and J. M. Pasteels, *Biochem. Syst. Ecol.*, 1982, **10**, 355.
[61] K. L. Mikolajczak, R. V. Madrigal, C. R. Smith, Jr., and D. K. Reed, *J. Econ. Entomol.*, 1984, **77**, 1144.

bean, *Phaseolus lunatus*, which contains linamarin and it has been shown that cyanide attracts it to feed.[62] Larvae of a Southern African butterfly, *Acraea horta*, were also shown to select only leaves of cyanogenic plant species in a choice situation. However, when crude fractions obtained from their normal host plant, *Kiggelaria africana*, were applied to filter paper disks, larvae preferred to eat discs treated with acyanogenic fractions rather than those containing the cyclopentenyl cyanoglycosides.[63] Some species which show a preference for cyanogenic plants may not suffer deleterious effects from other plant metabolites either.[64] Few situations are likely to be explicable solely by the presence or content of cyanoglycosides.

Thus, the role of cyanogens in relation to animal herbivores is neither clear nor consistent. Too many different responses exist, even within the comparatively few situations studied to allow general conclusions. It has always to be remembered though that there are many reports of animal fatalities covering several species of both animals and plants. These indicate that deterrence can be a weak effect.

In addition to animal herbivores, cyanogenic plants may be subject to attack by fungal pathogens. It has been indicated that pathogens of strongly cyanogenic plants possess the capacity to convert cyanide to formamide by means of cyanide hydratase (formamide hydrolyase),[65] a detoxication enzyme not found in animals. In *Stemphylium loti*, a fungal pathogen of *Lotus corniculatus*, cyanide induces the enzyme and the fungus also possesses a cyanide resistant pathway of respiration, a second defence against cyanide.[66] Cyanide resistant respiration is also found in *Microcyclus ulei*, a fungal pathogen of the rubber tree, *Hevea brasiliensis*, but this fungus does not possess cyanide hydratase.[67] In this particular plant–pathogen interaction cyanogenesis is a positive disadvantage to the plant because plants with high cyanogenic capacity are more susceptible to attack by *M. ulei*. Cyanide released in leaf damage restricts energy metabolism by the plant which decreases production of the protective phytoalexin scopoletin.[68]

These findings indicate the limitations of cyanide as a defence system for plants. Plants themselves may have too limited defence against their own chemical weapon. An organism with either a high capacity for cyanide detoxication or good cyanide resistant respiration will need to be deterred by other means.

Plants will have some capacity to cope with released cyanide because it can readily be incorporated into β-cyanoalanine by cyanoalanine synthase and

---

[62] L. B. Brattsten, J. H. Samuelian, K. Y. Long, S. A. Kincaid, and C. K. Evans, *Ecol. Entomol.*, 1983, **8**, 125.
[63] D. Raubenheimer, 'Cyanogenesis and the feeding preference of *Acraea horta* (L). (Lepidoptera: Acraeinae)', M.S. thesis, University of Capetown, 1987.
[64] J. T. Smiley and C. S. Wisdom, *Biochem. Syst. Ecol.*, 1985, **13**, 305.
[65] W. E. Fry and D. F. Myers, in 'Cyanide in Biology', ed. B. Vennesland, E. E. Conn, C. J. Knowles, J. Westley, and F. Wissing, Academic Press, London, 1981, p. 321.
[66] J. F. Rissler and R. L. Millar, *Plant Physiol.*, 1977, **60**, 857.
[67] R. Lieberei, *J. Phytopathol.*, 1988, **122**, 54.
[68] R. Lieberei, B. Biehl, A. Giesemann, and N. T. V. Junqueira, *Plant Physiol.*, 1989, **90**, 33.

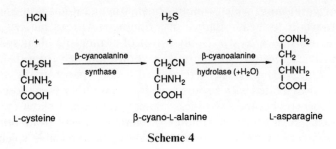

Scheme 4

thereafter, in many plants, into asparagine (Scheme 4).[69,71] Cyanoalanine synthase activity is usually higher in cyanogenic than acyanogenic plants[72] and this suggests that the enzyme is a major route for cyanide detoxification or that cyanogenesis has a role in general nitrogen metabolism, or both. It had been suggested at an early stage that cyanogens played a part in the nitrogen economy of plants[11] but this lacked immediate support. Recent findings point towards its validity in some instances. It is clear that cyanogens are metabolically active and are not inert end-products of metabolism. However, a question arises concerning a primary role in nitrogen metabolism because the aglycone is derived from an amino acid precursor; why is this not used more directly as the source of nitrogen? It would appear to be an inefficient diversion for the nitrogen to be transferred via a cyanogen intermediate unless there are advantages to be gained either from translocation or storage in such a form.

Studies with *Hevea brasiliensis* provide an example of nitrogen being transported in cyanoglycoside form. Linamarin in the seed endosperm is glycosylated to linustatin which is not a substrate for the plants own linamarase (β-glucosidase). In this diglucoside form it is transported to the growing parts of young seedlings where a diglucosidase releases the aglycone. Cyanide released from the aglycone is subsequently incorporated into asparagine via β-cyanoalanine.[73] Monoglucoside–diglucoside couples exist in other plants, e.g. both linamarin–linustatin and lotaustralin–neolinustatin in flax, *Linum usitatissimum*, whilst in many species of Rosaceae, prunasin is found in leaves and amygdalin in seeds. Analysis of different plant tissues at appropriate times may yield further examples of a role in transport of nitrogen.

Cyanolipids provide an example of nitrogen storage in seeds of *Ungnadia speciosa*.[74] The cyanolipids are completely consumed during development of the seedlings but without release of free cyanide provided that the plant is healthy. This implies conservation of the nitrogen and conversion to other compounds. Some cyanoglycosides are produced at the same time as the cyanolipids are being used but these could not account for more than a quarter

[69] Y. P. Abrol and E. E. Conn, *Phytochemistry*, 1966, **5**, 237.
[70] Y. P. Abrol, E. E. Conn, and J. R. Stoker, *Phytochemistry*, 1966, **5**, 1021.
[71] S. G. Blumenthal, H. R. Hendrickson, Y. P. Abrol, and E. E. Conn, *J. Biol. Chem.*, 1968, **243**, 5302.
[72] J. M. Miller and E. E. Conn, *Plant Physiol.*, 1980, **65**, 1199.
[73] D. Selmar, R. Lieberei, and B. Biehl, *Plant Physiol.*, 1988, **86**, 711.
[74] D. Selmar, S. Grocholewski, and D. S. Seigler, *Plant Physiol.*, 1990, **93**, 631.

of the cyanolipid nitrogen. Thus, there are clear examples of roles for cyanogens other than in defence and it seems probable that further examples of primary or, perhaps, multiple roles will be elucidated in the future.

6 Ecological Aspects

Cyanogenesis is not exclusive to plants. Many micro-organisms can produce or degrade cyanide.[1,75,76] Among animals, cyanogenesis appears to be restricted to arthropods, insects having received much recent attention.[77,78] Associations between diverse organisms that involve cyanogenesis appear to be common, *e.g.* many of the known cyanogenic insects are associated with plant families noted to include cyanogenic species. Ecological relationships are consequently suggested by these associations and are receiving considerable attention.

The proposed role for cyanogens in defence has been central to ecological investigations. The demonstration that cyanide was released from crushed tissues at all stages of the life-cycle of a moth, *Zygaena filipendulae*, that feeds on birds-foot trefoil[79] has been followed by a wide range of further studies that now encompass the genetics and biochemistry of cyanogenesis in plants and other organisms, of which studies on white clover, *Trifolium repens*, and birds-foot trefoil, *Lotus corniculatus*, appear to have been the most extensive.[12,50,52,80,81] In the plant genera *Lotus*, *Trifolium*, *Linum*, and *Manihot*, all important contributors of plant foods and all containing the linamarin and lotaustralin system, polymorphism has been shown. Certain features of their genetics or biochemistry may be similar, therefore, but differences in culture and location will introduce totally different ecological considerations in which cyanogenesis is only one feature. As food crops their distribution and extent of use can vary within comparatively short periods of time. For example, cassava is native to tropical America but was introduced to West Africa in the Sixteenth century, from where it spread to many other parts of Africa, and introduced to the Indian sub-continent and the Far East during the latter part of the Eighteenth century.[82] Its subsequent development has influenced the distribution of pests and diseases and cyanogenesis will have featured within these changes.[83]

[75] C. J. Knowles, *Bacteriol. Rev.*, 1976, **40**, 652.
[76] C. J. Knowles, in 'Cyanide Compounds in Biology', Ciba Foundation Symposium 140, Wiley, Chichester, 1988, p. 3.
[77] R. H. Davis and A. Nahrstedt, in 'Comprehensive Insect Physiology, Biochemistry and Pharmacology', ed. G. A. Kerkut and L. I. Gilbert, Pergamon Press, Oxford, 1985, Vol. 11, p. 635.
[78] A. Nahrstedt, in 'Cyanide Compounds in Biology', Ciba Foundation Symposium 140, Wiley, Chichester, 1988, p. 131.
[79] D. A. Jones, J. Parsons, and M. Rothschild, *Nature (London)*, 1962, **193**, 52.
[80] M. A. Hughes, in 'Cyanide in Biology', ed. B. Vennesland, E. E. Conn, C. J. Knowles, J. Westley, and F. Wissing, Academic Press, London, 1981, p. 495.
[81] M. A. Hughes, A. L. Sharif, M. A. Dunn, and E. Oxtoby, in 'Cyanide Compounds in Biology', Ciba Foundation Symposium 140, Wiley, Chichester, 1988, p. 111.
[82] R. D. Cooke and D. G. Coursey, in 'Cyanide in Biology', ed. B. Vennesland, E. E. Conn, C. J. Knowles, J. Westley, and F. Wissing, Academic Press, London, 1981, p. 93.
[83] J. Cock, R. MacIntyre, and M. Graham, Proceedings of the 4th Symposium of the International Society for Tropical Root Crops, CIAT, Cali, Columbia, International Development Research Centre, Ottawa, 1977.

Lima beans also contain linamarin but like many other beans they are noted for containing lectins and protease inhibitors in addition. Within the species, cyanogen content can vary widely with small, dark-seeded varieties generally containing much more than large, white-seeded varieties. Consequently the ecological contribution of the linamarin will be extremely variable. A range of toxins is also found in the bracken fern, *Pteridium aquilinum*. In both Europe[84] and America[85] relationships between the diversity and abundance of arthropods on bracken and its cyanogen content have been shown. Bracken contains polyphenols, thiaminase activity, and other toxins which would also be expected to contribute to its defence but even with such an array of chemical weapons, it has been argued that in some instances changes in plant structure during growth may become a more important determinant in colonization by some arthropods than chemical composition.[86]

Cyanide in soil and in water will also contribute to ecological development. Man's activities may be a major contribution to the cyanide in both within certain localities but natural sources make significant contributions also. Plant roots can exude cyanoglycosides into soil, and soil under cyanogenic plants may contain considerably more cyanide than under acyanogenic plants, with possible influence on soil micro-organisms.[87] It is extremely difficult to identify the sources of cyanide in water courses but plants near to them can certainly be included. They will contribute to seasonal variations that have been observed and the extent to which organisms are able to survive in the aqueous environment.[88,89]

Studies to date have revealed the complexity of ecological relationships in which cyanogenesis may be a contributory factor. Some of the known relationships between cyanogenic plants and other organisms suggest that cyanogenesis has been an important contributor to the development of particular ecological niches but the number of apparent exceptions and the range of natural variability that can be observed emphasize the need for further careful observation and new experimental approaches in this area.

7 Cyanide Detoxication

It has been known for almost a century that the major detoxication product of cyanide is thiocyanate. An enzyme catalysing this conversion was reported in 1933[90] and given the name rhodanese. The ending -ese denotes that it was named after the German name of the product and this unusual feature has probably contributed to the widespread retention of the trivial name for the

---

[84] J. H. Lawton, *Bot. J. Linn. Soc.*, 1976, **73**, 187.
[85] I. Schreiner, D. Nafus, and D. Pimentel, *Ecol. Entomol.*, 1984, **9**, 69.
[86] J. H. Lawton, in 'Diversity of Insect Faunas', Symposium Royal Entomology Society London 9, ed. L. A. Mound and N. Waloff, Blackwell, Oxford, 1978, p. 105.
[87] A. M. Dartnall and R. G. Burns, *Biol. Fert. Soils*, 1987, **5**, 141.
[88] H. Krutz, in 'Cyanide in Biology', ed. B. Vennesland, E. E. Conn, C. J. Knowles, J. Westley, and F. Wissing, Academic Press, London, 1981, p. 479.
[89] G. Leduc, in 'Cyanide in Biology', ed. B. Vennesland, E. E. Conn, C. J. Knowles, J. Westley, and F. Wissing, Academic Press, London, 1981, p. 487.
[90] K. Lang, *Biochem. Z.*, 1933, **259**, 243.

enzyme. In standard textbooks, rhodanese is accorded the pre-eminent place with respect to cyanide detoxication but this is an over-simplification. Although the systematic name is thiosulphate: cyanide sulphurtransferase, EC 2.8.1.1, it has broad substrate specificity and can transfer sulphane sulphur (divalent sulphur that, in its deprotonated form, is bonded only to other sulphur) to a wide range of thiophilic acceptors.[91] Thus, donors include thiosulphonates, polythionates, per- and polysulphides, and elemental sulphur in staggered 8-membered ring form, in addition to thiosulphate whilst acceptors include sulphite, sulphinates, and some mono- and dithiols in addition to cyanide. With such a range of possible substrates the *in vivo* contributions of alternative donors and the influence of alternative acceptors is far from clear.

Localization of the enzyme is important too. It is mitochondrial and substrates must be accessible. Cyanide can certainly gain access to mitochondria because it is known to inhibit enzymes located there. On the other hand sulphur donors may not gain easy access, although some transferable sulphur must be present within mitochondria in order for rhodanese to fulfil any physiological sulphur transfer function. This applies not only to cyanide detoxication, which is probably a secondary role, but also for a role in regulating the rate of mitochondrial respiration.[92,93]

Two further enzymes are able to generate sulphane sulphur.[91] 3-Mercaptopyruvate sulphurtransferase (EC 2.8.1.2), like rhodanese, has broad sulphur acceptor specificity. Cyanide is one acceptor, providing a second route for thiocyanate formation, while other acceptors generate sulphane sulphur and hence rhodanese substrates. The enzyme is found both within and without mitochondria, in the latter location possibly being linked to the physiological role of rhodanese. The other enzyme, thiosulphate reductase or thiosulphate: thiol sulphurtransferase (EC 2.8.1.3), cannot use cyanide as an acceptor. The only acceptors are, apparently, thiols but the persulphides thus formed could also be rhodanese substrates and so contribute to cyanide detoxication.

There are further ways in which thiocyanate might be formed, not involving enzyme catalysis. Cyanide can react directly with a form of sulphur that is bound by serum albumin.[94] Because blood offers the first protection against cyanide intoxication this reaction may be of considerable functional significance. Cyanide can also react non-enzymatically with di- or persulphides to give thiocyanates.

In view of the variety of possibilities for thiocyanate formation the quantitative significance of the rhodanese catalysed reaction is not clear and probably varies with cyanide load. What is now accepted, however, is that ample enzyme activity is available but that detoxication is limited by the availability of suitable sulphur donors at the required locations. Measurement of available sulphur presents considerable problems. Appropriate sulphur sources must

[91] J. Westley, in 'Cyanide Compounds in Biology', Ciba Foundation Symposium 140, Wiley, Chichester, 1988, p. 201.
[92] K. Ogata, X. Dai, and M. Volini, *J. Biol. Chem.*, 1989, **264**, 2718.
[93] K. Ogata and M. Volini, *J. Biol. Chem.*, 1990, **265**, 8087.
[94] R. Jarabak and J. Westley, *J. Biol. Chem.*, 1986, **261**, 10793.

either be acceptable enzyme substrates or be capable of reacting directly with cyanide under conditions that correspond to the physiological state. Recent measurements that make use of rhodanese appear to confirm that available sulphur is extremely limited and indicate that the total amount may correspond to the lethal dose of cyanide on a molar basis.[91] Alternative measurements based on cyanolysis give higher values but conditions for the reaction may lead to over-estimation.[95]

In addition to conversion to thiocyanate, some cyanide may be oxidized and converted to volatile products[96] and some may react directly with cystine to give 2-iminothiazolidine carboxylic acid which is then excreted in urine.[97] It can also react with hydroxycobalamin to give cyanocobalamin, thereby being prevented from inhibiting enzymes. Little cyanide can be excreted directly in urine. Small amounts may be eliminated by exhalation.[96] Taken together these alternatives make only a minor contribution to cyanide detoxication.

8 Effects in Animals and Man

The effects of cyanoglycosides are always considered to be due to the free cyanide that can be released from them. Little if any account has been taken of the carbonyl compound that is concomitantly derived from the aglycone. With respect to frequently ingested cyanoglycosides, linamarin gives rise to acetone which can arise during normal metabolism through decarboxylation of acetoacetate but amygdalin and prunasin among others will give rise to benzaldehyde which is itself mildly toxic. Nevertheless, when fatalities have occurred the symptoms recorded have been those of cyanide poisoning. This implies, in acute situations at least, that neither the carbonyl compounds nor the intact cyanoglycosides themselves exert significant toxicological effects.

Most cyanide in food can be expected to be present in glycosidic form. HCN itself is volatile (boiling point 26 °C), can readily diffuse through plant tissue, and is reactive and so can readily be eliminated or converted to other compounds. Cyanohydrins require acidic conditions for stabilization and these could exist in plant foods either from plant acids themselves or from acids produced during certain forms of processing such as fermentation. Nevertheless, results such as those obtained with cassava, support the conclusion that cyanoglycosides predominate and that non-glycosidic forms will seldom far exceed 10 % of the total cyanide potential.[47,48] It is important, therefore, to consider the fate of the intact cyanoglycosides.

The contribution that food plant glycosidases and α-hydroxynitrile lyases make to the liberation of cyanide is uncertain. Even if present in the fresh plant material, activity may have been eliminated as a result of storage or processing. On the other hand other plant foods, though not cyanogenic in themselves, could provide glycosidases capable of using cyanoglycosides as substrates.[13] It will be a common occurrence, therefore, for both enzymes and glycosides to be

---

[95] A. M. Buzaleh, E. S. Vasquez, and A. M. del C. Batlle, *Gen. Pharmac.*, 1989, **20**, 323.
[96] G. E. Boxer and J. C. Rickards, *Arch. Biochem.*, 1952, **39**, 7.
[97] J. L. Wood and S. L. Cooley, *J. Biol. Chem.*, 1956, **218**, 449.

consumed together. Some release of cyanide by this means can occur but conditions in the digestive tract do not favour rapid or extensive reaction.[98] All the glycosidases investigated have an acidic pH optimum (4.0–6.2); α-hydroxynitrile lyases similarly require slightly acidic conditions for maximum activity. Such conditions are hardly found in the monogastric digestive tract. It can be anticipated that substantial proportions of ingested cyanogens will not be catabolized by enzymes in food and this has been confirmed with intact glucosides administered orally to rats.[99,100] Some glucoside was excreted in urine unchanged showing both absorption of intact glucoside and avoidance of subsequent metabolism, but increased thiocyanate in blood was indicative of some release of cyanide. Absorption and excretion of glucoside vary considerably between individuals and so must the release of cyanide. Load of cyanogen and rate of absorption will influence the amounts reaching the large intestine which is clearly an important site for cyanide release through the metabolic capabilities of micro-organisms located there.[101] Although some intact glycoside can be excreted unchanged mammalian tissues contain β-glucosidase activities capable of releasing the aglycones which would spontaneously break down to give free cyanide. However, there is considerable variation between species in both the activity and tissue distribution of glycosidases and the quantitative contribution to the total release of free cyanide is unclear.[102,103] It seems probable that more is released by the intestinal microflora than by tissues since intravenous administration of a glycoside is tolerated well in comparison to oral administration.[104] Cyanide has been shown to inhibit pancreatic α-amylase activity but the significance of this for digestion is not yet clear.[105]

Release of cyanide from cyanoglycosides in the ruminant diet does not require the presence of plant enzymes. It has been long established that this can be achieved by ruminal micro-organisms, that glycoside breakdown occurs rapidly, as does disappearance of free cyanide from rumen contents.[106–108] More recently it has been shown that many individual strains of rumen bacteria are capable of hydrolysing cyanoglycosides, that rates of hydrolysis differ between individual glycosides, and that rates are also affected by the nature of

---

[98] J. E. Poulton, in 'Cyanide Compounds in Biology', Ciba Foundation Symposium 140, Wiley, Chichester, 1988, p. 67.
[99] D. C. Hill, in 'Cassava as Animal Feed', ed. B. Nestel and M. Graham, International Development Research Centre, Ottawa, 1977, p. 33.
[100] M. Sakata, A. Yoshida, C. Yuasa, K. Sakata, and M. Haga, *J. Toxicol. Sci.*, 1987, **12**, 47.
[101] G. W. Newton, E. S. Schmidt, J. P. Lewis, E. E. Conn, and R. Lawrence, *West. J. Med.*, 1981, **134**, 97.
[102] A. Freese, R. O. Brady, and A. E. Gal, *Arch. Biochem. Biophys.*, 1980, **201**, 363.
[103] J. Newmark, R. O. Brady, P. M. Grimley, A. E. Gal, S. G. Waller, and J. R. Thistlethwaite, *Proc. Natl. Acad. Sci. USA*, 1981, **78**, 6513.
[104] C. G. Moertel, M. M. Ames, J. S. Kovach, T. P. Moyer, J. R. Rubin, and J. H. Tinker, *J. Am. Med. Assoc.*, 1981, **245**, 591.
[105] V. A. Aletor, *Die Nahrung*, 1989, **33**, 457.
[106] I. E. Coop and R. L. Blakley, *N.Z. J. Sci. Technol., Sect. A*, 1949, **30(5)**, 277.
[107] R. L. Blakley and I. E. Coop, *N.Z. J. Sci. Technol., Sect. A*, 1949, **31(3)**, 1.
[108] I. E. Coop and R. L. Blakley, *N.Z. J. Sci Technol., Sect. A*, 1950, **31(5)**, 44.

the diet and time after feeding.[109-111] In view of the widespread potential of rumen bacteria to liberate cyanide it might be expected that considerable metabolism of cyanide would also be found but this appears not to be so. Rates of thiocyanate formation were low compared with rates of disappearance of cyanide in *in vivo* studies[109] and it is possible that some cyanide could be converted to formamide also but this has not yet been investigated. It seems probable that most of the difference between cyanide disappearance and thiocyanate formation is due to absorption of cyanide from the rumen. It is not known what limits the rate of thiocyanate formation but the supply of suitable forms of sulphur needs further investigation.

Because of the activity of the rumen microflora, ruminants may be more susceptible to poisoning by plant cyanogens than non-ruminants. Nevertheless, many cases of fatal poisoning of non-ruminants, including humans, have been recorded. In many instances natural variation in cyanogen content, inadequate processing, or lack of alternative foods may have contributed to ingestion of excessive amounts. A very different situation arose comparatively recently in North America where cyanogen preparations with the trade-names 'laetrile', 'nitrilosides, or 'vitamin B_{17}' existed, supposedly for the treatment of cancer.[112] The preparations were obtained from apricot pits. The composition of preparations was variable but amygdalin predominated although claims were made that the therapeutic constituent was the glucuronic acid derivative of prunasin. Not surprisingly fatalities resulted from overconsumption and use as an anticancer agent could not be supported.

Cases of acute poisoning appear to be most widespread among ruminants. With sorghum species the potential for cyanide can be high, in excess of 1 g HCN kg^{-1} dry matter in some instances, but sorghum has desirable attributes as a crop and so understanding of the factors affecting cyanogenesis and its potential for release of cyanide are important.[113] The content of dhurrin, the cyanoglucoside in all sorghum species, is influenced by age of the plant, genotype, and several environmental factors contributing to its growth. However, the rate of ingestion by animals ultimately determines the amount of cyanogen consumed and rate of ingestion will usually be greater for conserved material. It has been shown that comparatively little cyanide may be lost in the preparation of sorghum hay. Consequently, it has been advocated that only forage considered to be safe for grazing should be used for hay.[113] Silage making is preferable because the retention of moisture and acidic conditions allow more extensive release of cyanide by the action of the plant glucosidase. Similar considerations apply to other cyanogen containing forage crops. Arrowgrass (*Triglochin maritima*) has been held responsible for livestock losses in the United States[5] but not so in Britain where drought conditions are less likely to occur, a factor believed to influence cyanogen concentrations. In

[109] W. Majak and K.-J. Cheng, *J.Anim. Sci.*, 1984, **59**, 784.
[110] W. Majak and K.-J. Cheng, *Can. J. Anim. Sci.*, 1987, **67**, 1133.
[111] W. Majak, R. E. McDiarmid, J. W. Hall, and K.-J. Cheng, *J. Anim. Sci.*, 1990, **68**, 1648.
[112] V. Herbert, *Am. J. Clin. Nutr.*, 1979, **32**, 1121.
[113] J. L. Wheeler and C. Mulcahy, *Trop. Grassl.*, 1989, **23**, 193.

Africa, many grasses give positive tests for cyanide and stargrass (*Cynodon* spp.) has caused deaths under field conditions.[6] Although effects from ingestion of cyanogenic clovers and trefoils may be severe fatalities are unlikely, although some have been suspected.

The other area for concern is with browse plants, from which fatalities continue to be reported.[114] Often these instances arise in poorer countries where animal production systems are less developed and where the cyanogenic nature of the plants is less understood. One point that arises clearly from consideration of fatalities is that any bitterness in taste is insufficient to discourage consumption and any effect that cyanide release might have on intake is not sufficiently rapid to avoid drastic consequences.

The mechanisms of acute cyanide toxicity are still actively investigated, not particularly because of dietary concern but because of widespread industrial use of cyanide.[115] Inhibition of cytochrome oxidase by cyanide has been accepted as an important biochemical effect[116] but several other enzymes and metabolic pathways will be affected either directly or indirectly. It is thought that the main target organ is probably the central nervous system.[117] With a high dose of cyanide, natural respiration is stopped and cardiac arrest can occur. Antagonism of cyanide poisoning is directed towards sequestering cyanide and converting it to thiocyanate, thereby preventing it from reaching the most sensitive organs or enzymes.[117,118] The rapidity with which cyanide acts may not allow time for measures to be taken.

If acute toxicity is of particular concern with ruminants, it is humans that receive most attention with respect to chronic toxicity. Few cyanogenic plants are likely to be consumed regularly over extended periods but as a staple in tropical regions cassava is consistently consumed by many. It has been stated to be the major staple food for 40 % of the population of sub-Saharan Africa, the region in which it has given major cause for concern.[119]

Of several health disorders that have been linked to cassava the best understood is goitre which arises not from cyanide itself but from thiocyanate, its detoxication product. Thiocyanate is well known as an anti-thyroid agent because of the competitive nature of its interaction with iodine. Extensive investigations have shown that thiocyanate effectively increases the dietary requirement for iodine and that goitre with varying degrees of severity occurs where cassava is consumed and where iodine intake is low or marginal.[9,10,119] Consequently, hypothyroid effects can be avoided if measures are taken to increase iodine consumption. Cretinism is frequently observed in regions where goitre is prevalent. This, too, may arise from hypothyroidism during early

---

[114] V. S. Vihan and H. S. Panwar, *Indian J. Anim. Res.*, 1987, **21**, 53.
[115] B. Ballantyne and T. C. Marrs, 'Clinical and Experimental Toxicology of Cyanides', Wright, Bristol, 1987.
[116] D. Keilin, *Proc. R. Soc. London, Ser. B*, 1929, **104**, 206.
[117] J. L. Way, P. Leung, E. Cannon, R. Morgan, C. Tamulinas, J. Leong-Way, L. Baxter, A. Nagi, and C. Chui, in 'Cyanide Compounds in Biology', Ciba Foundation Symposium 140, Wiley, Chichester, 1988, p. 232.
[118] J. L. Way, *Ann. Rev. Pharmacol. Toxicol.*, 1984, **24**, 451.
[119] H. Rosling, 'Cassava Toxicity and Food Security', Tryck Kontakt, Uppsala, Sweden, 1987.

development and so cassava has been implicated in its occurrence. However, it is also found in cassava consuming regions where iodine intake is sufficient and where it does not accompany goitre.[9] This suggests that cassava consumption may impair intellectual development in some way other than or in addition to the anti-thyroid effects of the thiocyanate that arises from it. In seeking the mechanism of this effect it may be helpful to obtain much more detailed information on intakes of other micro-nutrients. There are several that might interact metabolically with cyanide and complex inter-relationships may occur.[120]

Micro-nutrient interactions may contribute to the neuropathies that have also been associated with cassava because the form of the disorder can differ from one region to another. In West Africa, tropical ataxic neuropathy is observed and deficiencies of protein, with particular reference to sulphur amino acid content, vitamin B_{12}, and riboflavin have been considered as contributory factors. Caeruloplasmin levels are found to be low in sufferers and this could implicate copper or vitamin A also.[121,122] Features common to other neuropathies have been considered in attempts to elucidate the causes of the disorders. These have included tobacco amblyopia (cyanide is a component of tobacco smoke), Leber's hereditary optic atrophy, Wilson's disease, and 'swayback' (the copper deficiency disease of sheep), among others but significant differences exist between all of these conditions.[123,124] Thus, the evidence for involvement of cyanide is strong but circumstantial. It is clear that ataxic neuropathies have multifactor aetiologies but involvement of cyanide seems certain.

The problems of investigating tropical ataxic neuropathy are immense. The condition is not found in the young; it is primarily associated with older sections of the population and may depend upon very long term consumption of cassava, during which time consumption of other micro-nutrients may have varied considerably. Analysis of blood samples or assessment of nutrient intake of existing patients might only be at best uninformative, at worst misleading. Furthermore, the human condition may not be reproduced in experimental animals. For example, cyanide can cause demyelination in rats and alter the myelin lipid spectrum[125] but in humans, changes occur mainly in the spinal cord rather than in the central nervous system. What is clear is that the disease is only found among poorer sections of the population and that it might be reduced or eliminated by changes in diet. Many who eat cassava are not affected by the disease. More efficient elimination of cyanide during processing,

[120] R. H. Davis, E. A. Elzubeir, and J. S. Craston, in 'Cyanide Compounds in Biology', Ciba Foundation Syposium 140, Wiley, Chichester, 1988, p. 219.
[121] B. O. Osuntokun, in 'Chronic Cassava Toxicity', ed. B. Nestel and R. MacIntyre, International Development Research Centre, Ottawa, 1973, p. 127.
[122] B. O. Osuntokun, in 'Toxicology in the Tropics', ed. R. L. Smith and E. A. Bababunmi, Taylor and Francis, London, 1980, p. 16.
[123] J. Wilson, in 'Chronic Cassava Toxicity', ed. B. Nestel and R. MacIntyre, International Development Research Centre, Ottawa, 1973, p. 121.
[124] J. Wilson, *Fund. Appl. Toxicol.*, 1983, **3**, 397.
[125] M. Wender, Z. Adamczewska-Goncerzewicz, J. Stanislawska, B. Knitter, D. Talkowska, and J. Pankrac, *Neuropat. Pol.*, 1986, **24**, 1.

which can be extremely variable, allied to increased intake of nutrients from other foods could result in elimination of the problem.

In East Africa, notably Mozambique, a different problem arises termed epidemic spastic parapesis.[119] It arises in dry seasons, when drought results in pronounced food shortage and cassava not only becomes even more dominant in availability but is consumed without the sun-drying normally used to reduce cyanide levels. The temporal nature of occurrences strongly implicate cassava as a causative factor. In comparison with tropical ataxic neuropathy it develops rapidly, is associated with short-term high cyanide intake rather than long-term lower intake, and is particularly prevalent among women and children, who may receive a low share of any additional foods that may be available. Low protein intake may exacerbate the disease although in experimental animals the interaction between dietary protein and cyanide has given inconsistent results.[126,127] Nevertheless, dietary protein intake has been implicated as a contributing factor in epidemic spastic parapesis and other nutrient deficiencies may contribute also. The damage to the spinal cord in this condition is permanent.

A particular type of diabetes has been noted to coincide with cassava consumption but it is not regarded as a major health problem in any specific area.[119] Another recent suggestion is that the thiocyanate formed from cyanide may enhance nitrosamine formation and relate to tumour incidence in human populations.[128] Chronic cyanide toxicity is not characteristic among other animals although it has been postulated that there is an increased risk of neonatal hypothyroidism and that certain neuropathies might be related to ingestion of cyanogens.[113]

9 Conclusion

Cyanogens have probably been responsible for more loss and suffering in animals, including man, than any other class of food toxin but the knowledge exists to make this a thing of the past. Foods containing cyanogens will continue to be important, economically and nutritionally, but plant breeding or varietal selection can be used to reduce initial cyanogen content and analytical procedures are available for risk assessment to be undertaken. Many of the foods are processed in some way before consumption and efficient processing can reduce cyanogen content to acceptable levels. Hence, the present need is for appropriate education, that will result in both reduced consumption of cyanogens and enhanced nutritional status among all exposed to this group of toxins.

---

[126] O. O. Tewe and J. H. Maner, *Food Chem.*, 1982, **9**, 195.
[127] J. V. Rutkowski, B. D. Roebuck, and R. P. Smith, *J.Nutr.*, 1985, **115**, 132.
[128] E. N. Manduagwu and I. B. Umoh, *Ann. Nutr. Metab.*, 1988, **32**, 30.

CHAPTER 10

Mycotoxins

BRIAN FLANNIGAN

1 Introduction

Among the filamentous fungi, or moulds, associated with crop plants there are many which produce secondary metabolites which are toxic to man and other animals. These toxic metabolites, or mycotoxins, are present in the mycelium, or network of branching filaments (hyphae), which colonizes senescing plant organs such as leaves, fruits, and seeds. They are also present in spores and may be excreted into the surrounding plant tissue. The illness or disease, referred to as mycotoxicosis, which is caused by ingestion of mycotoxin contaminated food commodities or feedstuffs may be acute or chronic and prove fatal, depending on the mycotoxin(s) involved. Certain toadstools, occasionally eaten in mistake for edible mushrooms, also contain toxins, *e.g.* the phalloidins and amanitoxins in *Amanita phalloides*. This toadstool, justifiably called Death Cap, is one of a number which in temperate countries every autumn cause illness or death among the unwary. For example, over the three years 1986–1988, in the Barcelona area there were 46 reported intoxications by toadstools, with four fatalities, three of which were caused by *A. phalloides*.[1] The toxins are not, however, referred to as mycotoxins, and toadstool poisoning is referred to as mycetism, rather than mycotoxicosis.

Since mycotoxins are associated with plants in the field and also stored plant produce, it is likely that they have affected man throughout agricultural history, but the first record of mycotoxicosis has been attributed to the second century Greek physician, Galen. He wrote that among people forced to eat stored barley and wheat during famine 'some die from a putrid or pestilential fever, others are seized by a scabby and leprosy-like skin condition'. From the Middle Ages on there were numerous fatal episodes of gangrenous and convulsive ergotism, but fortunately such episodes are rare nowadays. Although this mycotoxicosis, caused by consumption of cereals and bread containing the sclerotia of *Claviceps purpurea*, was widespread in Europe, the mycotoxins

---

[1] P. Sanz, R. Reig, J. Piqueras, G.Marti, and J. Corbella, *Mycopathologia*, 1989, **108**, 207.

which caused it were not isolated and identified as alkaloids until the Nineteenth century.

In the first half of the present century, well-documented evidence of mycotoxicoses caused by *Fusarium* spp. in cereals emerged from Russia. After a very rainy summer in 1923, people eating bread from that season's rye suffered from weakness, vertigo, headache, nausea, and vomiting. As well as this 'inebriant bread', in the Ukraine the use of 'inebriant linseed oil' in food also caused problems.[2] Septic angina or alimentary toxic aleukia (ATA) was another disease associated with *Fusarium*-contaminated grain. ATA occurred sporadically during the first three decades of the century in various parts of Russia, but appeared in endemic form from 1932 onwards. Especially during World War II and the postwar years until 1947, large numbers of people consuming grain which had overwintered under snow died from ATA, the symptoms of which included haemorrhage, necrotic angina, sepsis, and exhaustion of the bone marrow.[3]

However, research on mycotoxins did not really burgeon until after 1960, when 100 000 turkey poults in England died from turkey X disease, in which there is acute necrosis of the liver and hyperplasia of the bile duct. It was discovered that this disease and similar symptoms in other animals elsewhere had developed as a result of consumption of peanut meal. Detection of toxic factors and the fungus producing them, *Aspergillus flavus*, in peanut meal from several countries (as well as in other commodities) revealed that there was a worldwide problem.[4] When it was discovered that the toxins (named aflatoxins after *A. flavus*) were carcinogenic, research on an international scale intensified and since then a massive number of papers on aflatoxins has been published. The prevalence of aflatoxins in a wide range of commodities in developing countries in the Tropics, and in cottonseed, maize and peanuts in parts of US, is of considerable significance both economically and medically. Much research effort is therefore still expended on these toxins, but in recent years a considerable amount of research has been focused on mycotoxins produced by fusaria. Many other toxic metabolites have been isolated from fungi and the list continues to lengthen, but little is known about the natural occurrence of most of these mycotoxins, and consequently their true significance. This chapter will therefore devote itself primarily to considering certain major or well-researched mycotoxins, including aflatoxins, trichothecenes, zearalenone, and other *Fusarium* toxins.

2 Classification of Mycotoxins

Since setting out a full classification of mycotoxins which takes into account the wide chemical diversity of these relatively small molecules would be a lengthy process, it is perhaps sufficient here only to consider three broad categories of

[2] M. Dounin, *Phytopathology*, 1926, **16**, 305.
[3] A. Z. Joffe, in 'Mycotoxic Fungi, Mycotoxins, Mycotoxicoses', ed. T. D. Wyllie and L. G. Morehouse, Marcel Dekker, New York, 1978, Vol. 3, p. 21.
[4] J. M. Barnes, *J. Appl. Bact.*, 1970, **33**, 285.

(1)

(2)

(3)

(4)

(5)

(6)

(7)

(8)

(9)

(10)

(11)

mycotoxin. The categories are based on the intermediates from which these secondary metabolites are derived as the growth-related processes of primary metabolism are restricted. Firstly, there are polyketide-derived mycotoxins, formation of which requires acetyl coenzyme A that is normally involved in fatty acid synthesis during growth. These include the toxin patulin (1), which is derived from a tetraketide; citrinin (2) and ochratoxin A (3), from pentaketides; zearalenone (4), from a nonaketide; and aflatoxins (5–10), from a decaketide. The second category consists of terpene mycotoxins, which have mevalonic acid (derived from three molecules of acetyl coenzyme A) as a key intermediate. During growth, it is from this intermediate that the precursor of fungal sterols, isopentenyl pyrophosphate, is formed. This category includes around 80 sesquiterpene trichothecenes, including trichothecin (11). The third group consists of cyclic polypeptides and their derivatives. Their synthesis involves incorporation of amino acids which would otherwise enter into the structure of proteins. They include gliotoxin, an immunotoxin which is produced by the respiratory pathogen *Aspergillus fumigatus* and is formed from phenylalanine and serine, and the ergot alkaloids, in which a variety of amino acids are incorporated.

3 Aflatoxins

Aflatoxins are produced by *Aspergillus flavus* and *A. parasiticus*, and although there have been unsubstantiated reports of their production by various other moulds, it is generally held that these two species are the only aflatoxigenic moulds. They have a world-wide distribution and are particularly prevalent in tropical and subtropical regions, where they may colonize plant organs both in the field and under poor storage conditions after harvest. In temperate parts of the world, they tend to be regarded primarily as storage fungi, since they are seldom found under field conditions. A very wide range of plant products is known to be contaminated with aflatoxins, but the toxins are most often associated with peanuts (groundnuts), maize, rice, cottonseed, and Brazil, pistachio, and other nuts. Their biological activity has recently been reviewed by Palmgren and Hayes,[5] and their mode of action, together with that of other mycotoxins, by Hsieh.[6]

The aflatoxin molecule contains a coumarin nucleus linked to a bifuran and either a pentanone, as in AFB_1 (5) and the dihydro derivative AFB_2 (6), or a six-membered lactone, as in AFG_1 (7) and its corresponding derivative, AFG_2 (8). *A. parasiticus* produces all four of these aflatoxins, but toxigenic strains of *A. flavus* produce only AFB_1 and AFB_2. Two 4-hydroxylated derivatives of these last toxins have been found in peanuts and maize. These derivatives, AFM_1 (9) and AFM_2 (10), were first isolated from the milk of cows fed on aflatoxin-contaminated rations, but they have also been recovered from the meat, liver, kidneys, and urine of various animals, and from eggs. Various other minor aflatoxins are produced by *A. flavus* in culture,[5] and in the liver, and

[5] M. S. Palmgren and A. W. Hayes, in 'Mycotoxins in Food', ed. P. Krogh, Academic Press, London, 1987, p. 65.
[6] D. P. H. Hsieh, in 'Mycotoxins in Food', ed. P. Krogh, Academic Press, London, 1987, p. 149.

probably other organs, yet other derivatives are produced as a result of reduction, oxidative hydroxylation, O-demethylation, and epoxidation of AFB_1, the most potent and best researched of the aflatoxins.[5,6]

One of the metabolites of AFB_1 produced in the animal body is an 8,9-epoxide, which is formed by the action of a specific cytochrome P-450 monooxygenase and is considered to be the active form of AFB_1.[7] Electrophilic attack of the N-7 position of guanyl residues in DNA by the epoxide is likely to be the major cause of lesions which appear to account for the mutagenic and carcinogenic nature of AFB_1. The epoxide may also be hydrated to the dihydrodiol, which will form Schiff bases with the amino groups of bases. The hydrodiol binds to DNA *in vitro* and acts directly as a mutagen,[8] but it is held that it will be of much less importance *in vivo* because its high reactivity will cause it to react immediately with proteins at the locus of its production and thereby become unavailable for reaction with DNA. Sequestration of the dihydrodiol by proteins in this way could also be a first step in detoxification. In addition to attacking DNA, AFB_1 and its derivatives also bind to histones, ribosomal RNA, and proteins. A primary biochemical effect of aflatoxins in the target organ, the liver, is the inhibition of DNA synthesis, which occurs at concentrations below those which affect RNA and protein synthesis. This inhibition is caused by the modification of DNA template activity and inactivation of DNA polymerases and other necessary enzymes which arise from binding of AFB_1 and its metabolites to DNA and proteins.[6] Interference with RNA synthesis may not only block the subsequent synthesis of inducible enzymes, but also synthesis of various components of the immune system. In experimental animals, AFB_1 has been shown to cause complement deficiency, delayed production of interferon, and reduced levels of immunoglobulins IgA and IgG. It also may reduce the efficiency of phagocytes in clearing foreign material from the bloodstream. Dietary AFB_1 has been shown to significantly suppress the cell-mediated immune response in chickens, and has a residual effect on delayed-type hypersensitivity and phagocytic activity of reticuloendothelial cells.[9] Such damage to the defence mechanisms of the body would contribute to the observed reduction in resistance of animals to viral, bacterial, and *Candida* infections.[5,6]

Among other effects of aflatoxins at cellular level is the inhibition of energy production. AFB_1, AFG_1, and AFM_1 have all been demonstrated to inhibit cytochrome electron transport, and AFB_1 and AFM_1 have been found to uncouple oxidative phosphorylation. Mitochondrial permeability is also affected by AFB_1.[6]

Aflatoxins show both acute and chronic toxicity towards animals. The characteristic symptoms of acute aflatoxicosis are pathological changes in the liver, including enlargement, fat deposition, necrosis, and hyperplasia of the bile duct. Other effects observed are less specific, *e.g.* the loss in appetite,

---

[7] Y. Ueno, K. Ishii, Y. Omata, T. Kamataki, and R. Kato, *Carcinogenesis*, 1983, **4**, 1071.
[8] B. F. Coles, A. M. Welch, P. J. Hertzog, J. R. L. Smith, and R. C. Garner, *Carcinogenesis*, 1980, **1**, 79.
[9] S. K. Kadian, D. P. Monga, and M. C. Goel, *Mycopathologia*, 1988, **104**, 33.

lethargy, and wing weakness observed in turkey X disease, and even the gastrointestinal haemorrhage seen in some cases of aflatoxicosis in poultry. Experiments with small mammals have shown that there is considerable variation in resistance between species, and that there are differences according to sex, age, and size. LD_{50} values range from 1.0–1.4 mg AFB_1 kg^{-1} body weight in guinea pigs to 17.9 mg kg^{-1} in 150-g female rats. In addition to hepatic damage, other symptoms of acute toxicity which have been noted are congestion of lungs and intestinal submucosa, haemorrhagic necrosis of adrenals and spleen, necrosis of kidneys, and death/reabsorption or reduced growth in foetuses. Translocation of toxin occurs, as evidenced by later development of hepatic symptoms in progeny of rats consuming aflatoxin-contaminated feed during pregnancy. Among farm animals, poultry are particularly sensitive to aflatoxins; the extreme is the day-old duckling, with an LD_{50} of 0.5 mg kg^{-1}. Acute liver symptoms have been seen to develop in pigs fed on rations containing 0.28–0.81 mg AFB_1 kg^{-1}, and one animal fed at the highest rate died from gastrointestinal haemorrhage. In feeding trials, cattle on rations containing AFB_1 at 2 μg kg^{-1} have developed liver symptoms.

In humans, cases of acute hepatitis have been reported from Senegal, Uganda, India, and China.[10] Among the fatalities were more than 100 in Western India where the cause was consumption of maize containing 6.25–15.60 mg AFB_1 kg^{-1}. The daily AFB_1 consumption of those affected was estimated at 2–6 mg.[11] There is also evidence which can be taken to indicate that aflatoxins have a role in the pathogenesis of kwashiorkor, a deficiency disease associated with tropical areas of the world where the starchy staples contain only low levels of protein, e.g. maize, rice, or plantains, and are often contaminated with aflatoxins. Autopsy evidence has indicated aflatoxin-induced liver damage in kwashiorkor victims, but not in marasmus (both protein and energy deficiency) cases.[12] Recent evidence has established that aflatoxins accumulate in the body fluids and tissues and are only slowly eliminated.[13] While there is as yet insufficient evidence to say whether such accumulation is a cause, or only a consequence, of kwashiorkor, the case for the former is supported by the development of kwashiorkor symptoms in guinea pigs fed on aflatoxin-contaminated rations sufficient in protein.[14] In Czechoslovakia, New Zealand, Thailand, and US, aflatoxins have also been implicated in Reye's syndrome, a disease of children characterized by acute encephalopathy, hepatic enlargement, and elevated serum transaminase levels. However, no statistically significant differences in the aflatoxin content of blood and urine were observed in a study of Reye's syndrome patients and unaffected neighbours.[15] It appears

[10] WHO, 'Mycotoxins', Environmental Health Criteria 11, World Health Organization, Geneva, 1979, p. 68.
[11] K. A. V. R. Krishnamachari, R. V. Bhat, V. Nagarajon, and T. B. G. Tilak, Lancet, 1975, i, 1061.
[12] R. G. Hendrickse, Trans. R. Soc. Trop. Med. Hyg., 1984, **78**, 427.
[13] H. R. de Vries, S. M. Maxwell, and R. G. Hendrickse, Mycopathologia, 1990, **110**, 1.
[14] D. A. Long, Brit. Med. J., 1982, **285**, 1208.
[15] D. B. Nelson, R. Kimbrough, P. S. Landrigan, A. W. Hayes, G. C. Yang, and J. Benanides, Pediatrics, 1980, **656**, 865.

that several factors may be involved in this disease, including viral infection and a range of xenobiotic compounds, which may include aflatoxins.

Development of cancers due to chronic exposure of animals to aflatoxins in the diet has been demonstrated in a range of experimental animals, from trout and ducks to rats and monkeys. Although the liver is the primary target organ, tumours may also develop in the kidneys, pancreas, and other organs. The potency of AFB_1 and the other aflatoxins varies with the type of animal. For example, Hsieh concluded that in rats the potency of AFM_1 was two orders of magnitude lower than that of AFB_1,[16] but in trout the difference was only fivefold.[17] AFB_1 induces multicentric hepatocellular carcinoma, with the developing tumours sometimes becoming haemorrhagic and necrotic. In addition to carcinogenesis, chronic aflatoxicosis results in cirrhosis of the liver, reduced weight gain, and decreased resistance to infectious disease. Although trout, poultry, and rats are relatively sensitive to aflatoxin-induced carcinogenesis, in monkeys doses of 50–100 μg per day for periods up to six years are necessary for the development of tumours in the liver and other organs.[5]

Evidence of an association between aflatoxins in food and primary hepatocellular carcinoma (PHC) in humans living in various areas of central and southern Africa, Thailand, and Indonesia has been evaluated by WHO.[10] It was noted that the incidence of PHC was highest where concentrations of aflatoxins in food were greatest; the PHC incidence ranged from 1.2–13 cases per 100 000 people per year and the corresponding mean daily intake of aflatoxin estimated as 3.5–222.1 ng kg^{-1} body weight. As Krogh has pointed out, one limitation of these studies was that population, and not individual, daily intake of aflatoxin was considered.[18] They were also cross-sectional studies, although development of PHC takes a considerable time, and no consideration was given to the high incidence in these areas of two other risk factors, hepatitis B virus and alcohol consumption. Later studies linking aflatoxin intake and PHC development in the Philippines and Hong Kong are also considered to be flawed.

4 Mycotoxins Produced by *Fusarium*

The Genus *Fusarium*

Fusarium has always been an extremely important genus, because numbered among its species are phytopathogens which cause seedling blights, root rots, wilts, and cankers in both temperate and tropical areas, and have produced considerable losses in a variety of crops, including cereals, cotton, flax, tomatoes, and bananas. In addition to phytopathogenic species, there are others which colonize senescent plant organs saprophytically. Although noted largely as field fungi, some *Fusarium* spp. invade badly stored plant products too, *e.g.*

---

[16] D. P. H. Hsieh, in 'Trichothecenes and Other Toxins', ed. J. Lacey, Wiley, Chichester, 1985, p. 521.
[17] J. H. Canton, R. Kroes, M. J. van Logten, M. van Schothorst, J. F. C. Stavenutter, and C. A. H. Verhulsdonk, *Food Cosmet. Toxicol.*, 1975, **13**, 441.
[18] P. Krogh, *J. Appl. Bact.*, 1989, **67** (Supplement), 99S.

maize grain. It has also been recognized since the turn of the century that fusaria may be toxic to animals; disease in farm animals in Nebraska was attributed in 1904 to *F. moniliforme* invasion of grain used as feed. Since that time, numerous other disease episodes have been linked with *Fusarium* spp. and their toxins.[19] Clearly, fusarial trichothecenes and zearalenone are important and will be considered here, but there are other less well researched mycotoxins which also require our attention.

Trichothecenes

Trichothecenes are produced by a number of species in the genera *Acremonium* (*Cephalosporium*), *Cylindrocarpon*, *Dendrodochium*, *Myrothecium*, *Trichoderma*, and *Trichothecium*, but the widest range is produced by *Fusarium* spp., including a number of well-known phytopathogens. This class of compound takes its name from *Trichothecium*; trichothecin (11) was isolated from *T. roseum* and first described in 1949, but has only recently been isolated as a minor metabolite from one of the fusaria, *F. graminearum*.[20] The trichothecenes have a tetracyclic sesquiterpene skeleton which includes a six-membered oxane ring, an extremely stable epoxide group in the 12,13 position and a 9,10 olefinic bond (11). They have been classified into four groups. In Group A, positions 3, 4, 7, 8, and 15 may be substituted with hydroxyl or acoxyl functional groups. In Group B, positions 3, 4, 7, and 15 may be so substituted, but there is a carbonyl group at position 8. In Group C, there is a second epoxide function at the 7,8 position, and in Group D (the macrocyclic trichothecenes) positions 4 and 15 are linked by a di- or tri-ester bridge. As far as *Fusarium* is concerned, the toxins divide between Groups A and B. Diacetoxyscirpenol, or DAS, (12) and T-2 toxin (13) are two highly toxic members of Group A, and deoxynivalenol, or DON (14), is a commoner but less toxic member of Group B. Further examples of both groups are listed in Table 1.

Trichothecenes are potent inhibitors of protein synthesis in mammalian and other eukaryotic cells,[6] having an affinity for ribosomes and their 60S subunits. All toxic trichothecenes strongly inhibit the peptidyl transferase which is integral to the subunit, but some act at the initiation (I) step of the ribosomal cycle whilst others act on the elongation–termination (ET) steps. The I-type inhibitors include, in ascending order of potency, DAS, T-2, nivalenol, and fusarenon-X, but none is as potent as the macrocyclic trichothecene, verrucarin A, from *Myrothecium verrucaria*. DON, trichodermin, trichothecin, and verrucarol are among the ET-type inhibitors. In order of increasing effect overall on protein synthesis in mammalian cells, the most potent *Fusarium* trichothecenes are DON, DAS, diacetylnivalenol, T-2, nivalenol, and fusarenon-X. Substantial inhibition of RNA synthesis and blocking of most DNA synthesis also occur as secondary effects of inhibition of protein synthesis.

---

[19] W. F. O. Marasas, P. E. Nelson, and T. A. Tousson, 'Toxigenic *Fusarium* species—Identity and Mycotoxicology', Pennsylvania State University Press, University Park, Pennsylvania, 1984.

[20] S. Combrinck, W. C. A. Gelderblom, H. S. C. Spies, B. V. Burger, P. G. Theil, and W. F. O. Marasas, *Appl. Environ. Microbiol.*, 1988, **54**, 1700.

Table 1 *Some examples of trichothecenes produced by* Fusarium *spp.*

Group	Trichothecene	Substituents at position				
		3	4	15	7	8
A	Verrucarol	H	OH	OH	H	H
	Trichodermin	H	OAc	H	H	H
	Monoacetoxyscirpenol	OH	OAc	OH	H	H
	Neosolaniol	OH	OAc	OAc	H	OH
	HT-2 toxin	OH	OH	OAc	H	$OCOCH_2CH(CH_3)_2$
	Acetyl T-2 toxin	OAc	OAc	OAc	H	$OCOCH_2CH(CH_3)_2$
B	3-Acetyl deoxynivalenol	OAc	H	OH	OH	O
	Nivalenol	OH	OH	OH	OH	O
	Fusarenon-X	OH	OAc	OH	OH	O

It is acknowledged that inhibition of protein synthesis is the primary mechanism of toxicity, but the biological effects are diverse. Trichothecenes show acute toxicity to experimental animals, the toxicity varying considerably according to the trichothecene (Table 2).[21] For example, the LD_{50} for DON is more than 20 times higher than fusarenon-X in intraperitoneally injected mice. Symptoms shown in experimental animals include hypothermia; reduced respiratory rate; diarrhoea, apparently the result of the loss of plasma caused by the increased permeability in blood vessel walls; dermal inflammation and necrosis; vomiting, probably as a result of stimulation of the trigger gene in the brain stem (as a result of this effect DON is probably better known by its trivial name, vomitoxin); altered levels of important biogenic amines in brain tissue; cellular damage in the thymus, spleen, and bone marrow, with concomitant reductions in leucocytes and platelets (aleukia); and reductions in complement

---

[21] Y. Ueno, in 'Mycotoxins in Food', ed. P. Krogh, Academic Press, London, 1987, p. 123.

Table 2 *Comparison of LD_{50} values* (mg kg^{-1}) *of some* Fusarium *trichothecenes administered intraperitoneally to mice*

Trichothecene	LD_{50} (mg kg^{-1} body weight)
Verrucarin A[a]	0.5
Fusarenon-X	3.4
Nivalenol	4.1
T-2 toxin	5.2
HT-2 toxin	9.0
Neosolaniol	14.5
Diacetoxyscirpenol	23.0
Deoxynivalenol	70.0

[a] Macrocyclic trichothecene from *Myrothecium verrucaria*

formation and synthesis of immunoglobulins (IgA and IgM, but not IgG) and antibodies.[21]

Experiments with pure trichothecenes such as DAS, fusarenon-X, and T-2 do not reproduce all of the features of natural 'fusariotoxicosis', but it has been shown that T-2 at a level of 2 mg kg^{-1} in the diet can cause chronic to subacute lesions and death in dairy cattle. In pigs, DAS has caused haemorrhagic enteritis, and it has been reported that in trials 0.5–1.0 mg DON kg^{-1} feed has resulted in decreased feed consumption and weight gain. Some other reports indicate refusal of DON-contaminated maize containing this level of toxin, but it has also been reported that 10 mg DON kg^{-1} maize caused reduced feed intake without serious symptoms such as vomiting and haemorrhage occurring. Although it has been suggested that a dose of 0.1–0.2 mg DON kg^{-1} body weight is likely to cause vomiting in pigs, attribution of clinical symptoms to particular toxins, or particular doses, is rendered unsafe because mycotoxins do not occur singly in crops, *e.g.* grain containing DON often contains zearalenone and 15-acetyldeoxynivalenol,[22] or nivalenol.[23] However, it appears that cattle and poultry are more tolerant of DON and other trichothecenes than pigs.

As mentioned in the introduction to this chapter, ATA was an important natural mycotoxicosis of humans caused by consuming *Fusarium*-contaminated grain in Russia. The intoxication was attributed to extraordinarily high levels of T-2 and other trichothecenes produced by *F. sporotrichioides*, *F. poae*, and other fusaria present.[4] In an outbreak of ATA, the clinical symptoms which can be expected to develop are characterized by:

(1) a first stage involving a burning sensation resulting from inflammation of the mouth and fore-gut, emesis, diarrhoea, and abdominal pain;
(2) a second stage in which there is disturbance of bone marrow functions, with pronounced and progressive aleukia, lasting 3–4 weeks; and
(3) the sudden onset of a third stage with petechial haemorrhages on face,

---

[22] G. A. Bennett, D. T. Wicklow, R. W. Caldwell, and E. B. Smalley, *J. Agric. Food Chem.*, 1988, **36**, 639.

[23] T. Tanaka, A. Hawegawa, S. Yamamoto, U.-S. Lee, Y. Sugiura, and Y. Yeno, *J. Food Agric. Food Chem.*, 1988, **36**, 979.

head, trunk, and limbs, necrotic changes in the mouth, throat, and oesophagus which are accompanied by bacterial infections (septic angina), enlargement of the lymphatic glands, and parenchymatous hepatitis leading to jaundice.

About one-third of those who died of ATA in Russia did so as a result of constriction of the glottis (strangulation) caused by oedematous swelling. Among survivors, the rate of recovery depended on the degree of intoxication. The necrotic and haemorrhagic symptoms disappeared in about one month, but full recovery of the bone marrow functions took two months or longer.

Other natural toxicoses have been reviewed by various authors in a monograph on trichothecenes,[24] and by Ueno and Mirocha.[21,25] Among these were the serious outbreaks of mouldy corn (maize) toxicosis which occurred in the early 1960s in the midwestern states of the US. Symptoms in farm stock included feed refusal, diarrhoea, and reduction in milk production and weight gain, with massive haemorrhage in internal organs and death in some cases. DAS and T-2 toxin produced by *F. tricinctum* (*F. sporotrichioides*) were thought to be the cause of such haemorrhagic symptoms. Red mould disease (akakabi byo) has occurred sporadically in northern Japan and Korea and takes its name from the reddish colouration of wheat and barley kernels invaded by *F. graminearum* and other species in the field. Ingestion of the discoloured and shrivelled grain causes nausea, vomiting, congestion, and haemorrhage of the lungs, heart, and various other organs, and destruction of bone marrow in humans and animals. Akakabi byo is akin to the wheat scab which has frequently been recorded in US and became particularly widespread in Canada in 1980, 1981, and 1982. Since the affected wheat contained unusually large amounts of DON, it is fortunate that there have been no further serious outbreaks in Canada. Recently, however, it has been reported that an outbreak of disease in a large segment of the human population of the subtropical Kashmir Valley in India was caused by consumption of DON-contaminated wheat and wheat products.[26] Six of 37 *Fusarium* isolates from these materials produced DON in culture, four being *F. graminearum* and two *F. culmorum*.[27]

Zearalenone

The phenolic resorcyclic acid lactone, zearalenone (4), is also produced by a number of species of *Fusarium*. Since fusaria are commonly seed-borne, the co-occurrence of zearalenone with DON and other trichothecenes in grain is likely, particularly as some species produce both trichothecenes and zearalenone, *e.g. F. graminearum* and *F. culmorum*. Known trivially at first as F-2 and later as zearalenone, 6-(10-hydroxy-6-oxo-*trans*-1-undecenyl)-β-resorcyclic-

---

[24] Y. Ueno, 'Trichothecenes—Chemical, Biological and Toxicological Aspects', Kodansha, Tokyo, 1983.
[25] C. J. Mirocha, in 'The Applied Mycology of *Fusarium*', ed. M. O. Moss and J. E. Smith, Cambridge University Press, Cambridge, 1984, p. 141.
[26] R. V. Bhat, R. B. Sashidhar, Y. Ramakrishna, and K. L. Munshi, *Lancet*, 1989, **i**, 35.
[27] Y. Ramakrishna, R. V. Bhat, and V. Ravindranath, *Appl. Environ. Microbiol.*, 1989, **55**, 2619.

μ-lactone is not considered by some to be toxic in the usual sense, since it is primarily oestrogenic in its actions, and its LD_{50} values for different animals are very high, ranging from 2 to 10 g kg^{-1}.[28] Zearalenone behaves in a similar manner to an oestrogen in binding to receptor protein in the cytoplasm of cells to form a complex which is translocated to the nucleus and induces the synthesis of specific proteins. The natural hydroxy-derivative of zearalenone, α-zearalenol, has ten times the oestrogenic activity of zearalenone, but the β-isomer does not have this enhanced activity. As zearalenone reductases which hydroxylate zearalenone have been found in animal tissues,[29] it may be that α-zearalenol is the active form of zearalenone in animals.[6]

Zearalenone has little effect on poultry and, although it has activity against cattle, its most serious effects are seen in pigs. In prepubertal females, the mammae are enlarged, and the toxin shows a strong uterotrophic effect in causing oedematous swelling of the uterus and vulva (sometimes with prolapse of the vulva and rectum), as well as atrophy of the ovaries. In males, the mammae are correspondingly enlarged, and the testes atrophied.

Fusarochromanones

Fusarochromanone (TDP-1) was first described in 1986 as a water soluble toxin produced by strain *F. roseum* 'graminearum' (subsequently identified as *F. equiseti*). The compound (15) is a chromone derivative with an amino group at C-5 and a side chain at C-6.[30] It was found to cause tibial dyschondroplasia (TD) in chickens and reduced hatching of fertile eggs under experimental conditions. TD is a defect in ossification, primarily of the tibiotarsus, in which there is an uncalcified area of cartilage at the proximal end of the bone. It is common in broiler chickens and turkeys, and similar effects have been found in dogs, pigs, horses, and cattle. There is a variety of other causes for TD, however, ranging from environmental conditions to Thiram intoxication.[31] No link between the natural occurrence of fusarochromanone and TD has been established in US. However, the toxin has recently been detected in pelleted cereal/pea feed associated with TD in Denmark, and its production has also been confirmed in strains of *F. equiseti* previously isolated from Danish barley and peas.[32] However, in a survey of 62 isolates representing nine species of *Fusarium* from many areas of the world, only three isolates produced fusarochromanone in culture. All three were *F. equiseti*, but 28 others in this species did not produce the toxin. Fusarochromanone is also suspected of being involved in the human condition known as Kashin–Beck disease, where there is bone and joint deformation. It is a disease primarily of school-age children and is

---

[28] P. H. Hidy, R. S. Baldwin, R. L. Greasham, C. L. Keith, and J. R. McMullen, *Adv. Appl. Microbiol.*, 1977, **22**, 59.
[29] Y. Ueno, F. Tashiro, and T. Kobayashi, *Food Chem. Toxicol.*, 1983, **21**, 167.
[30] S. V. Pathre, W. B. Gleason, Y. W. Lee, and C. J. Mirocha, *Can. J. Chem.*, 1986, **64**, 1308.
[31] W. Wu, P. E. Nelson, M. E. Cook, and E. B. Smalley, *Appl. Environ. Microbiol.*, 1990, **56**, 2989.
[32] P. Krogh, D. H. Christensen, B. Hald, B. Harlou, C. Larsen, E. J. Pedersen, and U. Thrane, *Appl. Environ. Microbiol.*, 1989, **55**, 3184.

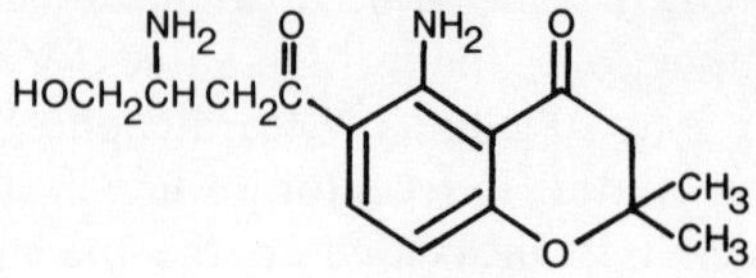

(15)

endemic to China, Korea, Siberia, and USSR, and an association with *Fusarium* contamination of wheat and maize has been reported in China.[33]

Two derivatives of fusarochromanone have been isolated from rice cultures of *F. equiseti*. TDP-2 is a C-3'-*N*-acetyl derivative,[33] and TDP-6 is an analog in which there is a hydroxyl group on C-3' and a methoxyl group on C-4'.[34] The biological activity of these derivatives has not yet been reported.

Moniliformin

The naturally occurring sodium or potassium salt of 1-hydroxycyclobut-1-ene-3,4-dione (16), which is known as moniliformin, was first described as a toxic metabolite of *F. moniliforme*, from which it took its name.[35] However, this strain (from maize in the US) lost its ability to produce the toxin in culture, and it was subsequently found that most strains of this species either produce only small amounts of moniliformin or none. A strain which was isolated from raw cotton in US and produced large quantities of the toxin in culture has now been identified as *F. proliferatum*, and highly productive isolates from millet in Nigeria and Namibia as *F. nyagamai*.[36,37] Moniliformin may be produced in quantity by *F. avenaceum* and *F. oxysporum*, and its production has also been noted in other species such as *F. culmorum* and *F. sporotrichioides*.[38]

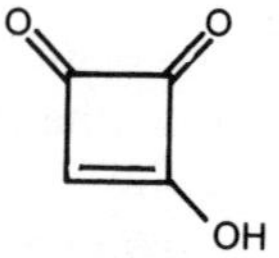

(16)

Moniliformin strongly inhibits mitochondrial oxidation of pyruvate to acetyl CoA and of *a*-ketoglutarate to succinyl CoA. Suppression of ATP-derived transmembrane transport is suggested by disturbance of intracellular osmo-

[33] W. Xie, C. J. Mirocha, R. J. Pawlosky, Y. Wen, and X. Xu, *Appl. Environ. Microbiol.*, 1989, **55**, 794.
[34] W. Xie, C. J. Mirocha, Y. Wen, and R. J. Pawlosky, *Appl. Environ. Microbiol.*, 1990, **56**, 2946.
[35] R. J. Cole, J. W. Kirkesey, H. G. Cutler, B. L. Doupnik, and J. C. Peckham, *Science*, 1973, **179**, 1324.
[36] W. F. O. Marasas, C. J. Rabie, A. Lubben, P. E. Nelson, and T. A. Tousson, *Mycologia*, 1988, **80**, 263.
[37] W. F. O. Marasas, P. E. Nelson, and T. A. Tousson, *Mycologia*, 1988, **80**, 407.
[38] P. M. Scott, H. K. Abbas, C. J. Mirocha, G. A. Lawrence, and D. Weber, *Appl. Environ. Microbiol.*, 1987, **53**, 196.

regulation and the resultant oedema in the myocardium and other tissues in experimental animals.[39] With an LD_{50} of 3.6 mg kg^{-1} body weight, which is one-tenth of that for rats, moniliformin is highly toxic to ducklings. Moniliformin has been found to be ten times more abundant in food from areas with a high incidence of human oesophageal cancer than in low-incidence areas,[40] but no causal role in the etiology of the disease has been demonstrated.

Fusarin C

Although moniliformin appears now to be at most a minor mycotoxin in *F. moniliforme*, this mould does produce a number of other mycotoxins in laboratory culture. *F. moniliforme* is ubiquitous, but it is particularly associated with tropical and subtropical regions, where it is the most abundant of the fusaria on maize. This is especially so in areas of southern Africa and China where there is a high incidence of oesophageal cancer,[40,41] and in China this species has been implicated in the formation of nitrosamines in maizemeal.[42,43] Extracts of *F. moniliforme* were found in 1979 to be strongly mutagenic in the Ames test against *Salmonella*,[44] and a role for *F. moniliforme* in the etiology of oesophageal cancer was envisaged. Culture material in the diet of rats was found to enhance nitrosamine-induced oesophageal cancer, and result in a higher incidence of hepatocellular carcinoma and overall tumour development.[45,46] Subsequently, a mutagenic compound, fusarin C (17), was isolated from an African strain, together with related nonmutagenic fusarins, A and D.[47] Strains from China, US, and Canada have also been found to produce fusarin C,[48-50] and the toxin has been extracted from both visibly *Fusarium* mouldered and 'healthy' kernels in a sample of maize.[51] This maize, from an oesophageal cancer area of Transkei, also contained moniliformin, DON, and zearalenone, and *F. graminearum* isolates from it produced the last two toxins and also fusarin C.[51] Recently it has been shown that the toxin can be produced on soyabean and other cereals by *F. moniliforme*.[50]

[39] P. G. Thiel, *Biochem. Pharmacol.*, 1978, **27**, 483.
[40] W. F. O. Marasas, F. C. Wehner, S. J. van Rensburg, and D. J. van Schalkwyk, *Phytopathology*, 1981, **71**, 792.
[41] C. S. Yang, *Cancer Res.*, 1980, **40**, 2633.
[42] M. Li, S. Lu, C. Ji, M. Wang, S. Cheng, and C. Jin, *Scientia Sinica*, 1979, **22**, 471.
[43] S. Lu, M. Li, C. Ji, M. Wang, Y. Wang, and L. Huang, *Scientia Sinica*, 1979, **22**, 601.
[44] L. F. Bjeldanes and S. V. Thomson, *Appl. Environ. Microbiol.*, 1979, **37**, 1118.
[45] S. J. van Rensburg, W. F. O. Marasas, W. C. A. Gelderbloom, P. G. Thiel, and C. J. Rabie, Proceedings of the Fifth International IUPAC Symposium on Mycotoxins and Phycotoxins, Vienna, 1982, p. 256.
[46] N. P. J. Kriek, W. F. O. Marasas, T. S. Kellerman, and J. E. Fincham, in Abstracts of the International Mycotoxin Symposium, Sydney, 1983, p. 12.
[47] W. C. A. Gelderblom, W. F. O. Marasas, P. S. Steyn, P. G. Theil, K. J. van der Merwe, P. H. van Rooyen, R. Vleggar, and P. L. Wessels, *J. Chem. Soc., Chem. Commun.*, 1984, 122.
[48] S. J. Cheng, Y. Z. Jiang, M. H. Li, and H. Z. Lo, *Carcinogenesis*, 1985, **6**, 903.
[49] J. M. Farber and G. W. Sanders, *Appl. Environ. Microbiol.*, 1986, **51**, 381.
[50] C. W. Bacon, D. R. Marijanovic, W. P. Norred, and D. M. Hinton, *Appl. Environ. Microbiol.*, 1989, **55**, 2745.
[51] W. C. A. Gelderblom, P. C. Theil, W. F. O. Marasas, and K. J. van der Merwe, *J. Agric. Food Chem.*, 1984, **32**, 1064.

(17)

In addition to causing mutation in bacteria, fusarin C has a mutagenic potency against rat liver preparations which approaches that of aflatoxin B_1 and sterigmatocystin.[51] However, it has not been demonstrated that fusarin C is a carcinogen. At lower concentrations than those required for mutagenesis, the toxin is immunosuppressive.[52] Immunosuppression has also been noted in poultry fed on maize infected with *F. moniliforme*.[50]

Fumonisins

Whilst the mutagen fusarin C does not appear to be involved in the carcinogenicity of *F. moniliforme*, isolation of a new group of toxins from a cancer-promoting extract of a Transkeian strain (MRC 826) was reported in 1988.[53,54] This strain was obtained from maize intended for human consumption and is hepatocarcinogenic in rats, and in horses can cause equine leukoencephalomalacia (ELEM), a fatal disease affecting co-ordination. ELEM is characterized by the development of random liquefactive (or malacic) lesions in the subcortical white matter of the cerebral hemispheres, with the cerebral cortex and the grey matter of the spinal cord also being affected. A family of structurally related mycotoxins, the fumonisins, were extracted from maize cultures of the mould. These are diesters of propane-1,2,3-tricarboxylic acid and 2-acetylamino- and 2-amino-12,16-dimethyl-3,5,10,14,15-pentahydroxy icosane as well as their C-10 deoxy analogues. The C-14 and C-15 dihydroxy groups are joined in ester formation with the terminal carboxy group of the propane-tricarboxylic acid.[53] One of the compounds, fumonisin B_1 (2-amino-12,16-dimethyl-3,5,10-trihydroxy-14,15-propane-1,2,3-tricarboxy icosane) has now been obtained from other strains in submerged cultures growing on a chemically defined medium,[55] as well as in corn cultures.[56] In addition to a cancer-promoting activity, fumonisin B_1 has a toxic effect on experimental rats,

[52] L. P. Chen and Y. H. Zhung, *J. Exp. Clin. Cancer Res.*, 1987, **6**, 15.
[53] S. C. Bezuidenhout, W. C. A. Gelderblom, C. P. Gorst-Allman, R. M. Horak, W. F. O. Marasas, G. Spiteller, R. Vleggaar, *J. Chem. Soc., Chem. Commun.*, 1988, 743.
[54] W. C. A. Gelderblom, K. Jaskiewicz, W. F. O. Marasas, P. G. Theil, R. M. Horak, R. Vleggaar, and N. P. J. Kriek, *Appl. Environ. Microbiol.*, 1988, **54**, 1806.
[55] M. A. Jackson and G. A. Bennett, *Appl. Environ. Microbiol.*, 1990, **56**, 2296.
[56] P. F. Ross, P. E. Nelson, J. L. Richard, G. D. Osweiler, L. G. Rice, R. D. Plattner, and T. M. Wilson, *Appl. Environ. Microbiol.*, 1990, **56**, 3225.

(18)

causing a significant reduction in weight gain compared with controls and insidious and progressive development of hepatitis.[54] This toxin has also been demonstrated to cause ELEM.[57] In an investigation prompted by numerous reports of ELEM and porcine pulmonary oedema syndrome during the 1989 harvest in US, fumonisin B_1-producing strains of *F. moniliforme* were isolated from maize and/or maize screenings in feed associated with outbreaks of these diseases. One feed also contained a strain of the related species *F. proliferatum* which produced this toxin.[56]

Wortmannin

There have been numerous reports of mycotoxicoses in which the symptoms have included haemorrhage and *Fusarium* has been implicated. Although there is some evidence that the trichothecenes T-2 toxin and DAS have been involved in some instances, there are other cases where these toxins do not appear to have been involved. In a recent investigation, various Brazilian isolates of the common phytopathogen *F. oxysporum* were found to cause haemorrhage in experimental rats.[58] However, only one of the 39 isolates of this species produced trichothecenes in rice culture, T-2 being the toxin found in quantity.[59] Acting on the observation that *Fusarium* isolates from cold regions are usually more toxic than those from tropical and subtropical regions, Abbas *et al.* examined rice cultures of 126 isolates (in five species) from cereal grains and soil in Arctic Norway for their ability to cause haemorrhage.[60] Some 69 isolates of *F. avenaceum*, *F. oxysporum*, and *F. sambucinum* caused haemorrhage in various organs of experimental rats, which could not be accounted for by the presence of trichothecenes or other known toxins. However, a haemorrhagic factor, H-1, was identified in most of these isolates. This factor was isolated and purified from one such isolate of *F. oxysporum* and found to be wortmannin (18).[61] This is an antibiotic compound first described in 1957 and previously

[57] N. P. J. Kriek, T. S. Kellerman, and W. F. O. Marasas, *Onderstepoort J. Vet. Res.*, 1981, **48**, 129.
[58] T. Kommendahl, H. K. Abbas, C. J. Mirocha, G. A. Bean, B. B. Jarvis, and M.Guo, *Phytopathology*, 1987, **77**, 584.
[59] C. J. Mirocha, H. K. Abbas, T. Kommendahl, and B. B. Jarvis, *Appl. Environ. Microbiol.*, 1989, **55**, 254.
[60] H. K. Abbas, C. J. Mirocha, and R. Gunther, *Mycopathologia*, 1989, **105**, 143.
[61] H. K. Abbas and C. J. Mirocha, *Appl. Environ. Microbiol.*, 1988, **54**, 1268.

isolated from *Penicillium wortmannii*, *P. funiculosum*, and *Myrothecium roridum*. Interestingly, none of the Brazilian isolates of *F. oxysporum* produced wortmannin in culture,[59] although isolates of *F. sambucinum* from New Zealand did.[62]

Chlamydosporol

A novel toxic metabolite recently isolated from *Fusarium* is chlamydosporol (19), or 5-hydroxy-4-methoxy-6,8a-dimethyl-6,7-dihydro-2H,8aH-pyrano[2,3-b]pyran-2-one.[63] This bicyclic 'lacto-ketal' was present as 6α and 6β isomers in rice cultures of *F. chlamydosporum* (= *F. fusarioides*) which had been isolated from rice in Malaysia,[64] and was subsequently also obtained from a fermentation of *F. tricinctum*.[63] Chlamydosporol is markedly less toxic than trichothecenes such as T-2 and DON in brine-shrimp and mammalian-tissue culture bioassays.[63] However, other strains of *F. chlamydosporum* have been found to produce neosolaniol monoacetate and moniliformin, and millet infected with this species has been reported to cause human mycotoxicoses in southern Africa.[65] In view of this, and the fact that synergy between metabolites may occur,[66] a fuller evaluation of chlamydosporol needs to be made.

5 Ochratoxin A

The nephrotoxic compound ochratoxin A, or OA (3), as the first described of a series of metabolites in which an isocoumarin moiety is linked to L-phenylalanine. The compounds take their name from *Aspergillus ochraceus*, since it was from a South African strain of this species that OA was first isolated. However, it is produced by other species closely related to *A. ochraceus* and by various *Penicillium* spp., including *P. cyclopium*, *P. purpurescens*, and *P. viridicatum*. Such penicillia are the principal sources of OA in stored cereals in temperate areas such as Europe and North America, although toxigenic strains of *A. ochraceus* are commonly isolated from crops in warmer climates. The range of

[62] U. Bosch, C. J. Mirocha, H. K. Abbas, and M. de Menna, *Mycopathologia*, 1989, **108**, 73.
[63] M. E. Savard, J. D. Miller, B. Salleh, and R. N. Strange, *Mycopathologia*, 1990, **110**, 177.
[64] B. Salleh and R. N. Strange, *J. Gen. Microbiol.*, 1988, **134**, 841.
[65] W. F. O. Marasas, P. E. Nelson, and T. A. Tousson, 'Toxigenic *Fusarium* species—Identity and Mycotoxicology', Pennsylvania State University Press, University Park, Pennsylvania, 1984, p. 70.
[66] P. F. Dowd, J. D. Miller, and R. Greenhalgh, *Mycologia*, 1989, **81**, 646.

plant commodities contaminated by OA includes all the major cereals, pulses, and soya-, coffee, and cocoa beans.[67]

OA affects protein synthesis by specifically and competitively inhibiting phenylalanyl-tRNA synthetase, the inhibition being reversed by L-phenylalanine.[6] There is a profound effect on the immune system; production of IgA, IgG, and IgM immunoglobulins and antibodies is reduced[6,68] and macrophage migration is inhibited.[69] The effect on antibody production appears to be at the level of protein synthesis, since phenylalanine counteracts the effect.[70] As a competitive inhibitor of membrane transport carrier proteins, OA affects mitochondrial respiration. Additionally, its uptake by mitochondria is energy dependent, leading to depletion of ATP, and it impairs oxidative phosphorylation. The net result is degeneration of these organelles. In the liver, synthesis of glycogen is reduced by a decline in glycogen synthetase activity and decomposition accelerated by enhanced phosphorylase activity, with the result that the glycogen level decreases and the blood glucose level increases. There is selective inhibition of synthesis of renal phosphoenolpyruvate carboxykinase,[71] and a consequent decrease in renal gluconeogenesis from pyruvate. It has been suggested that reduction in the activity of this enzyme (and γ-glutamyltranspeptidase) in biopsy specimens might be used as a marker for OA-induced disease in humans, renal biopsy having been successfully used in experiments with pigs.[67]

Susceptibility to acute intoxication with OA varies widely. The reported LD_{50} values for orally administered toxin range from 3.4 mg kg^{-1} body weight for hens and < 6 mg kg^{-1} for pigs to 28.0–30.3 mg kg^{-1} for male rats. The kidney is the principal target organ, renal failure being caused by tubular degeneration, but there may also be liver damage. Although OA is apparently not mutagenic, it is both teratogenic and carcinogenic, causing gross abnormalities in foetuses of experimental small mammals, and development of renal and hepatic tumours.[67]

Although porcine ochratoxicosis has been recorded in various European countries and US, it is of particular concern in Denmark and the extensive research of the late Palle Krogh and his colleagues on this problem is well documented.[67] In Denmark, porcine nephropathy is widespread and may reach epidemic proportions in seasons when conditions result in barley having a high moisture content. Although, OA is the main nephrotoxic mycotoxin in damp grain, it is often accompanied by another nephrotoxin, citrinin. In addition, viomellein, a heptaketide-derived quinone, is one of a number of other nephrotoxins produced by *P. viridicatum* which has been found in barley associated with porcine nephropathy. At slaughter, the reported prevalence rates of

[67] P. Krogh, in 'Mycotoxins in Food', ed. P. Krogh, Academic Press, London, 1987, p. 97.
[68] P. Dwivedi, R. B. Burns, and M. H. Maxwell, *Res. Vet. Sci.*, 1984, **36**, 104.
[69] A. C. Pier and M. E. McLoughlin, in 'Trichothecenes and Other Mycotoxins', ed. J. Lacey, Wiley, Chichester, 1985, p. 507.
[70] H.-D. Haubeck, G. Lorkowski, E. Kolsch, and R. Röschenthaler, *Appl. Environ. Microbiol.*, 1981, **41**, 1040.
[71] H. Meisner and P. Krogh, Proceedings of the Fifth International IUPAC Symposium on Mycotoxins and Phycotoxins, Vienna, 1982, p. 342.

nephropathy in pigs have ranged from 2.0–4.4 cases per 10 000. Only 27–39% of kidneys from the diseased animals contained OA residues, but it should be noted that the residual half-life of the toxin is 3–5 days in pigs. In Denmark, pig carcases are condemned if the OA concentration in the kidney exceeds 25 μg g^{-1}. As in pigs, nephropathy in poultry may have a number of causes, but OA appears to be an important factor.[67]

OA has been suggested as the cause of Balkan endemic nephropathy in humans, which occurs in rural populations in parts of Bulgaria, Romania, and Yugoslavia. It is a chronic, ultimately fatal condition in which there is marked reduction in the size of the kidneys, with changes similar to those seen in pigs intoxicated with OA. Earlier studies indicated that exposure to OA in food was greater in a high-prevalence area than in non-endemic areas of Yugoslavia. However, the presence of OA in blood did not correlate with contamination of food,[67] and the proportion of ochratoxigenic strains isolated from food did not differ between nephropathic and non-nephropathic areas of Croatia (37% *vs.* 35%).[72] Although attention has primarily been focused on the possible role of OA in this disease, it has recently been found that *P. aurantiogriseum* produces a water-soluble peptide or glycopeptide (*ca.* 1500 daltons) which is nephrotoxic in rats. *Penicillium* isolated from the staple cereal, maize, in a hyperendemic Yugoslavian village and identified as *P. aurantiogriseum*, were similarly nephropathic and acutely cytotoxic.[73]

6 Citrinin

First isolated in 1931 from *Penicillium citrinum*, citrinin (2) is produced by a number of penicillia, including the OA producers *P. purpurescens* and *P. viridicatum*. It is therefore not surprising that OA and citrinin have been found together in the barley samples from districts in Denmark where the incidence of porcine nephropathy is high. In addition to being found in cereals in North America and Europe, citrinin has been noted in maize from South-east Asia. With citreoviridin and other *Penicillium* toxins, it was also present in 'yellowed' rice, some 100 000 tonnes of which were imported from various parts of Asia by Japan between 1947 and 1954, and declared unfit for human consumption.

Citrinin is a less potent nephrotoxin than OA, LD$_{50}$ values for animals ranging from 35 to 67 mg kg^{-1} body weight. Its main role is probably in synergizing with OA in causing kidney damage,[67] and it also enhances the effect of OA in inducing development of renal tumours.[74]

7 Ergot Alkaloids

The ergots, or sclerotia, of *Claviceps purpurea* are tightly-packed masses of fungal mycelium which take the place of kernels in grasses and cereals, and mature as the grain ripens in the field. The purplish-black ergots, which contain

[72] Z. Cvetnic and S. Pepeljnjak, *Mycopathologia*, 1990, **110**, 93.
[73] S. E. Yeulet, P. G. Mantle, M. S. Rudge, and J. B. Greig, *Mycopathologia*, 1988, **102**, 21.
[74] M. Kanisawa, in 'Toxigenic Fungi—Their Toxins and Health Hazard', ed. H. Kurata and Y. Ueno, Kodansha, Tokyo, p. 245.

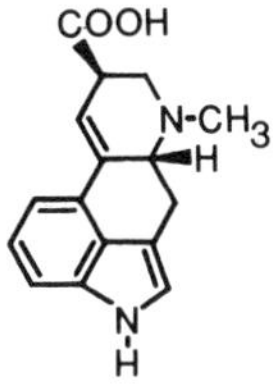

(20) 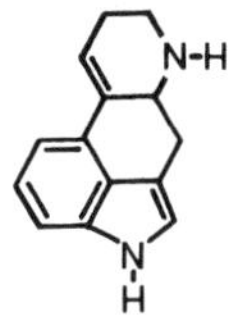   (21)

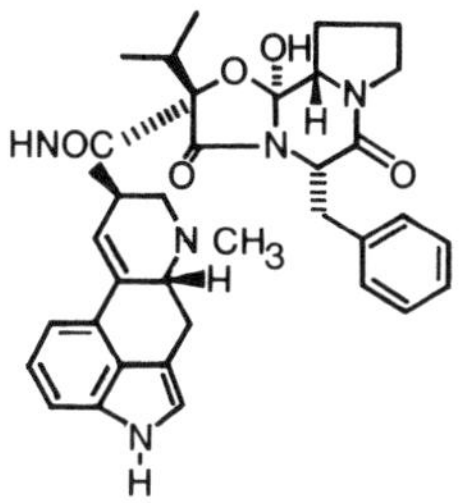

(22) 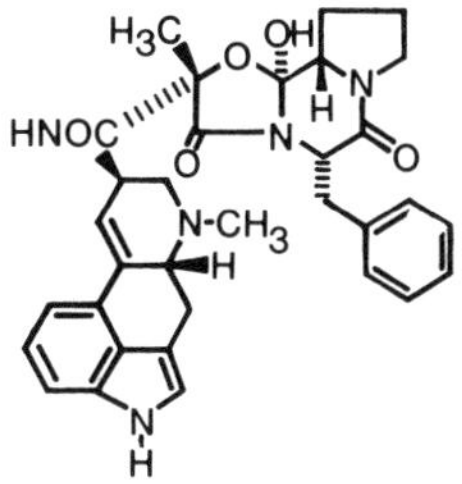   (23)

a variety of pharmacologically active compounds, may be roughly the same size and shape as the kernels of the host plant or they may be larger and protrude from the ear, as in the case of the cereal with which they are most often associated, rye. Whilst ergot is not as common in grain as in earlier times, it may still be found in small amounts annually, for example in various areas of Canada. Occasionally it has been present in sufficient quantities in barley and wheat to cause downgrading and serious financial loss to growers in Canada.[75]

Histamine and acetylcholine are among more than 100 substances which have been identified in sclerotia, but the most important of the physiologically important compounds are the alkaloids. These are 3,4-substituted derivatives of either lysergic acid (20) and its stereoisomer, isolysergic acid, or ergoline (21). There are considerable differences in the alkaloids present, both between different locations and between individual ergots, but the major components are often the lysergic acid derivatives ergocristine (22) and ergotamine (23).[76] Ergosine, ergocornine, α-ergokryptine, and ergometrine may also be prominent. Because of their pharmacological activity, there is extensive knowledge of the mode of action and effects of the ergot alkaloids, and this has been reviewed elsewhere in detail.[77]

In sheep, ergotamine tartarate administered orally at a daily rate of 1 mg kg^{-1} body weight resulted in death in 10 days,[78] and an acute LD$_{50}$ of 4 mg kg^{-1} has been determined for lysergic acid (the main alkaloid in the related species *C. paspali*).[79] The symptoms of ergotism in animals may be neurological and/or

[75] W. L. Seaman, in 'Mycotoxins: a Canadian Perspective', ed. P. M. Scott, H. L. Trenholm, and M. D. Sutton, National Research Council of Canada, Ottawa, 1985, p. 58.
[76] J. C. Young and Z.-J. Chen, *J. Environ. Sci. Health B*, 1982, **17**, 93.
[77] B. Berde and H. O. Schild, 'Ergot Alkaloids and Related Compounds', Springer-Verlag, Berlin, 1978.
[78] J. C. Greatorex and P. G. Mantle, *Res. Vet. Sci.*, 1973, **15**, 337.
[79] P. G. Mantle, P. J. Mortimer, and E. P. White, *Res. Vet. Sci.*, 1978, **24**, 49.

gangrenous; the former involve staggers, ataxia, tremors, and convulsions, and the latter necrosis and sloughing of the extremities due to the vasoconstrictant effect of the alkaloids. These symptoms are generally indicators of exposure to high doses of ergot alkaloids, but at lower doses (0.5 % ergot in the diet of cattle, 0.05 % in that of pigs) the only symptoms observed may be decreases in feed consumption and weight gain.[80] Many wild and pasture grasses are hosts to ergot fungi, and most cases of ergotism in cattle arise as a result of grazing on affected pasture, but hay may also be a cause. Since sheep are more selective grazers and avoid coarse flowering heads, cases of ovine ergotism are relatively rare. Convulsive ergotism and resultant sudden death of pigs are commoner than in cattle in US.[81] The most likely cause of these outbreaks is on- or inter-farm use of uncleaned, ergot-contaminated grain for feed,[76] but the incorporation of contaminated screenings in feeds may be another cause.

Fortunately, human ergotism is now extremely rare, but serious outbreaks still occur in developing countries. In one such outbreak in India, the cause was red millet contaminated by *C. fusiformis*.[82] The symptoms were less severe than would be expected if *C. purpurea* had been involved, and there were no fatalities. However, during one outbreak in Ethiopia in 1978, gangrenous ergotism developed in 93 people and 47 died.[83] The cause was sclerotia of *C. purpurea*, which had infested the abundant wild oat weeds in the local barley crop.

8 Phytotoxicity of Mycotoxins

Experimental studies have shown that various mycotoxins are phytotoxic. However, it should be noted that, in some cases, the plants used in tests have not been employed because the species producing the toxin is usually associated with them in Nature, but rather because they provide a convenient test system. A second point to be borne in mind is that in an appreciable number of cases, quoted in a review by Reiss,[84] the concentration necessary to produce a phytotoxic effect may be very high, *e.g.* 100 μg ml^{-1} or more. The experimental work tends therefore to indicate only the potential for phytotoxic effects in the presence of particular fungi and their toxins; it does not confirm that there is a connection between mycotoxins and disease symptoms in infected plants. For example, although dipping seedlings into culture broth of 642 isolates showed up a correlation between the degrees of toxicity to seedlings and to animals,[85] the contribution of phytotoxic trichothecenes and other mycotoxins in the pathogencity of particular *Fusarium* spp. for host plants has not been established. However, in the case of *F. graminearum* infection of maize, toxin

---

[80] H. B. Schiefer, in 'Mycotoxins: a Canadian Perspective', ed. P. M. Scott, H. L. Trenholm, and M. D. Sutton, National Research Council of Canada, Ottawa, 1985, p. 87.
[81] L. G. Morehouse, in 'Trichothecenes and Other Mycotoxins', ed. J. Lacey, Wiley, Chichester, 1985, p. 383.
[82] K. A. V. R. Krishnamachari and R. V. Bhat, *Indian J. Med. Res.*, 1976, **64**, 1624.
[83] T. Demeke, Y. Kidane, and E. Wuhib, *Ethiop. Med. J.*, 1979, **17**, 107.
[84] J. Reiss, in 'Mycotoxic Fungi, Mycotoxins, Mycotoxicoses', ed. T. D. Wyllie and L. G. Morehouse, Marcel Dekker, New York, 1978, Vol. 3, p. 119.
[85] A. Z. Joffe and J. Palti, *Mycopathol. Mycol. Appl.*, 1974, **52**, 209.

production appears to be related to the extent of fungal growth,[86] and in head blight of wheat caused by *F. culmorum* highly significant correlation has been observed between the DON content of kernels and the reduction in 1000-kernel weight and grain yield.[87]

A concentration of 100 μg AFB_1 ml^{-1} did not affect germination of lettuce or a range of crucifers, including cress, but it severely inhibited growth of seedlings. In peas, at lower concentrations AFB_1 is a synergist of the auxin β-indolylacetic acid (IAA), but at higher concentrations it is an antagonist to the growth-promoting activity of IAA. It has been found to inhibit chlorophyll formation in cowpea, formation of α-amylase and lipase in germinating cotton seed, and reproduction of mitochondria in root tissue of sweet potato. It also causes chromosomal aberrations in onion root cells.

The trichothecene DAS, stimulates root formation in cress at concentrations lower than 2.5 μg ml^{-1}. Its effects on stem growth and IAA are greater than those of AFB_1. T-2 toxin can totally inhibit germination of pea seeds at a concentration of 4 μg ml^{-1}, and causes dose-dependent wilting, necrosis, and reduction in fresh weight and length in seedlings. In soyabean, it has been found to inhibit growth promotion by the auxin 2,4-dichlorophenoxyacetic acid. At a concentration of 6 μg ml^{-1}, DON inhibits growth of tomato seedlings,[88] and at 0.3 μg ml^{-1} the toxin entirely prevents growth of coleoptile segments of particularly sensitive cultivars of wheat.[89] In barley, T-2, DAS, and DON have been observed to inhibit growth of the coleoptile and rootlets during experimental malting, and apparently also *de novo* synthesis of α-amylase and proteases.[90,91] Neosolaniol and isoneosolaniol (4,8-diacetoxy-12,13-epoxytrichothecene-3,15-diol) are also phytotoxic. The former is more potent than the latter, but less so than T-2 toxin, in inhibiting elongation of wheat coleoptiles.[92] In addition to the trichothecenes, another *Fusarium* toxin which is strongly phytotoxic is moniliformin.[35] It is also inhibitory to wheat coleoptile elongation, and when applied to tobacco seedlings caused necrosis, chlorosis, and distortion in individual leaves, and an abnormal rosette growth pattern. The oestrogenic *Fusarium* toxin, zearalenone, has been reported to inhibit growth of maize embryos,[93] and causes electrolyte leakage, reduced ATPase activity and other membrane phenomena in cereals and other plants.[94]

Other toxins mentioned earlier which are phytotoxic include patulin and citrinin. Wheat is particularly sensitive to patulin when it is applied at germination, during stem elongation, and at heading and flowering. Citrinin has been reported to cause wilting and leaf damage in beans, cotton, and sorghum. Like

[86] J. D. Miller, J. C. Young, and H. L. Trenholm, *Can. J. Bot.*, 1983, **61**, 3080.
[87] C. H. A. Snijders and J. Perkowski, *Phytopathology*, 1990, **80**, 566.
[88] A. Bottalico, P. Lerario, and A. Visconti, *Phytopathol. Mediterr.*, 1980, **19**, 196.
[89] Y. Z. Wang and J. D. Miller, *J. Phytopathol.*, 1988, **122**, 118.
[90] B. Flannigan, J. G. Morton, and R. J. Naylor, in 'Trichothecenes and Other Toxins', ed. J. Lacey, Wiley, Chichester, 1985, p. 171.
[91] S. F. D. Schapira, M. P. Whitehead, and B. Flannigan, *J. Inst. Brew.*, 1989, **95**, 415.
[92] R. J. Cole, J. W. Dorner, J. Gilbert, D. N. Mortimer, C. Crews, J. C. Mitchell, R. W. Winingstad, P. E. Nelson, and H. G. Cutler, *J. Agric. Food Chem.*, 1988, **36**, 1163.
[93] T. Brodnik, *Seed Sci. Technol.*, 1975, **3**, 691.
[94] A. Vianello and F. Macri, *Planta*, 1978, **143**, 51.

patulin, it has also been found to inhibit development of roots by epiphyllous buds of the succulent *Kalanchoe daigremontiana*.

9 Production of Mycotoxins

Production in the Laboratory

The usual conception of secondary metabolites is of molecules which are produced in submerged liquid culture when some nutrient, such as the nitrogen or phosphorus source, becomes limiting. Exponential growth of the mould can no longer be sustained and as a stationary phase is entered the secondary metabolites are synthesized. In laboratory fermentations, conditions can be manipulated so as to maximize production, and the metabolites recovered from both the mycelium and the culture broth. Again, if a solid-state fermentation is employed, in which the mould is usually grown on sterilized moistened grain, it is possible to optimize conditions for toxin synthesis. For example, yields of moniliformin as high as 33.7 g kg^{-1} maize have been achieved in laboratory culture.[95] Even allowing for variation introduced by differences in the cultural methods employed in separate studies, there are considerable qualitative and quantitative differences in toxin production between different strains of toxigenic species. Some strains are non-toxigenic in culture, *e.g.* many *A. flavus* strains do not produce aflatoxins. From laboratory studies employing different strains it can be seen that in some species a very wide range of toxins may be produced. That for *F. graminearum* includes the Group A trichothecenes calonectrin, isotrichodermin, monoacetoxyscirpenol, DAS, neosolaniol, T-2, and HT-2; the Group B trichothecenes trichothecin, DON, 3-acetyldeoxynivalenol, 15-acetyldeoxynivalenol, 3,15-diacetyldeoxynivalenol, nivalenol, and fusarenon-X; and the non-trichothecene toxins butenolide (4-acetoamido-4-hydroxy-2-butenoic γ-lactone) and zearalenone. Another widely distributed species producing an extensive range of toxins in culture is *F. crookwellense* (Table 3), which has been isolated in Australia, New Zealand, South America, US, Europe, and China from various cereals and other plants, or soil. The role of environmental factors on toxin production has also been investigated in laboratory experiments. For instance, the effect of water availability on synthesis has been determined for a range of toxins, including aflatoxin,[96] and of temperature on synthesis of aflatoxins,[96] T-2 toxin,[97] and DON and zearalenone in one strain of *F. graminearum*.[98]

[95] C. J. Rabie, W. F. O. Marasas, P. G. Thiel, A. Lübben, and R. Vleggaar, *Appl. Environ. Microbiol.*, 1982, **43**, 517.
[96] N. D. Davis and U. L. Diener, in Proceedings of the First US—Japan Conference on Toxic Microorganisms, ed. M. Herzberg, US Government Printing Office, Washington, DC, 1970, p. 43.
[97] H. R. Burmeister, *Appl. Microbiol.*, 1971, **12**, 739.
[98] R. Greenhalgh, G. A. Neish, and J. D. Miller, *Appl. Environ. Microbiol.*, 1983, **46**, 625.

Table 3 *Trichothecenes and other mycotoxins produced by* Fusarium crookwellense *under laboratory conditions*

	Type of culture		
	Liquid[c]	Rice, maize[d]	Maize[e]
Trichothecenes			
Fusarenon-X	−	+	+
Nivalenol	−	+	+
4,15-Diacetoxynivalenol[a]	+	−	−
8-Hydroxyisotrichodermin[a]	+	−	−
7-Hydroxyisotrichodermin[a]	+	−	−
7,8-Dihydroxyisotrichodermin[a]	+	−	−
3-Hydroxyapotrichothecene[a]	+[b]	−	−
4,15-Diacetoxy-7-deoxynivalenol	+	−	−
Isotrichodermin	+	−	−
8-Ketoisotrichodermin	+	−	−
Isotrichodermol	+	−	−
7-Hydroxyisotrichodermol	+	−	−
8-Hydroxyisotrichodermol	+	−	−
Culmorin	+	−	
Culmorone	+	−	−
Sambucinol	+	−	−
3-Deoxysambucinol	+	−	−
Sambucoin	+	−	−
Other mycotoxins			
Butenolide	+	−	−
Fusarin C	−	+	−
Zearalenone	+	+	+
Zearalenol	−	+[b]	+

[a] Most abundant trichothecenes, in descending order of abundance.
[b] α- and β-epimers.
[c] D. R. Lauren, A. Ashley, B. A. Blackwell, R. Greenhalgh, J. D. Miller, and G. A. Neish, *J. Agric. Food Chem.*, 1987, **35**, 884.
[d] P. Golinski, R. F. Vesonder, D. Latus-Zietkiewicz, and J. Perkowski, *Appl. Environ. Microbiol.*, 1988, **54**, 2147.
[e] A. Bottalico, A. Logrieco, and A. Visconti, *Mycopathologica*, 1989, **107**, 85.

Production in the Field

Although it is possible to study the interactions of important variables such as temperature and water availability (measured as the water activity, or a_w, of the substrate) on growth and toxin production in the laboratory, the situation in Nature is highly complicated, since so many variables are involved. One part of a plant organ such as a cereal grain, differs physically and nutritionally from another; conditions may be optimal for growth in one area, but growth may be severely restricted in another. Temperature and water availability vary diurnally and from one part of a plant to another. Other organisms present on the substratum also may exert an influence on toxin production. *Penicillium purpurogenum* and the rubratoxin it produces both enhance aflatoxin production by *Aspergillus parasiticus*. On the other hand, simultaneous inoculation of maize with *A. flavus* and *A. niger* may lead to significant reduction in the amount of aflatoxin which would otherwise be produced by *A. flavus* alone.[99] The application of pesticides and fungicides may also affect mycotoxin production, either by directly affecting the toxigenic fungi or by affecting competing microorganisms. For example, the fungicide triademenol, which inhibit sterol biosynthesis, has been shown to stimulate *in vitro* biosynthesis of aflatoxin by *A. flavus*.[100]

Investigation of the role of environmental and other factors in colonization and toxin production is most advanced in relation to contamination of peanuts, maize, and cottonseed by aflatoxigenic fungi.[101] It appears that *A. flavus* is adapted to the aerial and foliar environment and predominates over *A. parasiticus* on maize and cottonseed, whilst the latter seems better adapted to the soil environment and is therefore prominent in peanuts. The use of an extremely sophisticated control plot facility, in which soil temperature is among other environmental variables that can be controlled,[102] has shown that colonization by the mould and aflatoxin contamination are strongly influenced by the temperature around the maize cob, and of the soil around the geocarp (underground pod) of the peanut. The importance of maintaining adequate soil moisture levels, so that plants are not rendered susceptible to invasion as a result of drought stress, has been demonstrated for peanuts by use of this facility, and for maize by other field studies.[103] Drought results in elevation of the soil temperature, which promotes fungal growth and aflatoxin production in peanuts. However, drought also reduces the a_w which in turn leads to reduced production of stilbene phytoalexins by the seed.[104] Phytoalexin production is elicited by fungal invasion, and under favourable moisture regimes sufficient phytoalexin is normally produced by the seed to control growth of the fungus.

Aerial contamination is clearly the cause of colonization in maize and cottonseed; in maize, the route is via the senescing silks (with nutrients leaching from the pollen grains) to the glumes and kernels, and in cotton via nectaries and vascular tissue to the seeds.[105,106] The leaf scar may also be a point of entry in cotton, and insect or other damage also facilitates invasion and toxin production, not only in cottonseed, but also in maize and peanuts.[101]

Toxin production has also been examined in maize inoculated by toothpick with *F. graminearum* in the field at the mid-tassel stage.[86] Two chemotypes of *F. graminearum* have been differentiated: strains which produce nivalenol and fusarenon-X in culture, and those which produce DON and isomers of acetyldeoxynivalenone. Both chemotypes have been isolated in Japan,[107] but in

[99] D. T. Wicklow, B. W. Horn, and O. L. Shotwell, *Mycologia*, 1987, **79**, 679.
[100] P. Bayman and P. J. Cotty, *Mycol. Res.*, 1990, **94**, 1023.
[101] U. L. Diener, R. J. Cole, T. H. Sanders, G. A. Payne, L. S. Lee, and M. A. Klich, *Ann. Rev. Phytopathol.*, 1987, **25**, 249.
[102] P. D. Blankenship, R. J. Cole, T. H. Sanders, and R. A. Hill, *Oleagineux*, 1983, **38**, 615.
[103] G. A. Payne, D. K. Cassel, and C. R. Adkins, *Phytopathology*, 1986, **76**, 679.
[104] J. W. Dorner, R. J. Cole, T. H. Sanders, and P. D. Blankenship, *Mycopathologia*, 1989, **105**, 117.
[105] M. A. Klich, *Appl. Environ. Microbiol.*, 1990, **56**, 2499.
[106] H. E. Huizar, C. C. Berke, M. A. Klich, and J. M. Aronson, *Mycopathologia*, 1990, **110**, 43.
[107] M. Ichinoe, H. Kurata, Y. Sigiura, and Y. Ueno, *Appl. Environ. Microbiol.*, 1983, **46**, 1364.

North America and Argentina only the DON/acetyldeoxynivalenone chemotype has been reported.[108] In Antipodean, European and Chinese strains it is 3-acetyldeoxynivalenone which is the major isomer,[109] but the strain of *F. graminearum* used to inoculate the maize in this case was typical of North American strains in producing 15-acetyldeoxynivalenone.[109,110] Growth of the fungus appeared to reach a peak 6–7 weeks after inoculation. The DON concentration rose steadily to 580 p.p.m. at 7 weeks and then fell to 430 p.p.m. by 9 weeks; the 15-acetyldeoxynivalenone level (64 p.p.m.) at 2 weeks was similar to that of DON at that stage, but thereafter fell gradually to 20 p.p.m. This is somewhat different from liquid culture of this strain, where relative proportions of these toxins were reversed, very little DON being produced.[110] The toxins were also found in glumes and stalks, but it was not known whether they were translocated there or were synthesized *in situ*. The non-trichothecene toxins butenolide and zearalenone were also found in the ears. The former declined after 4 weeks, but the latter was present at less than 2 p.p.m. for most of the season and only reached *ca*. 10 p.p.m. in the last week. These observations clearly showed that there are considerable differences in the pattern of development for individual toxins produced by the same fungus, with the crop plant itself probably playing a part in determining trichothecene concentration, since leaf tissue preparations hydrolysed 15-acetyldeoxynivalenone to DON.[86]

Post-harvest Production

Since harvested products take with them into storage dormant mycelium and spores which have been produced in the field, and additional inoculum picked up during harvest and post-harvest operations, it is essential that the products should be rapidly and adequately dried before entering storage, which must also be efficiently managed if mould growth and toxin production are to be avoided. A considerable amount of aflatoxin contamination of peanuts occurs during the drying period if conditions are damp enough to allow fungal growth to continue. During storage, a succession of moulds can develop on, for example, cereals which have been inadequately dried or accidentally wetted in the store.[111] AFB_1 production has been found in maize stored at 24 °C or above, or when stored at a moisture content of 17.5 % or higher, and both aflatoxins and zearalenone may be produced as 'hot spots' develop.[112] Numerous studies have been made of the conditions of a_w and temperature under which toxigenic fungi grow and toxins are produced. These give some indication as to what might happen in bulk storage, but the methods of Abramson and his colleagues approach more closely the real situation by placing moistened lots of grain

---

[108] G. C. Faifer, M. S. de Miguel, and H. M. Goday, *Mycopathologia*, 1990, **109**, 165.
[109] C. J. Mirocha, H. K. Abbas, C. E. Windels, and W. Xie, *Appl. Environ. Microbiol.*, 1989, **55**, 1315.
[110] J. D. Miller, A. Taylor, and R. Greenhalgh, *Can. J. Microbiol.*, 1983, **29**, 1171.
[111] J. Lacey, *Int. Biodet.*, 1986, **22** (Supplement), 29.
[112] O. L. Shotwell, M. L. Goulden, R. J. Bothast, and C. W. Hesseltine, *Cereal Chem.*, 1975, **52**, 687.

within large bulks.[113,114] In this way, the production of OA (associated with *P. cyclopium* growth) and sterigmatocystin (associated with growth of *A. versicolor* and species in the *A. glaucus* group) was found to occur in barley with a moisture content of 20 % after 20 weeks.[113] Over the same time period, OA was found in wheat at 19 % moisture content, but no sterigmatocystin, although *A. versicolor* activity was higher.[114]

Production During Hydroponic Growth of Cereals

Further opportunities may arise for the growth of toxigenic fungi and production of mycotoxins when grain is used in the related hydroponic processes of malt production for the food, brewing, and distilling industries and production of fodder for livestock. In both processes, the grain is steeped until it is fully imbibed with water (moisture content around 45 %) and then set to germinate for a period of days, until an appropriate level of enzyme development and endosperm modification has been reached in malting, or until foliage has been produced in the fodder, *e.g.* 'barley grass'. One of the most hazardous moulds able to grow on the fully hydrated germinating grains is *Aspergillus clavatus*, the growth, physiology, and significance of which has been reviewed by Flannigan.[115] This mould, which is fortunately rarely encountered in the latest types of maltings, is the cause of the allergic respiratory disease, maltworker's lung. In addition to being allergenic, its mycelium and spores contain a range of mycotoxins, which includes patulin (1), ascladiol and cytochalasins E and K, as well as two neurotoxic toxins, tryptoquivaline and tryptoquivalone, and four compounds related to them.[115]

The optimum temperature for the growth of *A. clavatus* is 25 °C, and consequently increasing the malting temperature of barley from the 13–16 °C usual in UK leads to greatly enhanced growth.[116] As the ambient temperature during the malting of sorghum in South Africa can be 28 °C, *A. clavatus* can be one of the commonest moulds on this malt.[117] There have been scattered episodes of fatal toxicosis in livestock fed upon contaminated by-products of malting in Europe and South Africa, with the screenings (coleoptiles and rootlets) after kilning of malt usually being involved. Prominent among the symptoms of *A. clavatus* intoxication is neurological disorder, *e.g.* in a recent outbreak in cattle and sheep, ataxia was observed, and autopsy revealed degenerative changes in the central nervous system.[118] In addition to tryptoquivaline and tryptoquivalone, patulin (which is both toxigenic and carcinogenic when injected intraperitoneally into rats) is another toxin affecting the nervous system.

[113] D. Abramson, R. N. Sinha, and J. T. Mills, *Cereal Chem.*, 1983, **60**, 350.
[114] D. Abramson, R. N. Sinha, and J. T. Mills, *Mycopathologia*, 1990, **111**, 181.
[115] B. Flannigan, *Int. Biodet.*, 1986, **22**, 79.
[116] B. Flannigan, S. W. Day, P. E. Douglas, and G. B. McFarlane, in 'Toxigenic Fungi—Their Toxins and Health Hazard', ed. H. Kurata and Y. Ueno, Kodansha, Tokyo, 1984, p. 52.
[117] C. J. Rabie and A. Lübben, *S. Afr. J. Bot.*, 1984, **3**, 251.
[118] T. S. Kellerman, S. J. Newsholme, J. A. Coetzer, and G. C. A. van der Westhuizen, *Onderstepoort J. Vet. Res.*, 1984, **51**, 271.

Contamination of hydroponically germinated wheat by *A. clavatus* was the cause of deaths in cattle when this was used to provide forage during the dry summer of 1959 in France, and similarly contaminated sprouted maize has caused fatal tremorgenic mycotoxicosis of cattle in South Africa.[119] *A. clavatus* has also been implicated in intoxication of livestock by 'barley grass' in UK. Fusaria may also proliferate during malting and production of fodder from grain and *Fusarium*-contaminated fodder has been implicated in disease episodes, although corresponding malting by-products have not.

10 Detection and Quantification of Mycotoxins

Natural Occurrence of Mycotoxins

There is considerable geographical variation in the frequency and amounts of mycotoxin contamination, and superimposed upon this there is variation caused by differences in the weather from one year to the next. It is not possible here to do more than indicate the levels of important mycotoxins which have been encountered in some of the major crops. Aflatoxin has in exceptional cases been present at levels exceeding 200 mg kg^{-1} in cottonseed. More than one-third of the 60 % of aflatoxin-positive samples of peanuts imported into UK for animal feed and examined in 1980 contained more than 1 mg AFB_1 kg^{-1},[120] but in US the level in maize would not be expected to exceed 0.5 mg kg^{-1} in more than 3–4% of samples harvested in areas such as Virginia.

Among the *Fusarium* toxins, zearalenone and the trichothecenes DON and nivalenol are of worldwide distribution,[23] and zearalenone has been found at as high a level as 8 mg kg^{-1} in wheat in US, while DON has been detected at 20 mg kg^{-1} and nivalenol at 4.8 mg kg^{-1} in European maize. The highest concentrations in a recent survey of Dutch cereals, noted in a sample of feed-grade maize, were 0.68, 3.20, and 1.88 mg kg^{-1}, respectively.[121] Of the other important trichothecenes, T-2 has been found exceptionally at 25 mg kg^{-1} and DAS at 31 mg kg^{-1} in cereals. In visibly mouldered maize, as much as 16 mg kg^{-1} has been found. However, although some of these toxins are of worldwide distribution and the incidence of others may be high in certain areas, the concentration present in contaminated cereals is usually much lower than the extreme values quoted above. This also applies to the OA produced by some penicillia associated with grain, where the most extreme instance of contamination by this major toxin was 27.5 mg kg^{-1} in Danish barley.[67]

Sampling of Commodities

Clearly, on account of the potential health hazards, restrictions must be made on the level of mycotoxin accepted in commodities for consumption by live-

---

[119] J. S. Gilmour, D. M. Inglis, J. Robb, and J. McLean, *Vet. Rec.*, 1989, **124**, 133.
[120] A. E. Buckle, *Int. Biodet.*, 1986, **22** (Supplement), 55.
[121] T. Tanaka, S. Yamamoto, A. Hasegawa, N. Aoki, J. R. Besling, Y. Sugiura, and Y. Ueno, *Mycopathologia*, 1990, **110**, 19.

stock or humans. Since the decision to accept or reject a particular lot is based on the outcome of analysis of a sample from the lot in question, it is important to reduce error to the minimum. The error arises from the fact that there is not a normal distribution of toxin within any lot. In commodities such as peanuts or cereals, for example, probably only a very small proportion is contaminated by a particular toxin, and the level of toxin in the individual contaminated kernels may vary widely. The major component of the total error in testing for mycotoxins is in sampling the lot. Error arises if a bulk sample of insufficient size is amassed, if the number of subsamples comprising the sample is too few or if all parts of the lot are not adequately sampled. Errors in sample preparation, *i.e.* the division, comminution, and mixing involved in subsampling the original bulk sample for analysis, and analysis are minor components. A variety of methods is used for subdividing the sample, including coning and quartering, riffle division, and rotary cascade division, but it is possible that a method can be chosen which is not appropriate for a particular commodity. It has been found, for instance, that coning and quartering is not satisfactory for subdividing samples of palm kernel, although it is adequate for palm kernel cake.[122]

Since the greatest source of error is sampling, various sampling regimes have been developed in order to limit the error, and therefore the health risk to the consumer which arises from false acceptance (under-estimate of toxin level), and the risk of the producer suffering financially from false rejection (over-estimate of toxin level). Two sampling plans which are widely used are those of the Peanut Administrative Committee (PAC) and the Tropical Products Institute (TPI).[123,124] Both are based on the negative binomial distribution, but Jewers and his colleagues have observed that two Weibull distributions best describe aflatoxin levels when the sample number is small, and that a sequential sampling plan based on the Central Limit Theorem is needed.[122]

Bioassay for Mycotoxins

Bioassays can be used as a preliminary screen for toxicity in extracts prepared from commodities which are suspected to be contaminated with mycotoxin.[125] They may also be used to check on the toxicity of fractions as extracts are being fractionated for purification and subsequent identification of toxic elements. The use of plant bioassays has already been mentioned (Section 8), but microorganisms, animals, and mammalian tissue cultures have also been used.[126] Bacteria have proved to be of limited use, although agar-plate tests involving restriction of growth of certain *Bacillus* spp. have been used for aflatoxin and patulin. Bacteria cannot be used to assay trichothecenes and other mycotoxins,

[122] K. Jewers, R. D. Coker, B. D. Jones, J. Cornelius, M. J. Nagler, N. Bradburn, K. Tomlins, V. Medlock, P. Dell, G. Blunden, O. G. Roch, and A. J. Sharkey, *J. Appl. Bact.*, 1987, **67** (Supplement), 105S.
[123] J. W. Dickens, *J. Am. Oil Chem. Soc.*, 1977, **54**, 225A.
[124] K. Jewers, *R. Soc. Health J.*, 1982, **102**, 114.
[125] K. Mackenzie and J. Robb, *Int. Biodet.*, 1986, **22** (Supplement), 135.
[126] H. P. van Egmond, in 'Developments in Food Analysis Techniques—3', ed. R. D. King, Elsevier, London, 1984, p. 99.

but yeasts have been employed in assaying trichothecenes, since there is dose dependent inhibition of yeast growth by these toxins, and this can be measured in plate,[127] shake-flask,[90] and impedimetric assays.[128]

The ciliate protozoan, *Tetrahymena pyriformis*, has been used to assay for rubratoxin, whilst the larvae of the brine shrimp, *Artemia salina*, can be used in bioassays for trichothecenes, aflatoxins, and sterigmatocystin, and larvae of trout and other fish have also been used in bioassays. In these tests, the percentage of deaths in a given time is computed for dilutions of the given extract. Chick embryos and ducklings have also been widely used, and an early bioassay for T-2 and other trichothecenes depended on the dermatitic response of the shaved skin of the rabbit or rat.[126] In recent years, cell culture methods have been adopted for screening. In these methods tissue cultures of rat liver, baby hamster kidney, human epithelial, or other cells are exposed to the toxic extract and the effect on growth noted after 48 hours.[129] Cytotoxic effects are seen as inhibition of growth, detachment from the surface, or abnormal cell morphology. More recently in a more quantitative approach fluorescent flow cytometry has been used to assess the percentage viability of cells exposed to pure toxins or toxic extracts.[130] A limitation of tissue culture methods, and indeed other bioassays, is that they lack specificity; other substances co-extracted with mycotoxins may also be cytotoxic.[125]

Chemical Methods of Detection

For chemical identification and quantification of mycotoxins in commodities, the material is extracted with solvents appropriate to the chemical properties of the toxin.[126] Since lipids and other interfering substances are usually extracted by the organic solvents used, *e.g.* chloroform, it is necessary to carry out a column chromatographic clean-up step. Silica gel, aluminium oxide, Sephadex, or other materials are used in columns, and the interfering substances eluted prior to final elution and concentration of the mycotoxin(s). Alternatively, liquid:liquid extraction in separating funnels may be used in the clean-up.

A range of chromatographic techniques can be used to separate and identify mycotoxins. Various 'mini-column' methods have been used for aflatoxins, OA, zearalenone, and other toxins which are UV-fluorescent. The mycotoxin separates into one of a number of zones composed of different adsorbents, and its fluorescence can be compared with standards to give a semi-quantitative assessment of the mycotoxin status of the extract.[126] Thin-layer chromatography (TLC) has been a major tool in mycotoxin analysis, either employing a single solvent system to give separation in one dimension, or two systems for two-dimensional separation.[126] The concentrations of some toxins which are detectable by UV light, *e.g.* aflatoxins, can be computed by densitometric comparison

[127] K. T. Schappert and G. G. Khachatourians, *Appl. Environ. Microbiol.*, 1983, **45**, 862.
[128] M. O. Moss and G. K. Adak, *Int. Biodet.*, 1986, **22** (Supplement), 141.
[129] J. Robb and M. Norval, in 'Trichothecenes and Other Mycotoxins', ed. J. Lacey, Wiley, Chichester, 1985, p. 375.
[130] J. Robb, M. Norval, and W. A. Neill, *Lett. Appl. Microbiol.*, 1990, **10**, 161.

with standards, and comparison with standards can even be made by eye. Other mycotoxins are detected by spraying with specific reagents or charring with sulphuric acid. Cereals and animal feeds are often screened for a variety of commoner toxins using a multi-mycotoxin TLC method,[131] but the identity of mycotoxins present will usually require confirmation by another procedure.

One such confirmatory procedure is high performance liquid chromatography, which is also used for quantitation and has partially superseded TLC. Among the mycotoxins for which HPLC has been used are aflatoxins, OA, zearalenone,[23] fusarin C,[49] and fumonisins.[55] Although it may be possible to detect trichothecenes by HPLC after chemical derivatization,[126] gas–liquid chromatography (GC) is primarily used for these toxins. It is necessary to prepare volatile derivatives, which are detected by flame ionization detectors (FID) or electron capture detectors (ECD), the latter being sensitive and selective.[126] Frequently, the GC is coupled directly to a mass spectrometer (MS). GC/MS has been effectively used in conjunction with column chromatography and HPLC in analysing complex mixtures of trichothecenes (Table 3).

Immunochemical Methods of Detection

Immunochemical methods for mycotoxins have been developed in recent years, and have now become commercially available.[132] The methods have the particular advantage that, because of their specificity, the extensive clean-up procedures necessary for other analytical methods are not usually needed. Two methods have been used: enzyme-linked immunosorbent assay (ELISA) and radioimmunoassay (RIA).[126] For both methods, small mammals are used to raise antibodies to the mycotoxin, which is coupled to a protein because it is otherwise too small to induce antibody production. There are several variations of the ELISA method, but in a commonly employed procedure the pure mycotoxin or toxic extract is immobilized in the wells of a microtitre plate.[133] The antibody is added and allowed to bind to the mycotoxin. Unbound antibody is washed off, and a second antibody labelled with an enzyme and active against the anti-mycotoxin antibody is added. After incubation to allow binding and then removal of any unbound enzyme-linked antibody, a chromogenic substrate is added, and colour development is related to standards.[133] ELISA procedures have been developed for AFB_1,[126,132,133] OA,[126,133] sterigmatocystin,[133] T-2,[126,133-35] and a number of other toxins. In RIA, radioactively labelled mycotoxin and unlabelled mycotoxin (in the test extract) compete for active sites on a known amount of specific antibody. The radioactivity of the mycotoxin–antibody complex indicates the amount of unlabelled mycotoxin present. The lower the radioactivity, the more mycotoxin is present in the unknown. RIA techniques have been used for determination of AFB_1, AFM_1,

[131] D. S. P. Patterson and B. A. Roberts, *J. Ass. Off. Anal. Chem.*, 1979, **62**, 1265.
[132] A. A. G. Candlish, J. E. Smith, and W. H. Stimson, *Lett. Appl. Microbiol.*, 1990, **10**, 167.
[133] H. A. Kemp and M. R. A. Morgan, *Int. Biodet.*, 1986, **22** (Supplement), 131.
[134] S. Nagayama, O. Kawamura, K. Ohtani, J.-C. Ryu, D. Latus, L. Sudheim, and Y. Ueno, *Appl. Environ. Microbiol.*, 1988, **54**, 1302.
[135] R. Hack, E. Märtlbauer, and G. Terplan, *Lett. Appl. Microbiol.*, 1989, **9**, 133.

OA, and T-2 toxin.[126] As many toxins are structurally similar, related toxins may cross-react with antibodies, but the development of monoclonal antibodies has given increased specificity, so that it is now possible to detect as little as 5 pg mycotoxin per assay.[134,135]

11 Conclusion

Any realistic assessment of the present-day risk to human health posed by particular mycotoxins is hampered by the difficulty of quantifying exposure to mycotoxins in the diet. Nevertheless, it is perceived that they do pose a risk. For example, it has been calculated that in the US the risk of hepatic cancer in adults due to ingestion of AFB_1 in maize and peanut products is greater than from AFM_1 in milk and dairy products.[136] The fact that retrospective epidemiological studies of mycotoxin-associated diseases can be criticized in not taking into account other factors of possible relevance,[18] should also remind us that we have to take into account that toxins never occur alone in crops, and that synergy may occur between toxins. Special note should be taken of the fact that tests with caterpillars have shown that there can be powerful synergistic effects when one of the less toxic of the trichothecenes, DON, and other less toxic trichothecenes, *i.e.* culmorin, dihydrocalonectrin, and sambucinol,[66] are present together at naturally occurring concentrations.

Many mycotoxins are remarkably stable and are either not, or only partially, destroyed by cooking. Animal studies have shown that mycotoxin uptake from the alimentary canal into the tissues can be greatly reduced by the presence of additional fibre or clay minerals in rations.[137] It is possible to remove toxins from commodities by solvent treatment, and a variety of physical and chemical detoxification procedures can, or could, be used on feedstuffs.[137,138] Ammoniation, for example, is effective in detoxifying aflatoxin-contaminated oilseeds and cereals.[5,138] Breeding for resistance to fungal growth is being pursued in attempts to control aflatoxin development in maize, and contamination by DON and other *F. graminearum* mycotoxins in maize and wheat. However, careful attention to husbandry, drying, and storage will do much to prevent mycotoxin development and reduce health risks.

---

[136] D. P. H. Hsieh and B. H. Ruebner, in 'Toxigenic Fungi—Their Toxins and Health Hazard', ed. H. Kurata and Y. Ueno, Kodansha, Tokyo, 1984, p. 332.
[137] J. C. Young, in 'Mycotoxins: a Canadian Perspective', ed. P. M. Scott, H. L. Trenholm, and M. D. Sutton, National Research Council of Canada, Ottawa, 1985, p. 119.
[138] R. D. Coker, K. Jewers, and B. D. Jones, *Int. Biodet.*, 1986, **22** (Supplement), 103.

CHAPTER 11

Fibrous Polysaccharides

MARTIN EASTWOOD AND CHRISTINE A. EDWARDS

1 Introduction

Interest in the effects of dietary fibre on the gastrointestinal tract and metabolism, and hence human health, was stimulated by the claim that a number of Western diseases, *e.g.* ischaemic heart disease and colonic cancer, could be associated with increased industrialization, a greater processing of food, and consequently a decreased intake of dietary fibre. Evidence for this claim is still wanting, but there has developed a better understanding of the effects of plant cell walls, in particular polysaccharides, on human physiology, if not disease. Fibre is part of a family of complex carbohydrates. The complex carbohydrates have individual and diverse actions. As yet there is no logic based on chemistry which can identify or predict the biological action in the gastrointestinal tract of individual fibres. Each fibre is peculiar in its biological action, and is modified by its physical format and processing.

Dietary Fibre—Definition

The original definition of dietary fibre was of plant cell wall material which escaped digestion in the small intestine,[1] thereafter entered the colon, and the residue, after exposure to bacterial enzymes, passed on in the faeces. It is now appreciated that other dietary materials pass along the intestine in this manner for example starch, protein, and mucoproteins.[2] Therefore, there is much to be gained from classifying dietary polysaccharide polymers, exceeding 20 sugar residues, under an umbrella term of complex carbohydrates.[3]

The toxic and anti-toxic effects of the dietary complex carbohydrates are discussed in this chapter. Their action as toxicants are minimal but their effect

[1] H. Trowell, D. A. T. Southgate, T. M. S. Wolever, A. R. Leeds, M. A. Gassull, and D. J. A. Jenkins, *Lancet*, 1976, **i**, 967.
[2] M. Eastwood, in 'Diseases of the Colon Rectum and Anal Canal', ed. J. P. Kirsner and R. G. Shorter, Williams and Wilkins, Baltimore, 1988, p. 133.
[3] Report of the British Nutrition Foundations Task Force, 'Complex Carbohydrates in Foods', Chapman and Hall, London, 1990.

on the environment of the intestine and hence on the metabolism of other dietary and biliary secreted substances is profound.

2 Polysaccharides of the Plant Cell Walls (Chemical Structure and Role in the Plant)

Complex carbohydrates classified as dietary fibre that are indigestible by mammalian gastrointestinal enzymes are largely derived from and involved in the structure of the plant cell wall. In addition gums, mucilages, and seed galactomannans are used in the food industry and arabinoxylans from seed husks are extracted for therapeutic purposes, *e.g.* ispaghula. There are also storage polysaccharides which behave in a similar manner, *e.g.* resistant starch and inulin.[4-9] The amounts and types of polysaccharides in each cell wall are characteristic, not only of the plant but also the type of cell involved.

The complex carbohydrates forming dietary fibre are varied in chemical and physical structure. An important aspect of the plant cell wall is that water soluble polysaccharides interlock to form biological barriers which are water resistant. Primary cell wall, which is the principal contributor to dietary fibre, is 90 % polysaccharides and 10 % protein plus glycoprotein. The backbone of the plant cell wall, cellulose, is a polymer of linear β-(1,4)-linked glucose molecules, several thousand molecules in length. It occurs largely in a crystalline form in microfibrils, coated with a monolayer of more complex hemicellulosic polymers held tightly by hydrogen bonds. These are embedded in a gel of pectin polysaccharides. The cellulose microfibrils are coated with a layer of xyloglucans bound by hydrogen bonds, and this enables the insoluble cellulose to be dispersed within the wall matrix. Substitution of the hydroxyl group at C6 with xylose, as in xyloglucans, renders cellulose more soluble in alkali and water.

Other cells wall polysaccharides are the hemicelluloses and pectins.

The hemicelluloses have an important structural role in the cell wall. They are hydrogen bonded to the cellulose microfibrils, but may also have more complex regulatory roles. Hemicelluloses provide part of the true rigidity of the cell wall. The major hemicelluloses are xyloglucans, xylans, and β-glucans. The principal hemicellulose is xyloglucan a linear $(1\rightarrow4)\beta$-D-glucan chain substituted with xylosyl units, which may be further substituted to form galactosyl-$(1\rightarrow2)\beta$-D-xylosyl or fucosyl-$(1\rightarrow2)\alpha$-D-galactosyl-$(1\rightarrow2)\beta$-D-xylosyl units.

[4] R. R. Selvendran and A. V. F. V. Verne, in 'Dietary Fiber: Chemistry, Physiology, and Health Effects', ed. D. Kritchevsky, C. Bonfield, and J. W. Anderson, Plenum Press, New York, 1990, p. 1.
[5] R. Selvendren, *J. Cell Sci.*, 1985, (Supplement 2), 51.
[6] N. C. Carpita, in 'Dietary Fiber: Chemistry, Physiology, and Health Effects', ed. D. Kritchevsky, C. Bonfield, and J. W. Anderson, Plenum Press, New York, 1990, p. 15.
[7] D. P. Delmer and B. A. Stone, in 'The Biochemistry of Plants', ed. J. Preiss, Academic Press, Orlando, Florida, 1988, p. 373.
[8] H. N. Englyst and S. M. Kingman, in 'Dietary Fiber: Chemistry, Physiology, and Health Effects', ed. D. Kritchevsky, C. Bonfield, and J. W. Anderson, Plenum Press, New York, 1990, p. 49.
[9] M. M. Cassidy, S. Satchithanandam, R. J. Calvert, G. F. Vahouny, and A. R. Leeds, in 'Dietary Fiber: Chemistry, Physiology, and Health Effects', ed. D. Kritchevsky, C. Bonfield, and J. W. Anderson, Plenum Press, New York, 1990, p. 67.

Pectins are made up of a variety of polymers containing galacturonic acid. Besides polygalacturonic acid, a simple polymer composed of $(1\rightarrow 4)\alpha$-D-galactosyluronic acid units, there are more complex rhamnogalacturonans which are contorted rod-like polymers with repeating $(1\rightarrow 2)$-L-rhamnosyl-$(1\rightarrow 4)\alpha$-D-galactosyluronic acid units (rhm-galA). About one third of the rhamnosyl units are substituted. They form helical $(1\rightarrow 4)\alpha$-L-arabinosyl chains and highly branched chains with 2,5- and 3,5-linked arabinosyl branched residues. The ratio of rhamnose to galacturonic acid is 1:30–40.

Most pectic polysaccharides are highly methyl-esterified and slightly branched. They originate in the middle lamellae region. The pectins may act as biological glue, cementing cells together through ionic bonds. The precise function of pectins within the cell wall is unclear, but is closely associated with calcium. Most pectins are probably derived from the primary cell wall and appear to be soluble only after calcium ions are removed.

Pectic polysaccharides define the porosity of the cell walls of dicotyledons. The principal cross-linkage is provided by the helical $(1\rightarrow 4)\alpha$-D-galactosyluronic groups from adjacent polysaccharides and condensation with calcium converts soluble pectin into rigid 'egg-box' structures. Repeating rhm-galA units do not constitute the entire polymer but stretches of continuous galA units are connected to the rhamnose-rich regions to provide regions in which calcium cross-bridging may occur. Hydroxycinnamic acids such as ferulic and p-coumaric acid may be involved in cross-bridging through formation of ester bonds with neutral sugar side chains and these aromatic linkages may wire the pectins onto the hemicellulose matrix to provide two functional domains of pectin in the primary wall. The extent of calcium cross-bridging or esterification through aromatic linkages, degree of branching and size of neutral sugar side chains influence gel flexibility, porosity, and interaction with hemicellulosic polymers. Formation of the gel in the plant cell wall may depend on the control of calcium flux or levels in the wall. The ratio of calcium to magnesium in the cell walls of pea increases almost six-fold between the elongating and mature cells of the stem of the seedling. The bulk of the pectic polysaccharides of sugar beet primary cell wall are cross-linked to cellulose and are much less soluble in hot water than those of runner beans or onions.

Protein is also important in the walls of growing cells. Extensin, a structural hydroxyproline-rich glucoprotein is a major constituent in dicotyledons.

The glycoproteins within the cell wall may serve to provide extensive cross-linkage across the different polysaccharides components of the cell wall and may act to form a network with the cellulose microfibrils within a hemicellulose pectin gel.

Polysaccharide Composition and Distribution in Different Plant Tissues

The composition of the cell wall is dependent not only on the plant species but

Table 1 *Polymer composition of the cell walls of parenchymatous and lignified tissues of edible plant organs*[a]

	Fruits and vegetables		Cereal products	
	Parenchymatous	Lignified[b]	Parenchymatous	Lignified
Pectic polysaccharides	35–40	5	< 0.5	< 0.5
Cellulose	34	40	3–5	30–35
Hemicelluloses	10	25–30	80–85	45–50
Proteoglycans	5–10	} 5	} 5	5–10
Glycoproteins	5–10			—
Polyphenolics[c]	5	20–25	5	15

[a] The values are on a percentage dry weight basis and are approximate.
[b] The cell walls of lignified tissues of dicotyledons make only a small contribution to the DF content of plant foods.
[c] Polyphenolics include lignin and phenolic esters.

Derived from Selvendran and Verne, see ref. 4.

also on the tissue type.[10–13] The cell wall composition also depends on the maturity of the plant organ and harvesting, and to some extent post-harvest storage conditions. In addition to complex polysaccharides, non-carbohydrate components may influence the properties and physiological actions of the plant foods, *e.g.* lignin and phenolic esters in wheat bran, cutin, and waxy materials in leafy vegetables and suberin in roots and tubers. The polysaccharide components of a number of different plant tissues are shown in Table 1.

The parenchymatous tissues are the most important source of vegetable fibre. The vascular bundles and parchment layers of cabbage leaves, runner beans, pods, asparagus stems, and carrot roots are relatively immature and only slightly lignified on harvesting and digestion. The seeds of soft fruits such as strawberries contain very little dietary fibre. Lignified tissues are of greater importance in cereal sources such as wheat bran and oat products.

Cereals contain very little pectic substances but there is substantial arabinoxylan in wheat or β-glucan in barley and oats. The distribution of polysaccharides within the plant tissue also varies. Much of the β-glucan in oats is concentrated in the cells of the outer-most layer of the seeds, whereas the β-glucans in barley are more evenly distributed.

Some plant polysaccharides are harvested from the plant and purified before inclusion as processed food components or used therapeutically. These are more often soluble polysaccharides and include the following:

Guar gum (MW 0.25×10^6) a linear nonionic polymer and a seed galactomannan derived from the annual legume *Cyamopsis tetragonolobus* which contains 3–7% protein.

Gum karaya (MW 4.7×10^6) a dried exudate from *Sterculia urens Roxburgh* or from other *Sterculia* spp. It is a complex partially acetylated polysaccharide,

[10] K. Esau, in 'Anatomy of Seed Plants', John Wiley and Sons, 1960.
[11] K. Esau, *Hilgardia*, 1940, **13**, 175.
[12] D. Bradbury, I. M. Cull, and M. M. McMasters, *Cereal Chem.*, 1956, **3**, 329.
[13] J. A. Robertson, M. A. Eastwood, and M. M. Yeoman, *J. Sci. Food Agric.*, 1979, **30**, 388.

Table 2 *Classification of gum and mucilage additives and their acceptability*

'*ADI not specified*'
JECFA and EEC
  gum arabic
  gum tragacanth
  xanthan
  pectin
  amidated pectin
  galactomannans   [guar, locust bean (Carob) gum, tara gum]
  all modified celluloses
  carrageenans—but the carrageenan must be purified and not degraded in processing
  gellan—use limited to specific classes of product

EEC restricted classification but unrestricted by JECFA
  gum karaya   ADI 0–12.5 mg kg^{-1} body weight
                                       (EEC)

Restricted but under active re-assessment
  sodium alginate   ADI 0–50 mg kg^{-1} body weight
  propylene glycol alginate   ADI 0–25 mg kg^{-1} body weight

cylindrical in shape, and highly branched with interior galacturonorhamnose chains to which are attached galactose and rhamnose end groups. Glucuronic acid is also present.

Gum arabic (MW $0.5–1.5 \times 10^6$) with a complex acidic heteropolysaccharide of high molecular weight based on a highly branched array of galactose, arabinose, rhamnose, and glucuronic acid. Uronic acid residues tend to occur on the periphery of an essentially globular structure.

Gum tragacanth (MW $0.5–1 \times 10^6$) a dried exudate from *Astragalus gummifer Labillardiere* or other Asiatic *Astragalus* species. Gum tragacanth is a very complex gum with two major components, bassorin and tragacanthin. Bassorin swells but is not water soluble; tragacanthin is water soluble. Acid hydrolysis yields arabinose, fucose, galactose, glucose, xylose, and galacturonic acid.

Carrageenan, the generic term for chemically similar hydrocolloids obtained by aqueous extraction from certain members of the class Rhodophyceae (red seaweeds). Chemically these consist of the ammonium, calcium, magnesium, potassium, and sodium sulphate esters of galactose-3,6-anhydrogalactose copolymers. The relative proportions of actions existing in carrageenan in dependent upon processing. The prevalent copolymers are designated κ-, ι-, and λ-carrageenan. κ-Carrageenan is mostly the alternating polymer of D-galactose-4-sulphate and 3,6-anhydro-D-galactose; ι-carrageenan is similar except that the 3,6-anhydrogalactose is sulphated at carbon 2. There is a continuum of degree of sulphation between the κ and ι form. In λ-carrageenan, the alternating monomeric units are mostly D-galactose-2-sulphate (1,3 linked) and D-galactose-2,6-disulphate (1,4-linked).

These gums and mucilages are used as additives by food manufacturers, and therefore may contribute less than 2 % of a food.

Additives are classified by JECFA and EEC as either 'ADI not specified' or with a specified ADI (Acceptable Daily Intake). 'ADI not specified' is allocated

Table 3 In vitro *nutritional classification of starch*

Type of starch	Example of occurrence	Probable digestion in small intestine
Rapidly digestible starch	Freshly cooked starchy food	Rapid
Slowly digestible starch	Most raw cereals	Slow but complete
Resistant starch		
1 Physically inaccessible starch	Partly milled grain and seeds	Resistant
2 Resistant starch granules	Raw potato and banana	Resistant
3 Retrograded starch	Cooked potato, bread, and corn flakes	Resistant

if the regulatory committee is so satisfied that the substance is free of toxic properties that there is no need to express an upper limit on its use. For less safe substances, the recommended daily intake (over whole life times) is specified in a form such as $ADI = 0\text{--}10$ mg kg^{-1} body weight. A body weight of 60 kg is assumed for males/females of all ages. Table 2 shows the ADI classification of the gums and mucilages.

Starch, a ubiquitous storage polysaccharide, is an a-linked glucan and is the major carbohydrate of dietary food such as cereal grains and potatoes. The majority of dietary starch is susceptible to hydrolysis by salivary and pancreatic a-amylases, but, a proportion of dietary starch may escape digestion by human a-amylase.[8] This may be due either to physical inaccessibility or the inhibition of amylase activity by a rigid stereochemical structure caused during food processing (Table 3). This starch will pass undigested into the colon where it may act as a substrate for the bacteria and therefore have some of the biological properties of the undigestible plant cell wall polysaccharides described below.

Physical Properties of Fibre Important for Physiological Action

Plant polysaccharides from diverse sources differ in how they act in the gastrointestinal tract, but each may be likened to a water laden sponge passing along the intestine.[14] Physical properties that influence function along the gastrointestinal tract are a combination of the colligative properties of the water-soluble fibre components and the surface properties of the water-insoluble fibre components. These properties include water holding, viscosity, cation exchange, organic adsorption, gel filtration, and particle size distribution. These physical properties are interrelated. It has been suggested that dietary fibre can act as a form of ion exchanger and/or molecular sieve chromatography system throughout the gastrointestinal tract. The plant cell wall provides a water-soluble or insoluble matrix or phase through which the soluble contents of the gut can variably penetrate and also provide partition effects. The physiological actions that depend on these properties are discussed below.[14]

[14] M. Eastwood and W. G. Brydon, in 'Dietary Fibre, Fibre-depleted Foods and Disease', ed. H. Trowell and K. Heaton, Academic Press, London, 1985, p. 105.

Effect of Cooking and Ingestion on Cell Wall Structure

Cell wall polysaccharides have different solubility characteristics and the cell walls are degradable to varying degrees, depending on the structure and the conditions used. This is important when considering the action of cooking on cell wall structure and when comparing cooked and raw plant foods. An important function of insoluble fibres is to increase lumenal viscosity in the gut and although this is readily demonstrated for isolated polysaccharides it is not yet clear whether the soluble fibres in food have the same effect.[15]

Fresh uncooked apples undergo little cell sloughing on ingestion and mastication and the hydrochloric acid in the stomach only solubilizes a small portion of the pectins. Little hydrolysis of methyl-esterified pectins or ester cross-linked pectins occurs in the small intestine and so the pectins of raw apples contribute little to the viscous property of the digesta. Cooking the apples encourages cell sloughing[5] and hence significant proportions of the middle lamellae pectic polysaccharides are solubilized. These make the digesta more viscous.

Similarly, vegetables undergo structural change during cooking and mastication. The greatest difference between the intact carrot and the cooked fibre is in the amount of cellular disintegration. The cells in the intact carrot are each bounded by an intact cell wall; after cooking most, if not all, the cell walls have been ruptured and the cell contents lost. The carotenes are adsorbed to the cell wall and retained along with proteins. Although the cells are ruptured during cooking the general cellular network of the fibre remains intact.[10]

The grinding of foods before cooking and ingestion may also have pronounced effects on fibre action. Cell walls may be disrupted and the reduced particle size of some fibre preparations such as wheat bran may be less biologically effective.[16] The effect of other cooking processes, *e.g.* Maillard reactions, *etc.* are not kown.

3 Chemical Analysis and Measurement of Dietary Fibre

Dietary fibre is a term which encompasses several different types of complex carbohydrate. Each has its own structure and idiosyncratic action on the gastrointestinal tract. Present methods of analysis, however, measure the amount of dietary fibre in grammes and take little account of the chemical, physical, or physiological properties. There are basically two approaches to the analysis of dietary fibre — gravimetric and gas–liquid chromatography. Gravimetric methods measure by weighing an insoluble residue after chemical and enzymic solubilization of non-fibre constituents. Remaining protein is assayed and subtracted from the weight. These methods include the AOAC accepted method.[17]

---

[15] D. J. A. Jenkins, T. M. Wolever, A. R. Leeds, M. A. Gassull, P. Hausman, I. Diliwari, D. V. Goff, G. L. Metz, and K. G. M. M. Albreti, *Br. Med. J.*, 1978, 1392.
[16] K. W. Heaton, S. N. Marcus, P. M. Emmett, and C. H. Bolton, *Am. J. Clin. Nutr.*, 1988, **47**, 675.
[17] L. Prosky, N-G. Asp, I. Furda, J. W. De Vries, T. F. Schweizer, and B. Harland, *J. Assoc. Off. Anal. Chem.*, 1985, **68**, 677.

The gas–liquid chromatographic methods[18,19] involve the enzymic breakdown of starch and the separation of the low molecular weight sugars, acid hydrolysis to free sugars, derivatization to alditol acetates, and finally separation and quantitation of neutral monomers with GLC, together with determination of uronic acid and lignin. The GLC methods enable the nature of the carbohydrate to be determined in more detail, but still as yet give little information about the physical properties or the likely physiological action of the fibre.

4 Effects of Dietary Fibre Along the Gastrointestinal Tract

Complex carbohydrates affect the physiology of the gastrointestinal tract in a number of manners, which include:

1 nutrient absorption
2 sterol metabolism
3 caecal fermentation
4 stool weight.

Each fibre will influence these in a particular way and that peculiarity is altered by preparation and mode of cooking. There is little relationship between the results of quantitative chemical analysis and the effect that a fibre has along the gastrointestinal tract. There are now methods based upon water and solute immobilization techniques using dialysis bags and fermentation which enable a hierarchy of fibres to be established[20] in relation to the four fibre function modalities.

Mastication and Digestion Time

Diets which contain a substantial amount of complex carbohydrate content tend to be bulky, and require longer times for ingestion. The consumption of whole apple takes longer (17 minutes) than that required for puree (6 minutes) or apple juice (1.5 minutes) in equicaloric amounts.[21]

Gastric Emptying

Gastric emptying studies are bedevilled by problems of methodology. The physical nature of the gastric contents is as important as the chemistry of the components.[22] Most of the liquid component of a meal must empty before the

---

[18] H. N. Englyst and J. H. Cummings, *Analyst* (*London*), 1984, **109**, 937.
[19] O. Theander and J. H. Cummings, *Analyst* (*London*), 1984, **109**, 937.
[20] J. Adiotomre, M. A. Eastwood, C. A. Edwards, and W. G. Brydon, *Am. J. Clin. Nutr.*, 1990, **52**, 128.
[21] G. B. Haber, K. W. Heaton, D. Murphy, and L. F. Burroughs, *Lancet*, 1977, **ii**, 679.
[22] R. C. Heading, P. Tothill, G. P. McLaughlin, and D. J. C. Shearman, *Gastroenterology*, 1976, **71**, 45.

solids leave the stomach.[23] Solids leave the stomach more slowly than liquids. Isolated viscous fibres tend to slow the gastric emptying rate of liquids and disruptible solids.[24] It is almost certain, however, that gastric emptying for a fibre on its own will be quite different when taken along with other dietary constituents such as fat and protein. Likewise, it has been shown that the gastric emptying time for different fibre sources is variable.[14] Some low concentrations of viscous polysaccharides may even accelerate gastric emptying.[25]

Although dietary fibres such as bran and guar gum have a modest buffering capacity this does not appear to be effective in lowering gastric pH.[26]

Small Intestinal Absorption and Motility

Ingestion of both food as whole plant material, and food mixed with isolated viscous polysaccharides results in a reduction in nutrient absorption.[25,27] This is on the whole a reduction in rate rather than total amount absorbed, resulting, in the case of starch, in a much flatter post-prandial blood glucose profile.

In the case of whole plant material this appears to be due to the inaccessibility of nutrients within the cellular matrix of the plant. The effects on absorption can be decreased by grinding the food before cooking[28] or by thorough chewing[29]; both processes breaking open the cellular structure. In the case of isolated polysaccharides such as guar gum, the slowing of nutrient absorption appears to be a function of viscosity.[25,27] As discussed above, this may act to slow gastric emptying and hence the delivery of nutrients to the absorption sites in the small intestine. However, these polysaccharides still depress post-prandial hyperglycaemia even if gastric emptying is unaffected or indeed accelerated.[25] Moreover, glucose absorption from a perfused segment of human jejunum is reduced if guar gum is incorporated in the perfusate.[30]

An increase in lumenal viscosity would have several potential effects:

1 decreased interaction of enzyme and substrate
2 decreased movement of products to mucosa
3 decreased activation of mucosal receptors which stimulate secretion and motility and hormone release
4 decreased propulsion of fluid along the intestine.

---

[23] L. A. Houghton, N. W. Read, R. Heddle, M. Horowitz, P. J. Collins, B. Chatterton, and J. Dent, *Gastroenterology*, 1988, **94**, 1285.

[24] J. H. Meyer, Y. G. J. Elastoff, T. Reedy, J. Dressman, and G. Amidon, *Am. J. Physiol.*, 1986, **250**, 161.

[25] C. A. Edwards, N. A. Blackburn, L. Craigen, P. Davison, J. Tomlin, K. Sugden, I. T. Johnson, and N. W. Read, *Am. J. Clin. Nutr.*, 1987, **46**, 72.

[26] A. Rydning and A. Berstad, in 'Dietary Fibre Perspectives', Vol. 2, ed. A. R. Leeds and V. J. Burley, Libbey and Sons, London, 1990, p. 41.

[27] D. J. A. Jenkins, T. M. S. Wolever, A. R. Leeds, M. A. Gassull, P. Haisman, J. Dilawari, D. V. Goff, G. L. Metz, and K. G. M. M. Alberti, *Br. Med. J.*, 1978, **1**, 1392.

[28] K. O'Dea, P. Snow, and P. Nestel, *Am. J. Clin. Nutr.*, 1981, **34**, 1991.

[29] N. W. Read, I. McL. Welch, C. J. Austen, C. Barnish, C. E. Bartlett, A. J. Baxter, G. Brown, M. E. Compton, M. E. Hume, I. Storie, and J. Worldling, *Br. J. Nutr.*, 1986, **55**, 43.

[30] N. A. Blackburn, J. S. Redfern, M. Jarjis, A. M. Holgate, I. Hanning, J. H. B. Scarpello, I. T. Johnson, and N. W. Read, *Clin. Sci.*, 1984, **66**, 329.

These viscous polysaccharides appear to act as anti-motility agents resisting the actions of both mixing and propulsive intestinal contractions.[25,31]

The action of dietary fibre on pancreatic secretions is unclear. The decreased interaction between substrate and enzyme in the small intestine may result in a decreased rate of hydrolysis but there is usually a great excess of enzyme secreted by the pancreas. Decreased enzyme activity has been reported in patients with pancreatic insufficiency who were given a pectin or wheat bran containing meal. Direct inhibition of enzymes such as amylase, when a high legume diet is ingested, are probably due to enzyme inhibitors present in the plant material and are unrelated to the actions of polysaccharides. Studies investigating the action of dietary fibre on pancreatic secretions have shown a possible increase in pancreatic secretion after dietary fibre ingestion[32] and enzyme turnover in the lumen may also be slowed.[33]

In vitro dialysis studies have shown that viscous polysaccharides inhibit the action of pancreatic enzymes[34] and the movement of products.[27] This is probably due to a decrease in mixing rate[25] for small molecules, but diffusion may be reduced for large molecules such as micelles.[35]

An increased lumenal viscosity may also reduce the interaction of substrates with mucosal receptors and thus reduce activation of hormone and enzyme secretion mechanisms as well as motility. This may explain the acceleration of gastric emptying by some low concentrations of viscous polysaccharides.[25] However, the secretion of motilin was unaffected by guar gum[36] and gastrin secretion was in fact increased.[37]

Viscous polysaccharides tend to delay small bowel transit and this may be due to a resistance to the propulsive contractions of the intestine. Experiments in rats, however, suggest that most of this delay is due to change in ileal motility and that transit through the upper small intestine is little affected.[38,39] This may be because in the upper intestine the viscous properties are decreased by the large volume of intestinal secretions. The reduction in nutrient absorption in the upper small intestine may also result in an increased stimulation of the 'ileal brake'[40] resulting in the slowing of transit of a second meal. The addition of insoluble non-viscous fibre such as bran to food increases the time taken for the fasting motor pattern of motor activity to return to the small intestine after a meal,[41] generates aborally propagated clusters of contractions and may accelerate transit.[15]

[31] C. A. Edwards and N. W. Read, in 'Dietary Fibre Perspectives', Vol. 2, ed. A. R. Leeds and V. J. Burley, Libbey and Sons, London, 1990, p. 52.
[32] G. Isaksson, P. Lilja, I. Ludquist, and I. Ihse, *Digestion*, 1983, **27**, 57.
[33] L. P. Forman and B. O. Schneeman, *J. Nutr.*, 1980, 1992.
[34] G. Isaksson, I. Lundquist, and I. Ihse, *Gastroenterology*, 1982, **82**, 918.
[35] C. A. Edwards, I. T. Johnson, and N. W. Read, *Eur. J. Clin. Nutr.*, 1988, **42**, 307.
[36] D. R. Phillips, *J. Sci. Food Agric.*, 1986, **37**, 548.
[37] L. M. Morgan, J. A. Tredger, A. Madden, P. Kwasowski, and V. Marks, *Br. J. Nutr.*, 1985, **53**, 567.
[38] A. R. Leeds, in 'Dietary Fibre in Health and Disease', ed. G. V. Vahouney and D. Kritchevsky, Plenum Press, New York, 1982, p. 53.
[39] N. J. Brown, J. Worlding, R. D. E. Rumsey, and N. W. Read, *Br. J. Nutr.*, 1988, **59**, 223.
[40] N. W. Read, A. MacFarlane, R. Kusman, T. Bates, N. W. Blackhall, G. B. G. Farra, J. C. Hall, G. Moss, A. P. Morris, B. O'Neill, and I. Welch, *Gastroenterology*, 1984, **86**, 274.
[41] L. Bueno, F. Praddaude, J. Fioramonti, and Y. Ruckebusch, *Gastroenterology*, 1981, **80**, 701.

Sterol Metabolism

An important action of some fibres is to reduce the reabsorption of bile acids in the ileum. This may have important influences on serum cholesterol and also the amount and type of bile acid and fats reaching the colon. Bile acids may be trapped within the lumen of the ileum either because of a high lumenal viscosity or because they bind to the polysaccharide structure. A reduction in the ileal reabsorption of bile acid has several direct effects. The enterohepatic circulation of bile acids may be affected. In the caecum bile acids are deconjugated and 7α-dehydroxylated. In this less water soluble form bile acids are adsorbed to dietary fibre in a way that is affected by pH and is mediated through hydrophobic bonds. Alterations in the physical environments in the colon, *e.g.* pH, may increase the binding of bile acids to fibre and hence increase the loss of bile acid in the faeces.[42]

These so-called secondary bile acids may be reabsorbed and affect micellar properties in bile and alter the lithogenic properties of the bile.[43] It has been suggested that the inclusion of bran in the diet is likely to alter as a result of 7α-dehydroxylation in the colon, the amount of deoxycholic acid (3,12-dihydroxy cholic acid) in the colon and hence bile. The binding of bile acids to fibre or perhaps to bacteria may be important in the rate of loss of bile acids from the enterohepatic circulation. The consequence of this is that the enterohepatic pool is initially reduced. This may be renewed by increased synthesis of bile acids from cholesterol, reducing body cholesterol. Additionally, sequestration of bile acids and the micelles containing them may result in decreased fat absorption.[44] Ileostomy patients fed pectin have a greater ileal loss of fat.[45] This may result in the production of hydroxy fatty acids in the colon which may act as promotility agents.[46] It is known that some fibres in addition to altering the faecal bile acid excretion also alter faecal fat excretion. A problem relating to the concept of fibre adsorption being important in sterol homeostasis is that those fibres which are most potent in increasing faecal bile acid excretion, such as pectin, in fact are totally fermented and hence will not adsorb bile acids in the faeces.[14] Other fibres, *e.g.* gum arabic, are associated with a significant decrease in serum cholesterol without increasing faecal bile acid excretion. However the adsorption of a number of chemicals to fibre may alter their metabolism in the caecum and these have important consequences.[2]

An alternative mechanism for the effect of fibre on serum cholesterol is the action of propionic acid from fibre fermentation on liver cholesterol synthesis. Initial *in vitro* experiments have indicated that cholesterol synthesis by isolated

---

[42] M. A. Eastwood and D. Hamilton, *Biochim. Biophys. Acta*, 1968, **512**, 165.
[43] T. Low-Beer and E. W. Pomare, *Br. Med. J.*, 1975, **1**, 438.
[44] M. A. Eastwood and L. Mowbray, *Am. J. Clin. Nutr.*, 1976, **29**, 1461.
[45] I. Bosaeus, N. G. Carlsson, A. S. Sandberg, and H. Andersson, *Hum. Nutr. Clin. Nutr.*, 1986, **40C**, 429.
[46] R. C. Spiller, M. C. Brown, and S. F. Phillips, *Gut*, 1985, A1136.

hepatocytes is inhibited by propionic acid.[47,48] Whether this occurs *in vivo* at physiological concentrations is not clear.[49] Oral pectin but not caecally administered pectin decreases serum cholesterol.[50] Komai showed that 20 % pectin reduced plasma cholesterol in both conventional and germfree mice.[51] This indicates a small intestinal mechanism.

Bile acids are also potential carcinogen precursors and an increased load to the colon where they may be transformed by the colonic bacteria, especially if entrapped by a fermentable fibre, could theoretically lead to a greater cancer risk.

Colon Function

The colon may be regarded as two organs, the right side a fermenter, the left side affects continence.[2] The right side of the colon is involved in nutrient salvage so that dietary fibre, starch, fat, and protein are utilized by bacteria and the end products absorbed and used by the body. In addition to this the colon is part of the excretion system by the liver and biliary system. Chemicals of a molecular weight of more than 300–400 and poorly water soluble are excreted in the bile. They are often made water-soluble by chemical conjugation with sulphate, acetate, *etc.* and physically soluble by the detergent properties of bile acid. In the caecum these complex dietary residues, mucopolysaccharides secreted by the intestinal mucosa and biliary excretion products are fermented by the bacterial enzymes. Some of the end products are reabsorbed and hence an enterohepatic circulation is established. Others are excreted in faeces. Important in this are faecal bile acids which are a route whereby cholesterol is lost to the body. Fermentation in the caecum is also increased by dietary retrograde starch.[52]

The colonic flora is a complex ecosystem that contains mainly anaerobic bacteria which outnumber the facultative organisms at least 100:1. It is suggested that the colonic flora of a single individual is made up of possibly more than 400 bacterial species. The total bacterial count in the faeces is 10^{10} to 10^{12} colony forming units per millilitre. Despite the complexity of the ecosystem there is a remarkable stability of the microflora population. Although considerable variations in the microflora are found between individuals, studies in a single subject have generally shown that the microflora are stable over prolonged periods of time.[53] Whilst there is attraction in identifying individual bacteria because of the enormous range of bacterial species and the stability of

[47] J. W. Anderson and S. R. Bridges, *Proc. Soc. Exp. Biol. Med.*, 1984, **177**, 372.
[48] J. W. Anderson, D. A. Deakins, and S. R. Bridges, in 'Dietary Fibre: Chemistry, Physiology and Health Effects', ed. D. Kritchevsky, C. Bonfield, and J. W. Anderson, Plenum Press, New York, 1990, p. 339.
[49] R. J. Illman, D. J. Topping, G. H. McIntosh, R. P. Trimble, G. B. Storer, M. N. Taylor, and B-Q. Cheng, *Ann. Nutr. Metab.*, 1988, **32**, 97.
[50] F. Ahrens, H. Hagemeister, M. Pfeuffer, and C. A. Barth, *J. Nutr.*, 1986, **116**, 70.
[51] M. Komai and S. Kimura, *Nutr. Rep. Int.*, 1987, **36**, 365.
[52] H. Englyst and S. M. Kingman, in 'Dietary Fibre: Chemistry, Physiology and Health Effects', ed. D. Kritchevsky, C. Bonfield, and J. W. Anderson, Plenum Press, New York, 1990, p. 49.
[53] M. J. Hill, *Cancer Res.*, 1981, **41**, 3778.

the flora it is better to regard the caecal bacterial complex as an organ in its own right.

Enlargement of the caecum is a common finding when large quantities of an indigestible bulking agent is fed and this is now believed to be part of a normal physiological adjustment. Such an increase may be due to a number of factors; prolonged caecal residence, increased bacterial mass, increased bacterial products with an increased caecal weight causing caecal distension.[2]

The fermentation of fibre yields hydrogen, methane, and short chain fatty acids. Hydrogen is readily measured in the breath and has a diurnal variation with its nadir at mid-day and increases in the afternoon. Diverse sources of fibre influence the evolution of hydrogen in different ways. Disaccharides generate hydrogen more rapidly than trisaccharides which in turn evolve hydrogen more quickly than oligosaccharides. More complex carbohydrates may not be fermented as rapidly and may require induction of specific enzymes before they can be utilized. Methane is produced in the breath by only a proportion of the population. It would appear that the concentration of methane in the colon and spill over into the blood is all important.[54] Different sources of fibre yield varying amounts of short chain fatty acids, acetate, propionate, and butyrate. Acetate is a nutrient and it is possible that propionate affects cholesterol metabolism.[48] Butyrate may be important in colonic mucosal metabolism.[55]

Formation of Faeces, Water Absorption

One of the major functions of the colon is to absorb water and produce a plasticine-type of stool which can be readily voided at will from the rectum. The ileum contains a viscous fluid, the viscosity being created by mucus and water-soluble fibres. The molecular weight and the degree of cross-links and aggregation will affect the viscosity. If the viscosity increases to a certain point, peculiar to the constituent macromolecules, then a sol or hydrated carbohydrate complex will result. The sol will be coherent and homogeneous.

One to two litres of water per day pass into the caecum. Stool weight ranges from 50 to 250 g wet weight. The amount of stool excreted by an individual varies quite markedly from individual to individual and by that individual over a period of time.

There is an apparent concentration of caecal contents as a result of the absorption of water. This might be expected to create a gel. Faeces are not a gel, however, but a plasticine-like material, heterogeneous without viscosity and made up of water, bacteria, lipids, sterols, mucus, and fibre. In the caecum there is therefore a marked physical change, in part as a result of bacterial activity, in part by the presence of bacteria themselves. Such a plasticine-structure is lost in watery diarrhoea. The mechanism of this change, physiological or pathological, is unknown but many of the steps involved are described below.[14]

In the colon water is distributed in three ways.

[54] L. F. McKay, M. A. Eastwood, and W. G. Brydon, *Gut*, 1985, **26**, 69.
[55] W. E. W. Roeddiger, *Gastroenterology*, 1982, **83**, 424.

1 Free water which can be absorbed from the colon.
2 Water that is incorporated into bacterial mass.
3 Water that is bound by fibre.

Stool weight is dictated by time available for water absorption to take place through the colonic mucosa and the incorporation of water into the residue of fibre and the bacterial mass. Faeces consist of 75 % water, and dry weight being bacteria and residual fibre after bacterial fermentation and excreted compounds. The fibre matrix of faeces contains large amounts of bacteria, intermingled with small amorphous particles of food residue. The most important mechanism whereby stool weight increases is through the water holding of the unfermented dietary fibre. Wheat bran added to the diet increases stool weight in a predictable linear manner and decreases intestinal transit time.[56] The increment in stool weight is independent of the initial stool weight. Wholemeal bread, unless it is of a very coarse nature, has little or no effect on stool weight. The particle size of the fibre is all important. Coarse wheat bran is more effective than fine wheat bran.[57] The greater the water-holding capacity of the bran, the greater the effect on stool weight.[58] The effect of the water-binding by wheat bran is such that in addition to an increase in stool weight, other faecal constituents, bile acids, which in absolute amounts do not increase, are diluted and hence their concentration decreases.[59] Fibre may influence faecal output by the stimulation of colonic microbial growth as the result of ingestion of such fermentable sources as apple, guar, or pectin. This is an uncertain method as there is not always an increase in stool weight as a result of eating these fibres.[60,61] The exact mechanism is not known.

The increment in stool weight per gram of wheat bran varies in different populations. For control subjects increase in stool weight, whilst being dependent on the particle size of the bran, is generally of the order of 3 to 5 g wet stool weight per gram fibre. However in individuals with the irritable bowel syndrome and symptomatic diverticulosis the increment is of the order to 1 to 2 g wet weight per gram fibre. This suggests that there is a resistance to the movement of stool along the intestine in these situations.[56]

Dietary fibre also has a cation exchange property. As most of the fibres that have strong cation exchange capacity are fermented in the colon, it is unlikely that this cation exchange capacity is liable to increase faecal cation content.[62]

5 Dietary Fibre and Toxicology

The ingestion of the complex carbohydrates in dietary fibre does not have any direct toxic effects. A wide range of isolated gum and mucilages have been

[56] S. A. Muller-Lissner, *Br. Med. J.*, 1988, **296**, 615.
[57] A. N. Smith, E. Drummond, and M. A. Eastwood, *Am. J. Clin. Nutr.*, 1981, **34**, 2460.
[58] M. A. Eastwood, J. A. Robertson, W. G. Brydon, and D. MacDonald, *Br. J. Nutr.*, 1983, **50**, 539.
[59] M. A. Eastwood, J. R. Kirkpatrick, W. D. Mitchell, A. Bone, and T. Hamilton, *Br. Med. J.*, 1973, **4**.
[60] A. M. Stephen and J. H. Cummings, *Nature (London)*, 1980, **284**, 283.
[61] M. A. Eastwood, W. G. Brydon, and D. M. W. Anderson, *Am. J. Clin. Nutr.*, 1988, **296**, 615.
[62] A. A. McConnell, M. A. Eastwood, and W. D. Mitchell, *J. Sci. Food Agric.*, 1974, **25**, 1457.

shown to be safe to ingest in a variety of trials.[63-66]

Apart from very few reported cases of intestinal obstruction, the effects of dietary fibre on the toxicology of the human are indirect. The greatest impact is through the slowing of absorption in the small intestine or by increased bacterial activity in the colon.

Intestinal Obstruction

Intestinal obstruction by a bolus of fibre may rarely occur in the oesophagus and small intestine. When the oesophageal lumen is narrowed for any cause, the ingestion of food with a fibrous consistency, particularly meat but also vegetable, is liable to become impacted. Impaction may even occur with soft food, particularly with an elderly person with abnormal oesophageal motility and luminal narrowing.

Obstruction of the small intestine by a bolus is particularly likely to occur in patients who have undergone gastric surgery that accelerates gastric emptying. The commonest food indicated are orange segments whose pith has not been removed. Exceptionally intestinal bolus colic occurs with a normal pylorus after eating a large amount of fibrous food such as nuts. Dried fruit is a special risk as it may swell in the intestine.

Strictures of the small intestine due to Crohn's disease or other causes are usually considered a contrary indication of eating high fibre foods. Nevertheless, provided food is masticated appropriately there seems to be no special risk from high fibre foods in these individuals. In fact one authority has suggested that an enhanced carbohydrate diet may be advantageous to such patients.[67] In the only case reported of a self induced obstruction[68] the patient was psychiatrically disturbed, had a history of laxative abuse, and was eating 200 g of All Bran daily.

On the other hand, volvulus of the sigmoid colon is associated with populations who eat a high fibre diet. This is found in African and Asian populations and associated with an exceptionally long and bulky bowel and a long mesentery. It has been suggested that the colon becomes abnormally elongated because of its bulky content. It is also possible that the ingestion of maize starch with a high retrograde starch content, results in unusually high gas production leading to a twisting of the pedicle of such an intestine.[69] It is interesting that in occupied European countries during World War II when the fibre content of the diet increased due to food shortages, there was a paralleled increased incidence of volvulus of the colon.

[63] D. M. W. Anderson, M. A. Eastwood, and W. G. Brydon, *Food Hydrocolloids*, 1986, **1**, 37.
[64] M. A. Eastwood, W. G. Brydon, and D. M. W. Anderson, *Food Additives Contaminants*, 1987, **4**, 17.
[65] D. M. W. Anderson, W. G. Brydon, and M. A. Eastwood, *Food Additives Contaminants*, 1988, **5**, 237.
[66] M. A. Eastwood, W. G. Brydon, and D. M. W. Anderson, *Am.J. Clin. Nutr.*, 1986, **44**, 51.
[67] K. W. Heaton, J. R. Thornton, and P. A. Emmett, *Br. Med. J.*, 1979, **9ii**, 764.
[68] J. Y. Cairn and W. F. Day, *Br. Med. J.*, 1979, **9**, 1249.
[69] Report of the Royal College of Physicians, in 'Medical Aspects of Dietary Fibre', Pitman Medical, London, 1980.

It has been suggested that dietary fibre has an advantageous influence on the prevention of the development of gallstones, type II maturity onset diabetes, appendicitis, colonic cancer, colonic diverticulosis, and haemorrhoids. Other conditions which have been suggested to be prevented are even more speculative, *e.g.* hiatus hernia, varicose veins, and coronary artery disease.[70]

Carrageenan and Colitis

It has been demonstrated that extracts of various red seaweeds in drinking water caused ulcerative disease of the colon in several animal species, *e.g.* guinea pigs, rabbits, rats, and mice. The extracts contained carrageenan, a sulphated polysaccharide of high molecular weight 100 000–800 000.[71]

Degradation of the carrageenan produced a degraded product with a molecular weight of less than 30 000 but retaining its original sulphate content and polyionic properties. The degraded product was found to be more ulcerogenic than the native or undegraded carrageenan. Other high molecular weight sulphated products such as sodium lignosulphonate caused some lesions in the colons of animals. The ulcers are found at a concentration of 5 % of degraded carrageenan and the lesions are found first in the caecum and later involve the more distal parts of the large bowel, including the rectum, in the guinea pig and rabbit. Microscopically, the ulceration is largely confined to the mucosa and associated with an acute or sub-acute inflammatory cellular infiltrate in the marginally ulcerated areas. Of particular interest in relation to ulcerative colitis, crypt abscesses and cystic dilatation of the gland are present, particularly in more acute reactions. In common with man, clinical features include weight loss, amelioa, diarrhoea, occult blood, and sometimes mucus in the faeces. On the other hand no association between these degraded carrageenans and ulcerative colitis have been found. For example, Bonfils[72] observed no incidence of ulcerative colitis in 200 patients receiving degraded carrageenan in the treatment of peptic ulcer. In countries in which red seaweed are included in the diet, such as Japan and the Pacific Islands, there is no indication that the incidence of ulcerative colitis differs from that of other countries. Nevertheless, it is highly significant that disease of the colon can be caused by the ingestion of sulphated products of high molecular weight at least for some animal species. If this was a drug it would be banned from human consumption. When the aetiology of ulcerative colitis and Crohn's disease is finally elucidated, it will be interesting to recollect these experiments.

Sequestration of Minerals and Vitamins

Complex carbohydrates, particularly those that possess uronic and phenolic acid groups or sulphated residues such as pectins and alginates may bind

---

[70] D. P. Burkitt and H. C. Trowell, in 'Refined Carbohydrate Foods and Disease', Academic Press, London, 1975.
[71] R. Marcus and J. Watt, *Lancet*, 1969, **ii**, 489.
[72] S. Bonfils, *Lancet*, 1970, **ii**, 314.

magnesium, calcium, zinc, and iron.[62,73,74] Nonpolysaccharide components of plant cell walls such as phytates,[75-77] silicates,[78] and oxalates [79,80] also chelate divalent cations. Many of the fibre preparations used to study the effect of unpurified plant material on mineral availability contain phytate and some researchers consider phytate to be the major factor in mineral binding by plant material.[81,82] However, there is still some binding of minerals by plant material when phytate binding has been excluded.[83] Sugar beet has been shown to enhance zinc and iron absorption in rats.[84] The binding of minerals may be reduced by acid, protein,[76,77] and agents such as ascorbate, citrate and EDTA.[76,85,86]

The absorption of both water-soluble and fat-soluble vitamins may also be reduced by fibre ingestion, either because they are trapped within the plant matrix, or because they are bound by specific fibre components, or, in the case of fat-soluble vitamins, because lipid absorption is reduced.

The absorption of carotene from vegetables was higher when the vegetable was finely chopped or homogenized.[87] In addition, experimental pellagra was more easily induced with diets containing whole corn than with diets containing unenriched milled corn[88] indicating that niacin was trapped within the plant matrix. The availability of vitamin B_6 from fibre-rich foods, such as cereals and wholemeal bread, is less than that from nonfibrous food such as milk and fish.[89-91] Vitamin B_{12}-depleted rats fed cellulose or pectin excreted more radiolabelled vitamin B_{12} that rats on fibre-free diets,[92] and wheat bran and pectin in the faecal excretion of vitamin B_6.[93,94] Nicotinic acid bound in cereal

[73] J. G. Reinhold, B. Faradji, P. Abadi, and F. Ismail-Beigi, *J. Nutr.*, 1976, **106**, 493.
[74] H. H. Sandstead, J. M. Munoz, R. A. Jacob, L. M. Klevay, S. J. Reck, G. M. Logan, F. R. Dintzis, G. E. Inglett, and W. C. Shuey, *Am. J. Clin. Nutr.*, 1978, **31**, S180.
[75] J. G. Reinhold, K. Nasr, A. Lahimgarzadeh, and H. Hedayati, *Lancet*, 1973, **ii**, 283.
[76] L. Hallberg, *Scand. J. Gastroenterol.*, 1987, **22** (Supplement 129), 73.
[77] B. Sandstrom, *Scand. J. Gastroenterol.*, 1987, **22** (Supplement 129), 80.
[78] G. V. Vahouny, in 'Nutritional Pathology—Pathobiochemistry of Dietary Imbalances', ed. H. Sidransky, Marcel Dekker Inc., 1985, p. 207.
[79] L. V. Avioli, *Arch. Intern. Med.*, 1972, **129**, 345.
[80] J. L. Kelsay, *Am. J. Gastroenterol.*, 1987, **82**, 983.
[81] D. Oberleas and B. F. Harland, in 'Zinc Metabolism: Current Aspects in Health and Disease', A. R. Liss, New York, 1977, p. 11.
[82] N. T. Davies, in 'Dietary Fibre: Current Development of Important to Health', ed. K. W. Heaton, Newman Publishing, London, 1978, p. 112.
[83] J. G. Reinhold, G. J. Salvador, and P. Garzon, *Am. J. Clin. Nutr.*, 1981, **34**, 1384.
[84] S. S. Fairweather-Tait and J. A. Wright, *Br. J. Nutr.*, 1990, **64**, 547.
[85] J. L. Kelsay, in 'Dietary Fibre Basic and Clinical Aspects', ed. G. V. Vahouny and D. Kritchevsky, Plenum Press, New York, 1986, p. 361.
[86] T. Hazell and I. T. Johnson, *Br. J. Nutr.*, 1987, **57**, 223.
[87] W. B. Van Zeben and T. F. Hendricks, *Int. Z. Vitaminforsch.*, 1948, **19**, 265.
[88] G. A. Goldsmith, J. Gibbens, W. G. Unglaub, and O. N. Miller, *Am. J. Clin. Nutr.*, 1956, **4**, 151.
[89] J. E. Leklem, T. D. Schultz, and L. T. Miller, *Fed. Proc.*, 1980, **39**, 558.
[90] J. L. Kelsay, in 'Dietary Fibre in Health and Disease', ed. G. V. Vahouny and D. Kritchevsky, Plenum Press, New York, 1982, p. 91.
[91] H. Kabir, J. E. Leklem, and J. T. Miller, *J. Nutr.*, 1983, **113**, 2412.
[92] R. W. Cullen and S. M. Oace, *J. Nutr.*, 1978, **108**, 640.
[93] L. B. Nguyen, J. F. Gregory, and B. L. Damron, *J. Nutr.*, 1981, **111**, 1403.
[94] A. S. Lindberg, E. Leklem, and L. T. Miller, *J. Nutr.*, 1983, **113**, 2578.

fibre could be released by alkaline hydrolysis or treatment with lime water.[90] Cereal fibre had no effect on the absorption of folate[95,96] or riboflavin.[90]

The faecal excretion of vitamin A was increased in humans fed on a mixed fibre diet[90] compared with a low-fibre diet supplemented with β-carotene to contain equivalent amounts of vitamin A. The results of studies on the effect of pectin on vitamin A metabolism in animals have produced conflicting results.[97-99] Surprisingly, Kasper[100] found that 15 g of pectin given to humans with a large dose of vitamin A in a formula meal results in higher plasma levels of vitamin A than the same amount of vitamin A given with the meal alone. Erdman[99] found that there was an inverse relationship between the methoxyl content of pectin and β-carotene utilization in chicks suggesting that the reduction in β-carotene utilization was due to the viscous nature of the high methoxyl pectins. Vitamin E bioavailability in rats was also decreased by pectin ingestion,[101] but unaffected by wheat bran.[102]

The reduction in absorption of minerals and vitamins could, in theory, have adverse nutritional consequences, particularly in poor people in developing countries where diets may be marginal in micro-nutrients but high in fibre, in children in whom zinc uptake may be deficient,[103] and in patients with anorexia who may be consuming a diet containing large amounts of dietary fibre with very little other nutrients.[104] Experiments in rats have shown that viscous polysaccharides intensified the symptoms of B_{12} depletion in rats.[92,105] However, normal Western diets contain levels of minerals and vitamins in excess of daily requirements and some of the minerals trapped in the lumen of the small intestine may be released and absorbed in the colon if the fibre is fermented.[103] Mineral balance studies have indicated that in people on nutritionally adequate diets, the ingestion of mixed high fibre diets or dietary supplementation with viscous polysaccharides is unlikely to cause mineral deficiencies,[85,106] and most researchers conclude that there are no important deleterious effects of fibre ingestion on vitamin bioavailability.[78,90,93,94,107,108]

[95] K. A. Ristow, J. F. Gregory, and B. L. Damron, *J. Nutr.*, 1982, **112**, 750.
[96] P. M. Keagy, B. Shane, and S. Oace, *Am. J. Clin. Nutr.*, 1988, **47**, 80.
[97] E. C. Hwang, P. Griminger, and H. Fisher, *Nutr. Rep. Int.*, 1975, **11**, 185.
[98] W. E. J. Phillips and R. L. Brien, *J. Nutr.*, 1970, **100**, 289.
[99] J. W. Erdman, G. C. Fahey, and C. B.White, *J. Nutr.*, 1986, **116**, 2415.
[100] H. Kasper, U. Rabast, H. Fassl, and F. Fehle, *Am. J. Clin. Nutr.*, 1979, **32**, 1847.
[101] E. E. Schaus, B. O. de Lumen, F. I. Chow, P. Reyes, and S. T. Omaye, *J. Nutr.*, 1985, **115**, 263.
[102] W. A. Behrens, R. Madere, and R. Brassard, *Nutr. Res.*, 1986, **6**, 215.
[103] W. P. T. James, in 'Medical Aspects of Dietary Fibre', ed. G. A. Spiller and R. M. Kay, Plenum Press, New York, 1980, p. 239.
[104] O. Sculati, G. Giampiccoli, L. P. Morandi, and D. Vecchiati, *Scand. J. Gastroenterol.*, 1987, **22** (Supplement 129), 278.
[105] R. W. Cullen and S. M. Oace, *Fed. Proc.*, 1980, **39**, 785.
[106] K. M. Behall, D. J. Scholfield, K. Lee, B. S. Powell, and P. B. Moser, *Am. J. Clin. Nutr.*, 1987, **46**, 307.
[107] T. D. Schukltz and J. E. Leklem, *Am. J. Clin. Nutr.*, 1987, **46**, 647.
[108] E. Wiske, T. F. Schweizer, and W. Feldheim, in 'Dietary Fibre: Chemical and Biological Aspects', ed. D. A. T. Southgate, K. Waldron, I. T. Johnson, and G. R. Fenwick, Royal Society of Chemistry, Cambridge, Special Publication No. 83, 1990, p. 203.

Effect of Complex Carbohydrates on Drug Absorption and Metabolism

Ingestion of dietary fibre may affect drug absorption in two ways. Firstly, if a soluble fibre reduces absorption in the small intestine by reducing gastric emptying or inhibiting mixing in the small intestine then the rate of absorption of the drug and therefore the time to reach effective dose may be delayed. Viscous polysaccharides have been shown to delay the absorption of paracetamol.[109] Secondly, if the drug enters the enterohepatic circulation any bacterial metabolism of the drug may be altered by a fermentable fibre and thus the half life of a drug may be increased or decreased. An example of this is digoxin. Digoxin has a narrow therapeutic range, and is passively absorbed in the small intestine and so should be subject to the influence of agents that slow gastric emptying or decrease small intestinal absorption. Digoxin is also reduced to an inactive metabolite in the colon so an inactive metabolite will be absorbed. If antibiotics are given at the same time as the digoxin, the serum concentrations of digoxin increase.[110]

Several studies have investigated the effects of dietary fibre on digoxin. Bran given with digoxin capsules had a small delaying effect on digoxin absorption but this was not clinically significant.[111] Guar gum in a liquid meal did not significantly affect digoxin absorption after 5 days of guar supplementation[112] and only slightly decreased the rate of absorption of a single dose.[113] The bioavailability of amoxycillin, however was decreased in subjects from underdeveloped countries ingesting a high fibre diet.[114] The bioavailability of penicillin was also reduced when ingested with guar gum.[113] Soluble viscous fibres guar and konjac mannan reduced the plasma level of sulphonylurea achieved after ingestion of glibenclamide[115] but in another study guar gum had no effect on glipizide absorption.[116] Wheat bran decreased the absorption rate of the anti-coagulant phenprocoumon.[117] Thus drug absorption may be affected by dietary fibre but the effect depends on the nature of the fibre as well as the drug. This may be particularly important in drugs with a narrow therapeutic range or where the patient is taking soluble fibre therapeutically such as in diabetes.

Colonic Bacterial Enzymes

The bacteria in the colon produces an 'organ' of intense metabolic activity. This activity is mainly reductive, and may be involved in not only the deactivation of

[109] S. Holt, R. C. Heading, D. C. Carter, L. F. Prescott, and P. Tothill, *Lancet*, 1979, **i**, 636.
[110] J. Linderbaum, D. G. Rind, V. P. Butler, D. Tse-Eng, and J. R. Saha, *New Eng. J. Med.*, 1981, **305**, 789.
[111] B. F. Johnson, S. M. Rodin, K. Hock, and V. Shekar, *J. Clin. Pharmacol.*, 1987, **27**, 487.
[112] W. Creutzfeldt, *Z. Gastroenterol.*, 1982, **3**, 164.
[113] R. Huupponen, P. Seppala, and E. Isalo, *Eur. J. Clin. Pharmacol.*, 1984, **26**, 279.
[114] M. Lutz, J. Espinoza, A. Arancibu, M. Araya, I. Pacheco, and O. Brunser, *Clin. Pharmacol. Therap.*, 1987, **42**, 220.
[115] K. Shima, H. Ikegami, A. Tanaka, A. Ezaki, and Y. Kumahara, *Nutr. Rep. Int.*, 1982, **26**, 297.
[116] R. Huupponen, S. Karhuvaara, and P. Seppala, *J. Clin. Pharmacol.*, 1985, **28**, 717.
[117] N. R. Kitteningham, S. Mineshita, and E. E. Ohnhaus, *Klin. Wochenschr.*, 1985, **63**, 537.

ingested toxins but also the production of carcinogens and co-carcinogens. This is in contrast to the liver which is oxidative. The range of metabolic transformations that the intestinal flora perform on ingested compounds are shown in Table 4. The major enzymes involved in these activities include azoreductase, nitrate reductase, nitroreductase, β-glycosidase, β-glucuronidase, and methyl-mercury-demethylase.

Table 4 Range of biochemical transformation by intestinal bacteria

Hydrolysis	Aromatization	Reduction
Glucuronides	Ethereal sulphates	Carbon–carbon double bonds
Esters	Sulphamates	Nitro-acid A20 bonds
Amides	Glycosides	N-oxides, N-hydroxy compounds
		Carboxyl groups
		Alcohol, phenols
		Arsonic acid
Degradation		Synthesis
Decarboxylation		Esterification
Dealkylation		Acetylation
Deamination		Formation of nitrosamines
Dehalogeneration		

The effect of dietary fibre on the activity of these enzymes has been extensively reviewed by Rowland.[118,119] The action of a dietary fibre appears to be dependent on its fermentability. Fibres which are highly fermentable such as pectin increase the activity of most of these enzymes whereas poorly fermented fibres such as cellulose have no effect or decrease activity.[118,119] The effect of dietary fibre on these enzymes may be due to several mechanisms. Most obviously, fibres which are fermented may increase the size of the bacterial population and therefore the enzyme activity. Increased bacterial metabolism may reduce luminal pH and change absorption or the toxic activity of a compound, and finally the increased bulk of the intestinal contents with either water or solid material may dilute enzymes and thus reduce activity.

Nitrate Reductase

Bacterial reduction of nitrate to nitrite is associated with formation of genotoxic N-nitroso compounds.[120] These products may be of importance in the achlorhydric stomach but do not seem to be important in the large intestine.[119] Animals fed pectin containing diets, however, were more susceptible to nitrate-induced methaemoglobinaemia.[121]

---

[118] I. Rowland and A. K. Mallett, in 'Dietary Fibre: Chemistry, Physiology, and Health Effects', ed. D. Kritchevsky, C. Bonfield, and J. W. Anderson, Plenum Press, New York, 1990, p. 195.
[119] A. K. Mallett and I. R. Rowland, in 'Role of the Gut Flora in Toxicity and Cancer', Academic Press, London, 1988, p. 347.
[120] I. R. Rowland, *Toxicol. Pathol.*, 1986, **16**, 147.
[121] A. Wise, A. K. Mallett, and I. R. Rowland, *Xenobiotica*, 1982, **12**, 111.

Nitroreductase

Pectin increased the reduction of aromatic and heterocyclic compounds. 2,6-Dinitrotoluene, which binds covalently to hepatocyte DNA repair assays, was administered to animals on 0 %, 5 %, or 10 % pectin diets. The pectin fed rats had significantly more hepatic binding than the control animals and this appeared to be related to increased levels of nitroreductase.[122] Nitrobenzene produces methaemoglobinaemia and degenerative changes in the testes, liver, and spleen.[123] Rats fed pectin and treated with nitrobenzene showed marked methaemoglobinaemia whereas control rats did not.[124]

β-Glycosidase

The β-glycosidase activity of the colonic flora may be responsible for the activation of flavonoids. These polyphenolic compounds are common in plants as glycosidic conjugates. They are large hydrophilic molecules and therefore escape absorption in the small intestine. In the colon they are hydrolysed and transformed to genotoxic agents. For example, faecal cultures produce aglycone quercetin from rutin or quercitin. These toxic compounds may also inhibit or induce mammalian enzymes which enhance the action of mutagens.[125,126]

β-Glycosidase is also responsible for the generation of the toxic product methylazoxymethanol (MAM) from cycasin. Cycasin induces tumour formation at many sites in the body. MAM may also enter the enterohepatic circulation and be detoxified in the liver to be reactivated in the colon by β-glucuronidase.[127]

β-Glucuronidase

Bacterial β-glucuronidase may be responsible for the release of toxic substances that have been previously detoxified by the liver by conjugation with glucuronic acid prior to excretion in the bile. The toxin then reenters the enterohepatic circulation. This increases the carcinogenic activity of agents such as dimethylhydrazine (DMH).

Azoreductase

The azoreductase activity of the intestinal flora may release highly reactive intermediates from azo dyes which react with proteins and nucleic acids.

The action of fibre on the activity of these enzymes may be species-dependent and animal studies do not always indicate effects in man.[118]

---

[121] A. Wise, A. K. Mallett, and I. R. Rowland, *Xenobiotica*, 1982, **12**, 111.
[122] J. D. DeBethizy, J. M. Shenil, D. E. Rickett, and T. E. Harm, *Toxicol. Appl. Pharmacol.*, 1983, **69**, 369.
[123] D. E. Rickent, J. A. Bond, R. M. Long, and J. P. Chism, *Toxicol. Appl. Pharmacol.*, 1983, **67**, 206.
[124] R. S. Goldstein, J. P. Chism, J. M. Sherill, and T. E. Hamm, *Toxicol. Appl. Pharmacol.*, 1984, **75**, 547.
[125] K. W. Wattenberg and J. L. Leong, *Cancer Res.*, 1970, **30**, 1922.
[126] R. L. Sousa and M. A. Marletta, *Arch. Biochem. Biophys.*, 1985, **240**, 345.
[127] A. K. Mallett and I. R. Rowland, *Dig. Dis.*, 1990, **8**, 71.

6 Reduced Toxicity

High fibre intake is reported to reduce the toxicity of certain detergents, antioxidants, and heavy metals. Results of studies, mainly in rats, are summarized in Table 5.[128-144] The effects on Amaranth toxicity may also be related to the bulking capacity of the fibre. Takeda and Kiriyama[139] compared the effects of a range of dietary fibres with different physical properties and showed that those with the highest swelling volume or water holding capacity had the greatest effect. Methylmercury toxicity may be reduced by the activity of methylmercury demethylase.

7 Colonic Carcinogenesis

The role of dietary fibre in the cause or prevention of cancer in humans is very unclear. The original hypothesis that low dietary fibre intake was responsible for the high colonic cancer occurrence in industrialized countries was based on the observation that these were the populations with a low fibre intake. However, a low fibre intake is usually accompanied by a high fat intake and there may be many other contributing factors that are more important than fibre. The results of experimental studies in animals designed to investigate whether dietary fibre has a beneficial or harmful effect on carcinogenesis have produced confusion and conflicting results.[145] Firstly the wide variety of plant materials that make up dietary fibre make it very misleading to consider dietary fibre as a whole when looking at its effects on any part of the gut but most especially in the colon. Here fermentability and the resultant effects on the gut bacteria may be the critical property of a polysaccharide in determining its action. Second in studies of induced carcinogenesis, the animal models vary in not only their age and sex but also the length of the diet, the dose of the fibre,

---

[128] D. W. Woolley and L. D. Krampitz, *J. Exp. Med.*, 1943, **78**, 333.
[129] B. H. Ershoff, *Proc. Soc. Exp. Biol. Med.*, 1954, **87**, 134.
[130] B. H. Ershoff, *Proc. Soc. Exp. Biol. Med.*, 1957, **95**, 656.
[131] B. F. Chow, J. M. Burnett, C. T. Ling, and L. Barrows, *J. Nutr.*, 1953, **49**, 563.
[132] B. H. Ershoff and H. J. Hernandez, *J. Nutr.*, 199, **69**, 172.
[133] B. H. Ershoff and W. E. Marshall, *J. Food Sci.*, 1975, **40**, 357.
[134] B. H. Ershoff, *J. Nutr.*, 1960, **70**, 484.
[135] T. Kimura, H. Furuta, Y. Matsumoto, and A. Yoshida, *J. Nutr.*, 1980, **110**, 513.
[136] T. Kimura, H. Imarura, K. Hasegawa, and A. Yoshida, *J. Nutr. Sci. Vitaminol.*, 1982, **28**, 483.
[137] B. H. Ershoff, *Proc. Soc. Exp. Biol. Med.*, 1963, **112**, 362.
[138] B. H. Ershoff and E. W. Thurston, *J. Nutr.*, 1974, **104**, 937.
[139] H. Takeda and S. Kiriyama, *J. Nutr.*, 1979, **109**, 388.
[140] B. H. Ershoff, *Proc. Soc. Exp. Biol. Med.*, 1977, **107**, 822.
[141] R. H. Wilson and F. DeEds, *Arch. Ind. Hyg. Occup. Med.*, 1950, **1**, 73.
[142] T. M. Paul, S. C. Skoryna, and D. Waldron-Edward, *Can. Med. Assoc. J.*, 1966, **95**, 957.
[143] B. Momcilovic and N. Cruden, *Experentia*, 1981, **37**, 498.
[144] I. R. Rowland, A. K. Mallett, J. Flynn, and R. J. Hargreaves, *Arch. Toxicol.*, 1986, **59**, 94.
[145] L. R. Jacobs, in 'Dietary Fibre: Chemistry, Physiology, and Health Effects', ed. D. Kritchevsky, C. Bonfield, and J. W. Anderson, Plenum Press, New York, 1990, p. 389.

Table 5 Antitoxic effects of dietary fibre in rats and mice

Toxic substance	Symptom	Fibre	Effect	Animal	Ref.
glucoascorbic acid	Vitamin C deficiency weight loss	alfalfa grass	reduced or removed toxic effect	mice rats	128 129 130
Detergents					
Tween 60		soyabean meal alfalfa psyllium wheat bran	reduced effect	rats	131 132 133
Tween 20 polyoxyethylene stearate sorbitan monolaurate	inhibited growth	alfalfa meal or residue not juice cellulose	inhibited or abolished effect	rats	134 132
Various detergents including Tween 20, 60 SDS SLS	↑exfoliation intestinal cells ↓disaccharidase	edible burdock	prevented		135 136
Antioxidant					
DBH 2,5,di-t-butyl hydroquinone	↓weight gain	stock diet vs. semi-purified	reduced effect increased survival		137

Food colours				
Amaranth ⎫				138
Tartrazine ⎬	↓ weight gain	10% cellulose pectin alfalfa	rats	139
Sunset yellow brilliant blue red no. 40 ⎭		edible burdock carrot root powder blond psyllium		140
sodium cyclamate	↓ weight gain	stock diet	rats	Ershoff 1972
		agar, roast bean gum, Aaga, guar karaya, alginate, tragacanth, celery, lemon peel, psyllium, carrot, alfalfa		133
Heavy metals				
Cadmium	cadmium deposit	6% fibre	rats	141
Strontium 89 45	accumulate in blood and bone	calcium alginate	rats	142
Calcium 47		cellulose		143
Methylmercury	deposits in blood, brain, intestine	wheat bran	mice	144

and the type, dose, and route of administration of the carcinogen. Also they may bear little resemblance to the events leading to carcinoma in man.

The hypotheses for the action of dietary fibre fall into two main groups:

1 the consequence of bacterial fermentation
2 physical properties of the polysaccharide.

Bacterial Fermentation

When a fibre is fermented it has direct actions—production of short chain fatty acids, reduced pH, and indirect actions—increased bacterial enzymes, such as β-glucuronidase and the bile acid 7-dehydroxylase, as a result of increased bacterial numbers.

Short Chain Fatty Acids

Dietary fibre is fermented to short chain fatty acids, chiefly acetic, propionic, and n-butyric, carbon dioxide, hydrogen, and methane. The production of short chain fatty acids has several possible actions on the gut mucosa. Firstly they may act as cellular fuels. All of the short chain fatty acids are readily absorbed by the colonic mucosa, but only acetic acid reaches the systemic circulation.[146] Butyric acid is metabolized before it reaches the portal blood, propionic acid is metabolized in the liver. This would suggest that butyric acid is being used as a fuel by the colonic mucosa and *in vitro* studies of isolated cells have indicated that the short chain fatty acids and butyric acid in particular are the preferred energy sources of colonic cells.[147] It has been suggested that short chain fatty acids are essential for the health of the mucosa. More recently it has been shown that short chain fatty acids are potent stimulants of cellular proliferation not only in the colon but also in the small intestine and this appears to be mediated by a blood borne factor.[148] Increased cellular proliferation would superficially appear to indicate increased opportunity for the action of carcinogens. Indeed, this may be one explanation for the effect of fermentable fibres increasing tumour formation in animal models such as the DMH model. However most of these studies are performed on starved intestinal cells within an empty lumen. It has not been clearly demonstrated that an increase in short chain fatty acids above the usually recorded levels has any further effect on cellular proliferation. However, diets containing fermentable fibre have been shown to stimulate cellular proliferation,[149] but this may be due to other factors such as bile acid concentration or pH.[145] Conversely the short chain fatty acids have also been reported to have anti-mutagenic properties. Propionic acid has been shown as a

[146] J. H. Cummings, E. W. Pomare, W. J. Branch, C. P. E. Naylor, and G. T. MacFarlane, *Gut*, 1987, **28**, 1221.
[147] W. E. W. Roediger, *Gastroenterology*, 1982, **83**, 424.
[148] T. Sakata, *Br. J. Nutr.*, 1987, **58**, 95.
[149] R. R. Goodlad, W. Lenton, M. A. Ghatei, T. E. Adrien, S. R. Bloom, and N. A. Wright, *Gut*, 1987, **28**, 171.

potent mitogen for human colonocytes *in vitro*.[150] Butyric acid stimulated differentiation of cell clones of human colonic cancer cells *in vitro*,[151] though in some studies butyric acid has been shown to enhance carcinogenesis.[152] The influence of diet on colonic cancer aetiology is difficult to understand. The use of artificial carcinogens to produce colonic cancers in animals may have no relevance to the human disease. New approaches are required before the role of fibre in such a complex process can be understood.

Reduced pH

The reduction in colonic pH seen after ingestion of some dietary fibres,[153] may be beneficial as the enzymes responsible for bile acid transformation (see below) have a high pH optimum.[154] A low pH would also reduce the solubility of bile acids and fatty acids reducing their putative mitogenic and carcinogenic effects. The faecal pH of colorectal cancer patients may be higher than normal people.[155,156]

Bacterial Enzymes

Increased bacterial numbers induced by ingestion of fermentable fibre is associated with an increase in the activity of certain enzymes which are thought to be related to carcinogenesis. β-Glucuronidase and β-glycosidase have been discussed above. 7-α-Dehydroxylase is the enzyme thought to be responsible for the transformation of bile acids to cancer promoters. The role of bile acids and this enzyme in cancer development is far from clear. Faecal bile acid output is increased by several types of dietary fibre but this does not seem to happen in patients with cancer. The relationship seems to be more complex with perhaps ratios of deoxycholic and lithocholic acid being more important.[157] Though relationship between this ratio and stool weight is to be found in a normal unaffected population and may at the most be a weak association.[158]

Physicochemical Properties

Part of the original hypothesis was that fibres would increase colonic content turnover and bulk and thus dilute carcinogen concentration. For some fibres such as wheat bran which increases faecal bulk without increasing bile acid

---

[150] E. Friedman, C. Lightdale, and S. Winawer, *Cancer Lett.*, 1988, **43**, 121.
[151] C. Augeron and C. I. Laboisse, *Cancer Res.*, 1984, **44**, 3961.
[152] H. J. Freeman, *Gastroenterology*, 1986, **91**, 596.
[153] J. R. Lipton, D. M. Coder, and L. R. Jacobs, *J. Nutr.*, 1988, **118**, 840.
[154] M. J. Hill, *Am. J. Clin. Nutr.*, 1974, **27**, 1475.
[155] I. A. MacDonald, G. R. Webb, and D. E. Mahoney, *Am. J. Clin. Nutr.*, 1978, **31**, 233.
[156] J. R. Thornton, *Lancet*, 1981, i, 1081.
[157] R. W. Owen, M. H. Thompson, M. J. Hill, M. Wilpart, P. Mainguet, and J. Roberfroid, *Nutr. Cancer*, 1987, **9**, 67.
[158] W. G. Brydon, M. A. Eastwood, and R. A. Elton, *Br. J. Cancer*, 1988, **57**, 635.

output this may be true. But many others that increase faecal bulk also increase the output of bile acids and steroids and thus do not dilute.[159]

These fibres must trap the bile acids in the small intestine by binding or physical entrapment and by increasing colonic bile acid they may indeed increase the exposure of the colon to carcinogens.

The measurement of total bile acid concentration in faeces may not be totally appropriate. Some of this bile acid may be unavailable for interaction with the mucosa. The concentration of carcinogens or co-carcinogens in the aqueous phase of the colonic contents may be more appropriate although much more difficult to assess.

The role of fibre in the aetiology of colon cancer, if any, is complex. There may be only a circumstantial epidemiological relationship. Any explanation has to account for the greater prevalence of colonic cancer in the distal colon, *i.e.* descending, sigmoid, and rectum. This excludes fermentation as a direct aetiological cause, as the caecal fermentor would be more vulnerable if this was the case. However the relatively sterile small intestine rarely develops malignancy. Fermentation products, for example butyric acid (believed by some to be a stabilizer of colonic mucosal turnover, whatever that may mean) decreases in amount from the right to the left colon, rendering the left colon potentially deficient in such nutrients. This could explain the geographical differences.

Non-fermentable, water binding fibres would dilute putative carcinogens, especially if the stool, static in the distal colon, is expanded by held water. The role of fibre in inhibiting the genesis of colonic cancer is therefore mediated through at least two mechanisms. Genetic predisposition and even the possibility of viral causes are important considerations.

8 Conclusion

In general the diverse effects of individual dietary fibre appear to be neutral or beneficial. Fibre, a sub-set of dietary complex carbohydrates, modulate the progression of intestinal contents and absorption in the upper gastrointestinal tract. In the lower gastrointestinal tract fibre alters bacterial mass and metabolism and may increase faecal bulk and water content. The only area of possible toxic effects is that of increases in bacterial enzymes such as β-glucuronidase and β-glycosidase when fermentable fibres are ingested. The significance of these enzymes in human disease has not been established and needs further study. In addition, there is speculation for pro-cancer effects of fermentable fibre. Most of the fibre in our diet is of cereal origin and is poorly fermented. The significance of the fermentable fibre that we ingest in promoting cellular proliferation and carcinogenesis is probably minute but still needs consideration and investigation.

[159] M. J. Hill and F. Fernandez, in 'Dietary Fibre: Chemistry, Physiology, and Health Effects', ed. D. Kritchevsky, C. Bonfield, and J. W. Anderson, Plenum Press, New York, 1990, p. 339.

CHAPTER 12

Saponins

G. ROGER FENWICK, KEITH R. PRICE, CHIGEN TSUKAMOTO, AND KAZUYOSHI OKUBO

1 Introduction

Saponins, which derive their name from their ability to form stable, soap-like foams in aqueous solutions are a diverse and chemically-complex group of compounds which occur naturally in plants, and to a lesser extent in marine animals (*e.g.* starfish). More than one hundred plant families contain saponins, which generally occur as complex mixtures, varying in nature and amount according to plant part, physiological age, and environment.

Inspection of the natural product literature reveals considerable information on saponins available prior to 1960. Subsequently there has been a tremendous increase in work in this area, evidence of which can be found by the regular reviews and updates of the literature (see references 1–18 in reference 1). The complex natures of the saponin structures, their water solubility and ability to bind strongly to other components of the plant matrix, as well as their susceptibility to partial or complete hydrolysis during isolation and workup may all be cited as factors delaying the advance of knowledge about the chemical and biological properties of this group of natural products. The development of techniques for the isolation, separation, and purification of water-soluble plant constituents, and of complementary chemical, enzymatic, and physicochemical methods for structural elucidation has had a major effect on the development of research in this area. An additional spur has been the identification of diverse physiological and pharmacological properties amongst isolated saponins. Such findings raised the possibility of exploiting saponins and saponin-rich plant extracts as pharmacological and veterinary products, health foods, and herbal medicines and led to extensive research programmes on, for example, ginseng and liquorice.

In comparison, knowledge about the nature, amount, and effect of saponins in plants intended for human foods and animal feedingstuffs can only be described as fragmentary. That progress has been made can be seen by comparing

Table 1 *Historical development of research into soyasaponins*

1923	Studies on 'soaps' in soyabean (S. Muramatsu, *J. Chem. Soc. Jap.*, 1923, **44**, 1035).
1929	Studies on soyabean saponin (T. Sumiki, *Bull. Agric. Chem. Soc. Jap.*, 1929, **5**, 27).
1974	Soyasaponin I–III structures (I. Kitagawa *et al.*, *Chem. Pharm. Bull.*, 1974, **22**, 3010).
1982	Soyasaponin I–III structures revised (I. Kitagawa *et al.*, *Chem. Pharm. Bull.*, 1982, **30**, 2294).
1985	Soyasaponin A_1, A_2 structures (I. Kitagawa *et al.*, *Chem. Pharm. Bull.*, 1985, **33**, 598 and 1069).
1987	Soyasaponin IV structure (J. Burrows *et al.*, *Phytochem.*, 1987, **26**, 1214).
1988	Acetylated soyasaponin structures (I. Kitagawa *et al.*, *Chem. Pharm. Bull.*, 1988, **36**, 2819). (T. Taniyama *et al.*, *Chem. Pharm. Bull.*, 1988, **36**, 2829).
1991	Group E structures (M. Shiraiwa *et al.*, *Agric. Biol. Chem.*, 1991, **55**, 911).
1991	*Glycine soja* saponin structures (C. Tsukamoto *et al.*, *Abstr. Ann. Meeting Jap. Soc. Biosci., Biotechnol., Agrochem.*, 1991, abstr. 2Qa4).

the initial review of the subject, by Oakenfull in 1981,[2] with those of Price *et al.*[1] and Oakenfull and Sidhu[3]; an indication of the rate of progress in this area is shown in Table 1. Despite the importance of the soybean as a global commodity, and its major role as both a human food and animal feedstuff, it was only as recently as 1982 that the three mono-desmosidic saponins were finally characterized, the structures of the first bis-desmosides being established three years later. Subsequent progress has been facilitated by the advances in chemical techniques referred to above; the stimuli to such research have included the suggestion that dietary saponins may lower plasma cholesterol levels (thereby offering protection against coronary artery disease) and may increase the permeability of the intestinal mucosa. Considerable interest has also resulted from the recent findings linking saponins to plant protection and disease resistance.

The present review describes the sources of saponins in crop plants, their structures, levels in crops, and the effects of processing. However, as with all secondary metabolites it is the *significance* of their presence in human foods and animal feedingstuffs which is important, and later sections will address the biological effects of saponins as well as discussing their role in crop protection. In covering such an area, the authors are mindful of the recent comprehensive coverage by Price *et al.*[1] and Oakenfull and Sidhu.[3] Particular attention will thus be paid to four classes of crop plant, soyabean (*Glycine max*), alfalfa (*Medicago sativa*), quinoa (*Chenopodium quinoa*), and liquorice (*Glycyrrhiza glabra*). Together these serve to exemplify the diversity in structure and biological activity of saponins and point to future areas of research.

[1] K. R. Price, I. T. Johnson, and G. R. Fenwick, *CRC Crit. Rev. Food Sci. Nutr.*, 1987, **26**, 27.
[2] D. Oakenfull, *Food Chem.*, 1981, **7**, 19.
[3] D. Oakenfull and G. S. Sidhu, in 'Toxicants of Plant Origin, Volume II', ed. P. R. Cheeke, CRC Press Inc., Boca Raton, 1989, p. 97.

Table 2 Saponins in plants used as foods and feedingstuffs

Aglycone	
Triterpenoid	Steroid
Soyabean	Oats
Beans	Capsicum peppers
Peas	Aubergine
Alliums	Tomato (seed)
Tea	Alliums
Spinach	Asparagus
Sugar beet	Yam
Quinoa	Fenugreek
Liquorice	Ginseng
Alfalfa	
Sunflower	
Horse chestnut	
Ginseng	

2 Chemical Structures and Sources

Whilst all saponins contain an aglycone (sapogenol, sapogenin) linked to one or more sugars or oligosaccharide moieties, they fall naturally into two groups depending upon whether the aglycone is triterpenoid or steroidal. In cultivated crops the former group is predominant, although steroidal saponins are commonly found in plants used as herbs or for their 'health-giving' properties.[1] The main crops containing saponins are listed in Table 2.

The complexity of saponin structure (and thereby the diversity of biological activities) has its basis not only in the variability of aglycone structure and position(s) of attachment of the glycosidic moieties, but also in the nature of the latter. Aglycones are generally linked to D-galactose (gal), L-arabinose (ara), L-rhamnose (rham), D-glucose (glc), D-xylose (xyl), D-mannose (man), and D-glucuronic acid (glcUA) some of which may be acetylated; chain lengths of 2–5 saccharide units are most frequent. The oligosaccharide chains are generally linear, although branching is not uncommon. Detailed coverage of the saponins of food, forage, and herbal plants has been reported recently[1] and the interested reader is directed to this review and the references contained therein.

At this early juncture it is necessary to digress to discuss the nomenclature of saponins. Given the complexity of these molecules within both the (oligo)saccharide and aglycone moieties it is unrealistic to expect adherence to systematic rules of nomenclature. Consequently many saponins are named in a trivial manner which serves to specify the botanical origins of the saponins, thus, for example, in addition to soyasaponins one finds reference to phaseollosides A–E (from *Phaseolus vulgaris*), azukisaponins (from azuki beans, *Vigna angularis*), asparasaponins I and II and officinalisnins I and II (from *Asparagus officinalis*), and glycyrrhizin (from liquorice, *Glycyrrhiza glabra*). Whilst such trivial nomenclature has obvious advantages, especially in an area where the location and role of individual saponins is of interest to plant breeders, plant pathologists, food technologists, and nutritionists, it may lead to confusion and has the potential for duplication. This is particularly the case with the

Table 3 *Nomenclature given to the soyasaponins by the research groups of Kitagawa (Osaka) and Okubo (Tohoku)*

Structure	Kitagawa	Okubo
A group—acetylated[a]		
glc-gal-glcUA-A-ara-xyl(2,3,4-triAc)	acetylA4	Aa
glc-gal-glcUA-A-ara-glc(2,3,4,6-tetraAc)	acetylA1	Ab
rham-gal-glcUA-A-ara-glc(2,3,4,6-tetraAc)	—	Ac
glc-ara-glcUA-A-ara-glc(2,3,4,6-tetraAc)	—	Ad
gal-glcUA-A-ara-xyl(2,3,4-triAc)	acetylA5	Ae
gal-glcUA-A-ara-glc(2,3,4,6-tetraAc)	acetylA2	Af
ara-glcUA-A-ara-xyl(2,3,4-triAc)	acetylA6	Ag
ara-glcUA-A-ara-glc(2,3,4,6-tetraAc)	acetylA3	Ah
A group—deacetylated[a]		
glc-gal-glcUA-A-ara-xyl	A4	deacetylAa
glc-gal-glcUA-A-ara-glc	A1	deacetylAb
gal-glcUA-A-ara-xyl	A5	deacetylAe
gal-glcUA-A-ara-glc	A2	deacetylAf
rham-gal-glcUA-A[b]	—	—
ara-glcUA-A-ara-xyl	A6	deacetylAg
ara-glcUA-A-ara-gly	A3	deacetylAh
B group		
glc-gal-glcUA-B	V	Ba
rham-gal-glcUA-B	I	Bb
rham-ara-glcUA-B	II	Bc
gal-glcUA-B	III	Bb
ara-glcUA-B	IV	Bc
E group[c]		
glc-gal-glcUA-E		Bd
rham-gal-glcUA-E		Be

[a] Sugar chains on left of A are linked to 3-O and those on right to 22-O.
[b] Soyasaponin isolated by Curl *et al.* and given the name soyasaponin A_3.
[c] Soyasapogenol E contains a ketone function at C-22.

soyasaponins, which have been extensively investigated by Japanese groups led by Kitagawa[4-6] and Okubo.[7-9] Table 3 shows that the same compounds have been separated and identified by both groups, but that there is no agreement over the nomenclature. In this paper the authors are using the nomenclature of Okubo; it is to be hoped that before long a wider agreement could be reached on nomenclature so that potential confusion in the literature could be minimized.

The structures of typical soyasaponins are shown in Figure 1; these are not only confined to soyabeans but may also be found in other beans and peas, and in alfalfa. Soyasaponins may be subdivided into three groups, (A, B, and E

[4] I. Kitagawa, M. Saito, T. Taniyama, and M. Yoshikawa, *Chem. Pharm. Bull.*, 1985, **33**, 1069.
[5] I. Kitagawa, H. K. Wang, T. Taniyama, and M. Yoshikawa, *Chem. Pharm. Bull.*, 1988, **36**, 153.
[6] T. Taniyama, M. Yoshikawa, and I. Kitagawa, *Yakugaku Zasshi*, 1988, **108**, 562.
[7] M. Shimoyamada, S. Kudo, K. Okubo, F. Yamauchi, and K. Harada, *Agric. Biol. Chem.*, 1990, **54**, 77.
[8] S. Kudou, I. Tsuizaki, T. Uchida, and K. Okubo, *Agric. Biol. Chem.*, 1990, **54**, 1341.
[9] M. Shiraiwa, F. Yamauchi, K. Harada, and K. Okubo, *Agric. Biol. Chem.*, 1990, **54**, 1347.

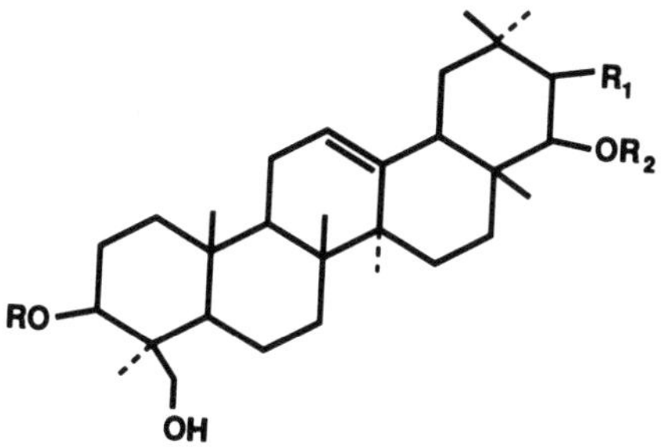

Figure 1 *Structures of group A and B saponins*

R	R_1	R_2
Glc(1→2)gal(1→2)glcUA-	-OH	2, 3, 4-tri Ac.xyl(1→3)ara-
Glc(1→2)gal(1→2)glcUA-	-OH	2, 3, 4, 6-tetra Ac.glc(1→3)ara-
Rham(1→2)gal(1→2)glcUA-	-OH	2, 3, 4, 6-tetra Ac.glc(1→3)ara-
Glc(1→2)ara(1→2)glcUA-	-OH	2, 3, 4, 6-tetra Ac.glc(1→3)ara-
Gal(1→2)glcUA-	-OH	2, 3, 4, 6-tetra Ac.glc(1→3)ara-
Ara(1→2)glcUA-	-OH	2, 3, 4, 6-tetra Ac.glc(1→3)ara-
Ara(1→2)glcUA-	-OH	2, 3, 4-tri Ac.xyl(1→2)ara-
Glc(1→2)gal(1→2)glcUA-	-H	-H
Rham(1→2)gal(1→2)glcUA-	-H	-H
Rham(1→2)ara(1→2)glcUA-	-H	-H
Gal(1→2)glcUA-	-H	-H
Ara(1→2)glcUA-	-H	-H

respectively), depending upon whether the aglycone is olean-12-en-3β,22β,24-tetraol (soyasapogenol A), olean-12-en-3β,24-triol (soyasapogenol B), or olean-12-en-3β,24-diol-22-one (soyasapogenol E). If the former aglycone contains two ether-linked sugar chains (attached to positions 3 and 22) it is termed a *bis-desmoside*; in soyabeans, soyasapogenol B contains sugars attached to position 3 alone and is thus a *mono-desmoside*. Differences in chemical properties between mono- and bis-desmosides may be exploited to facilitate their separation and isolation. Group E saponins (Bd and Bc, Table 3), in the purified form are rather unstable, being readily converted to the group B compounds, Ba and Bb. The former compounds are, however, stable in the plant and in crude plant extracts, even over long periods of storage. The group B and, especially, group A saponins (acetylated and deacetylated) have presented considerable challenge for the natural product chemist.[1] Kitagawa's research group at Osaka were the first to suggest the presence of partially acetylated saponins in soybean seeds; these workers considered, however, that it would not be possible, even using X-ray analysis and NMR to determine the exact position of the acetyl groups. For this reason an initial alkaline treatment was included in their isolation procedures, thereby removing acetyl groupings from the native saponins.

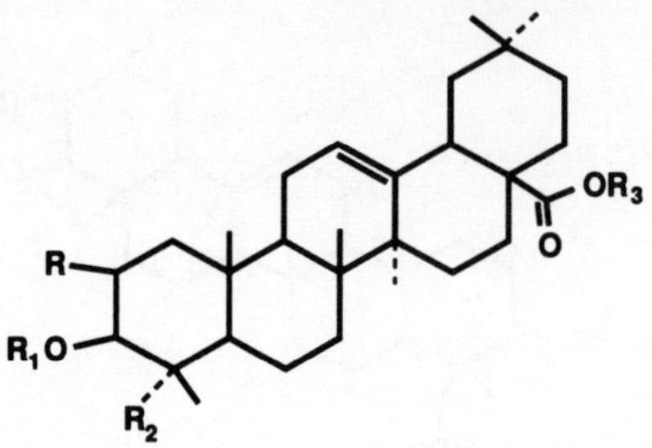

Figure 2 *Structures of typical saponins found in alfalfa*

R	R_1	R_2	R_3
-H	Glc(1→2)ara-	-CH$_2$OH	-H
-H	Ara(1→2)glc(1→2)ara-	-CH$_2$OH	-H
-OH	Glc-	-COOH	Ara(1→2)rham(1→4)-xyl-
-OH	Glc-	-COOH	Glc-
-OH	GlcUA-	-COOH	Ara(1→2-rham(1→4)-xyl-

Independently Okubo and colleagues had obtained evidence linking the occurrence of undesirable bitter and astringent characteristics in soyabeans, its products and extracts to the presence of partially acetylated group A saponins; at that time it was not possible to fully characterize these compounds. The importance of these acetylated species was emphasized when studies showed that soyasaponins A1–6 did not possess bitterness or astringency. It thus appeared that the presence of partially acetylated sugar moieties was a necessary requirement for such sensory characteristics and a programme of structural analysis was embarked upon[10,11] which resulted in the group A bisdesmosidic saponins from American and Japanese soyabeans being shown to contain acetyl groupings on the terminal position of the C-22 oligosaccharide chain (soyasaponins Aa–Ah, Table 3). Most soyabean varieties contain either soyasaponins Aa or Ab and have thus been divided into Aa- and Ab-lines, respectively. Recently the Ab-line soyabeans have been analysed in detail by high performance liquid chromatography and thin layer chromatography. Soyasaponin Ab was the major component, with saponins Ac and Af also being present. There was, however, no evidence for the presence of soyasaponin A1 (the deacetylated product of Ab). Interestingly soyasaponin A2 (the deacetylated product of Af) was detected indicating that the nature and occurrence of saponins in soyabeans is more complex than thought to be the case even a year ago. The most recent development has been the identification of new saponins in *Glycine soja* (Table 1).

Alfalfa (lucerne) is a valuable source of vegetable protein for temperate climates; the aerial parts may be used as a forage crop or extracted to provide leaf protein concentrate, LPC. Soyasaponins are found in the seed of alfalfa but

[10] I. Kitagawa, T. Taniyama, Y. Nagahama, K. Okubo, F. Yamauchi, and M. Yoshikawa, *Chem. Pharm. Bull.*, 1988, **36**, 2819.
[11] T. Taniyama, Y. Nagahama, M. Yoshikawa, and I. Kitagawa, *Chem. Pharm. Bull.*, 1988, **36**, 2829.

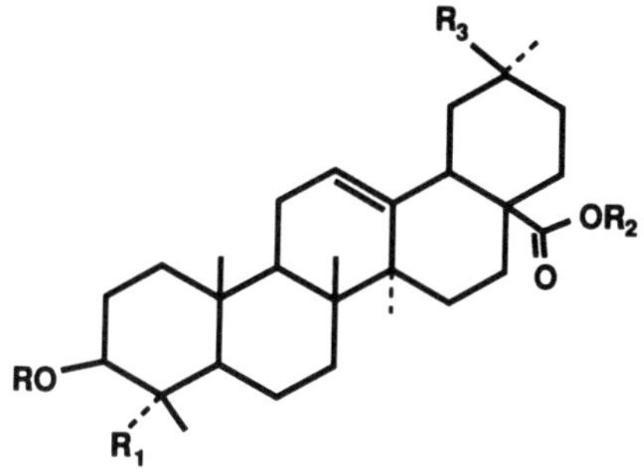

Figure 3 *Structures of typical saponins found in quinoa*

R	R_1	R_2	R_3
Ara-	-CH$_2$OH	Glc-	-CH$_3$
Glc(1→3)ara-	-CH$_2$OH	Glc-	-CH$_3$
Glc(1→3)ara-	-CH$_2$OH	Glc-	-CH$_3$
Ara-	-CH$_2$OH	Glc-	-COOCH$_3$
Glc(1→3)ara-	-CH$_2$OH	Glc-	-COOCH$_3$
Glc(1→3)gal-	-CH$_2$OH	Glc-	-COOCH$_3$
Glc(1→2)glc(1→3)ara-	-CH$_3$	Glc-	-COOCH$_3$
H-	-CH$_3$	-H	-COOCH$_3$
Glc(1→2)glc(1→3)ara-	-CH$_3$	-H	-COCH
Glc(1→2)glc(1→3)ara-	-CH$_3$	Glc-	-CH$_3$
H-	-CH$_3$	-H	-CH$_3$
Glc(1→2)glc(1→3)ara-	-CH$_2$OH	-H	-COOCH$_3$
-H	-CH$_2$OH	-H	-COOCH$_3$
GlcUA-	-CH$_3$	-Glc-	-CH$_3$
GlcUA-	-CH$_2$OH	-Glc-	-CH$_3$
-H	-CH$_2$OH	-H	-CH$_3$
Xyl(1→3)glcUA-	-CH$_3$	-Glc-	-CH$_3$
Xyl(1→3)glcUA-	-CH$_2$OH	-Glc-	-CH$_3$

co-occur with additional saponins in the roots and above-ground parts. These additional saponins contain aglycones possessing carboxylic acid groups, notably hederagenin (olean-12-en-3β,23-diol-28-oic acid) and medicagenic acid (olean-12-en-2β,3β-diol-23,28-dioic acid).[1] Kitagawa *et al.*[12] have identified soyasaponin Be (dehydrosoyasaponin I) in the aerial parts of American alfalfa, together with glycosides of azukisapogenol (olean-12-en-3β,24-diol-20α-carboxylic acid). Unpublished investigations[13] on aerial parts of Polish alfalfa have revealed glycosides of zanhic acid (olean-12-en-2β,3β,16β-triol-23,28-dioic acid). Typical alfalfa saponins are shown in Figure 2, many having been isolated and characterized by Massiot and colleagues[14,15] and by joint work between the Institute of Food Research in the UK and the Institute of Soil Science and Plant

---

[12] I. Kitagawa, T. Taniyama, T. Murakami, M. Yoshihara, and M. Yoshikawa, *Yakugaku Zasshi*, 1988, **108**, 547.
[13] W. Oleszek, K. R. Price, and G. R. Fenwick, manuscript in preparation.
[14] G. Massiot, C. Lavaud, L. LeMen-Olivier, and G. van Binst, *J. Chem. Soc., Perkin Trans.*, 1988, 3071.
[15] G. Massiot, C. Lavaud, D. Guillaume, and L. LeMen-Olivier, *J. Agric. Food Chem.*, 1988, **36**, 902.

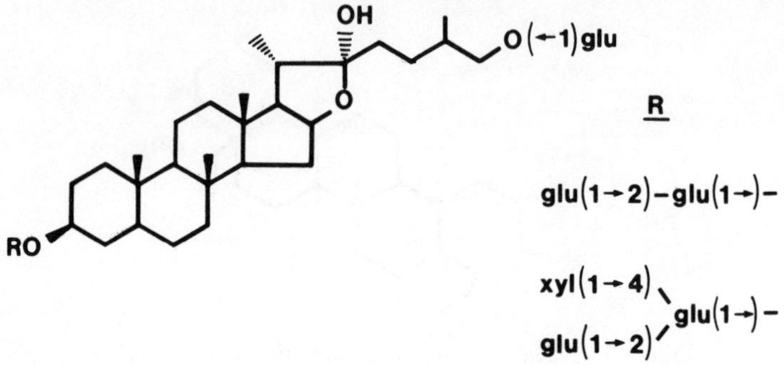

Figure 4 *Structures of asparasaponin I, II found in asparagus*

Cultivation at Pulawy in Poland.[16] The latter workers have also been engaged in detailed studies of the variation of saponin content with *Medicago lupulina*.[17,18]

Quinoa is a native South American crop domesticated by the Incas and which subsequently formed a staple food of this Andean civilization. The outer layers of the grain contain bitter saponins which have to be removed prior to human consumption. In South America this removal is achieved by washing; however, as quinoa is attracting attention in other, less climatically-favourable countries, alternative techniques are being developed, notably abrasion and decortication. Recent investigations have shown the germ of quinoa grain to contain saponins possessing oleanolic acid (olean-12-en-3β-ol-28-oic acid), hederagenin, and phytolaccagenic acid (olean-12-en-3β,23-diol-28-oic-30-oate methyl ester). The structures of the saponins characterized by Mizui *et al.*[19] are shown in Figure 3. Subsequently these workers have separated and identified seven further oleanane saponins from the same source.[20] In parallel investigations, guided by shrimp-brine toxicity screening and bitterness assays, McLaughlin and co-workers[21,22] reported five saponins (quinosides A–E) and characterized two. In contrast to the Japanese workers and others,[23] McLaughlin and co-workers were unable to detect the presence of saponins containing aglycones other than hederagenin and oleanolic acid.

Studies by Kawano *et al.*[24,25] have revealed steroidal saponins to be present in

[16] W. Oleszek, K. R. Price, I. J. Colquhoun, M. Jurzysta, M. Ploszynski, and G. R. Fenwick, *J. Agric. Food Chem.*, 1990, **38**, 1810.
[17] W. Oleszek, M. Jurzysta, P. Gorski, S. Burda, and M. Ploszynski, *Acta Soc. Bot. Pol.*, 1987, **57**, 119.
[18] W. Oleszek, K. R. Price, and G. R. Fenwick, *J. Sci. Food Agric.*, 1988, **43**, 289.
[19] F. Mizui, R. Kasai, K. Ohtani, and O. Tanaka, *Chem. Pharm. Bull.*, 1988, **36**, 1415.
[20] F. Mizui, R. Kasai, K. Ohtani, and O. Tanaka, *Chem. Pharm. Bull.*, 1990, **38**, 375.
[21] W. W. Ma, P. F. Heinstein, and J. L. McLoughlin, *J. Nat. Prod.*, 1989, **52**, 1132.
[22] B. N. Meyer, P. F. Heinstein, M. Burnouf-Radosevich, N. E. Delfel, and J. L. McLaughlin, *J. Agric. Food Chem.*, 1990, **38**, 205.
[23] C. L. Ridout, K. R. Price, M. S. Dupont, M. L. Parker, and G. R. Fenwick, *J. Agric. Food Chem.*, 1991, **54**, 165.
[24] K. Kawano, K. Sakai, H. Sato, and A. Sakamura, *Agric. Biol. Chem.*, 1975, **39**, 1999.
[25] K. Kawano, H. Sato, and A. Sakamura, *Agric. Biol. Chem.*, 1977, **41**, 1.

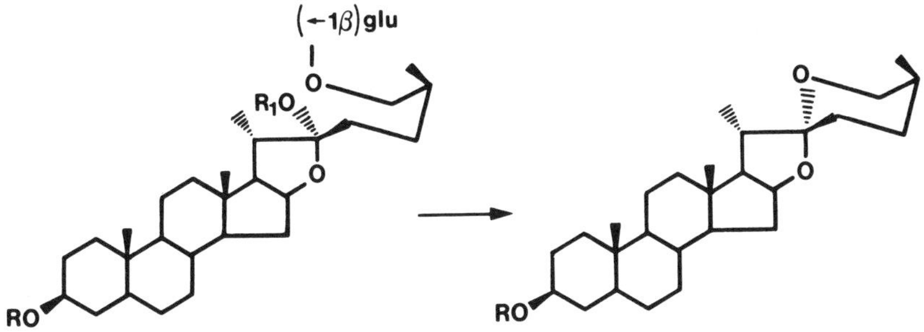

Figure 5 *Interconversion of furostan–spirostan structures*

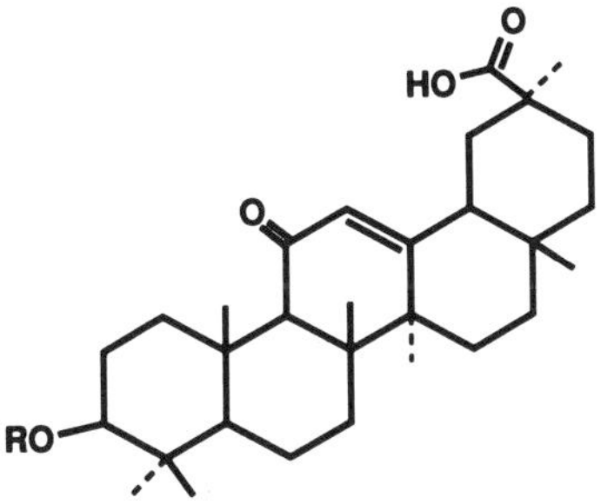

Figure 6 *Structure of glycyrrhetinic acid (upper), glycyrrhizin (lower) from liquorice*

asparagus (*Asparagus officinalis*). Typical of the compounds isolated are officinalisnin I and II, which contain the 5β-furostan-3β,22α,26-triol aglycone, and asparasaponin I and II, containing yamogenin (furost-5-en-3β,22α,26-triol) (Figure 4). Excessive bitterness in white shoots and bottom cuts of asparagus is a particular problem in the processing industry; it is of interest that when the asparasaponins were examined, only the latter compound was found to be bitter. By analogy with other plants, it would be expected that the C-26 glucose moiety of these molecules may be cleaved by enzymic or bacterial action to form spirostan derivatives (Figure 5). A series of additional steroidal saponins (asparagosides A–I) have been separated by Goryanu and co-workers (see reference 1) and the majority have been fully characterized.

Liquorice is one of the most widely investigated of the economically-important medicinal plants.[26] The roots and rhizome of liquorice contain large amounts of a mixture of saponins, primarily glycyrrhizin; this saponin (Figure 6), which contains the aglycone, glycyrrhetinic acid (olean-12-en-3β-ol-11-one-30-oic acid) is unusual in two respects. Firstly, it contains a β-D-glucuronopyranosyl(1→2)β-D-glucuronopyranose moiety and, because of this, has an intensely sweet taste, reportedly 50 times sweeter than sugar. Consequently glycyrrhizin has been used as a natural sweetening agent; selection of raw materials, monitoring of extraction processes, and quality of glycyrrhizin-

[26] G. R. Fenwick, J. Lutomski, and C. Nieman, *Food Chem.*, 1990, **38**, 119.

containing confectionery, chewing tobacco, and health products thus requires a method of quantitative analysis. The second unusual feature of glycyrrhizin, the presence of a conjugated dienone in the aglycone, provides a characteristic chromophore which has been exploited in spectrophotometric and, more recently, HPLC methods of analysis (see below). Recently Kitagawa et al.[27,28] have described five new oleanene saponins from liquorice of Chinese origin and isolated two sweet saponins, structurally related to glycyrrhizin (so-called apioglycyrrhizin and araboglycyrrhizin) from *Glycyrrhiza inflata*. A patent filed by Japanese workers[29] describes the isolation of an analogue of glycyrrhizin (possessing a terminal galacturonic acid moiety) from an unspecified *Glycyrrhiza* source; the compound is claimed to possess three times the sweetness of glycyrrhizin.

3 Methods of Extraction, Separation, and Structural Elucidation

Saponins are generally isolated by extraction with organic solvents in a rather non-specific manner. Lipids and pigments may be removed by exhaustively extracting the dried or freeze-dried powdered plant tissue with acetone or hexane. Subsequent extraction of the residue with methanol removes saponins, together with low molecular weight species such as sugars, phenolics, glycosides, oligosaccharides, and flavonoids. The use of cholesterol, ammonium or sodium salts to fractionate saponins into two classes was once much used,[1] but tends now to be superseded by chemical and chromatographic procedures. Thus separation of mono- and bis-desmosides has been achieved by extraction with methanol and water respectively, or by partitioning between butanol:water mixture. Oleszek[30] has described a simple method for isolation and separation of saponins from alfalfa extracts; the saponins are selectively eluted from a C_{18} solid-phase column with aqueous methanol. Eluants containing 50–60 % methanol removed only medicagenic acid glycosides, with 70–100 % methanol being needed to elute hederagenin and soyasapogenol glycosides. Paper, electrophoretic, ion exchange, and adsorption column chromatographic methods have largely been overtaken by droplet counter current chromatograph, high performance liquid chromatography, and so-called 'flash'[1] chromatography; in the latter cases the use of reversed-phase column packings has proved especially effective.

Droplet counter current chromatography (DCCC)[1] has the advantage that since no solid packing is used, there are no losses due to absorption. The method has been exploited with particular success by Hostettmann and co-workers in Lausanne. The technique is not, however, well suited to saponins

[27] I. Kitagawa, J. L. Zhou, M. Sakagami, T. Taniyama, and M. Yoshikawa, *Chem. Pharm. Bull.*, 1988, **36**, 3710.
[28] I. Kitagawa, M. Sakagami, F. Hashiuchi, J. Zhou, M. Yoshikawa, and J. Ren, *Chem. Pharm. Bull.*, 1989, **37**, 551.
[29] M. Uchida, M. Wada, H. Nakamura, S. Ishihara, N. Mano, and T. Komoda, *Chem. Abstr.*, 289, **110**, 133965e.
[30] W. Oleszek, *J. Sci. Food Agric.*, 1988, **44**, 43.

containing acidic moieties. Centrifugal thin layer chromatography has been used with limited success and it is unlikely that the method will have general applicability until problems associated with the use of reversed-phase supports are overcome.[1] At present, the method appears to be effective for low molecular weight, neutral saponins containing one or two simple sugars. 'Flash' chromatography[31]—a simple and inexpensive hybrid of conventional column chromatography and HPLC—has been found to be an effective means of purifying and separating saponins and other glycosides. As an example, soyasaponin Bb may be isolated from pea seed (*Pisum sativum*) by applying a methanol extract (redissolved in water) to a column of reversed-phase octadecylsilane bonded to silica gel. The column was washed with water (to remove oligosaccharides) and eluted with methanol. Further purification was effected by chloroform:methanol (1:1 by volume) elution of normal phase silica gel.[32]

In the majority of cases, combinations of the above, and other, techniques have proved most effective and it is likely that some 'tailoring' will be necessary to suit individual needs. This is especially so when, as in the case of alfalfa or quinoa, the saponin mixture contains aglycones of diverse chemical structure. The use of mineral acid during saponin isolation should be avoided; not only may such use lead to partial, or complete, hydrolysis but in some cases artefacts have been shown to occur as a result of modification of the aglycone structure.[33,34]

Until recently the structural elucidation of saponins relied on their hydrolysis to yield oligosaccharide(s) and the aglycone, after which the structures of the separate moieties were determined independently (reference 1 for general coverage). To minimize the possibility of artefact formation, a range of hydrolytic chemical, physicochemical, and enzymatic methods have been developed. Particular interest has attached to methods having a particular structural requirement—for example, treatment with acetic acid/pyridine or lead tetraacetate[12] is specific for sugars which contain a glucuronide moiety; if an oleanane aglycone is attached directly to a uronic acid residue efficient cleavage has been achieved by UV irradiation in methanol. The oligosaccharides are identified by conventional methods, the individual sugars being compared with standards and quantified by gas chromatography or high performance liquid chromatography.[35] Positions of linkage are usually identified by gas chromatography following alditol acetate formation or permethylation. The nature of the aglycone–sugar linkage is normally determined enzymatically or by inspection of ^{1}H or ^{13}C NMR spectra. A wide range of chromatography and physicochemical methods has been employed for the structural elucidation of aglycones, these include infrared spectroscopy, ^{1}H and ^{13}C NMR, mass spectrometry, optical rotatory dispersion, circular dichroism, and X-ray crystallography. The assignment of triterpene stereochemistry has been facil-

---

[31] K. R. Price, C. L. Curl, and G. R. Fenwick, *Food Chem.*, 1987, **25**, 145.
[32] K. R. Price and G. R. Fenwick, *J. Sci. Food Agric.*, 1984, **35**, 887.
[33] O. Tanaka and R. Kasai, *Prog. Chem. Org. Nat. Prod.*, 1984, **46**, 1.
[34] K. R. Price, G. R. Fenwick, and M. Jurzysta, *J. Sci. Food Agric.*, 1986, **37**, 1027.
[35] J. Kikuchi, K. Nakamura, O. Nakata, and Y. Morikawa, *J. Chromatogr.*, 1987, **403**, 319.

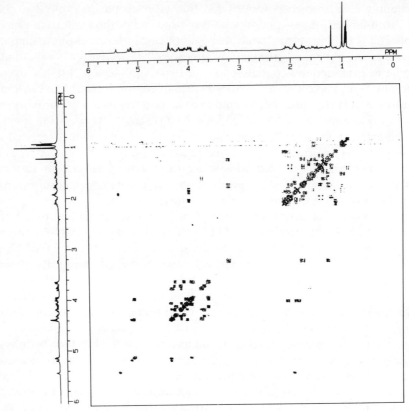

Figure 7 *2D Phase sensitive double quantum filtered COSY spectrum of soyasapogenol B taken at 400 MHz*

itated by lanthanide-induced shift and internuclear double resonance techniques.[36] In recent years double resonance and other sophisticated techniques have been employed for the analysis of saponin structure either directly, *i.e.* without a requirement for prior hydrolysis or derivatization[37] (Figure 7), or following peracetylation which has been shown to reduce problems of peak broadening.[38]

Conventional (EI, CI) mass spectrometry was of limited use for saponin analysis because of its need for sample volatilization. Consequently it was limited to providing supportive information (molecular weight, sugar fragmentation) for low molecular weight saponins which could be effectively converted into the permethyl- or peracetyl-derivatives. In recent years, however, the area has been increasingly dominated by newer ionization techniques (including field-, plasma-, and flash desorption as well as fast atom

[36] R. L. Baxter, K. R. Price, and G. R. Fenwick, *J. Nat. Prod.*, 1990, **53**, 298.
[37] A. Penders, C. Delande, H. Perpermans, and G. van Binst, *Carbohydr. Res.*, 1989, 109.
[38] G. Massiot, C. Levaud, D. Guillaume, L. LeMen-Olivier, and G. van Binst, *Chem. Ind.*, 1986, 1485.

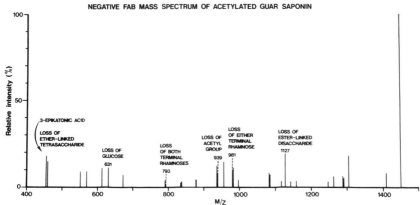

Figure 8 *Fast-atom bombardment mass spectra of acetylated guar saponin (Curl et al., Phytochemistry, 1986, 25, 11, 2675). Spectrum shows the fragmentation of a saponin, MW = 1452, m/z 1451, and the subsequent losses of an ester linked disaccharide m/z 1127, a terminal rhamnose m/z 981, from an ester linked tetrasaccharide, an acetyl group m/z 939, both terminal rhamnoses from the tetrasaccharide m/z 7903, a glucose m/z 631 and finally loss of a glucuronic acid to leave the aglycone, 3-epikatonic acid m/z 455*

bombardment). These enable the direct analysis of intact saponins (Figure 8). Schulten and co-workers[39] have compared field desorption and fast atom bombardment procedures and have emphasized the importance of the interaction of the sample and the glycerol matrix in the latter. The complementary nature of the information which can be obtained from inspection of the positive and negative ion fast atom bombardment mass spectra has been discussed by Fraisse et al.[40] whilst Price and co-workers[16,23,41] have applied such techniques to studies of saponins from legumes, grains, quinoa, and alfalfa. Hattori et al.[42] have utilized high performance liquid chromatography–fast atom bombardment mass spectrometry for the screening of saponins in crude oriental drugs. The method shows clear potential although specific clean-up procedures may be necessary for individual samples.

4 Analytical Methods

The method used for the analysis of saponins will, to a considerable extent, depend upon the type of information required. Thus plant breeders selecting for low saponin alfalfa or quinoa have used screening procedures based upon foam production, haemolytic activity, or inhibition of the growth of *Trichoderma viride*.[1,3] Such methods are calibrated against the effects of crude saponin mixtures or purified material, and as such should be seen as semiquan-

---

[39] H-R. Schulten, S. B. Singh, and R. S. Thakur, *Z. Naturforsch., Teil C*, 1984, **39**, 201.
[40] D. Fraisse, J. C. Tabet, M. Becchi, and J. Raynaud, *Biomed. Environ. Mass Spectrom.*, 1986, **13**, 1.
[41] K. R. Price, J. Eagles, and G. R. Fenwick, *J. Sci. Food Agric.*, 1988, **42**, 183.
[42] M. Hattori, Y. Kawata, N. Kakiuchi, K. Matssura, and T. Namba, *Shoyakugaku Zasshi*, 1988, **42**, 228.

titative methods based upon biological or chemical properties. They provide little, if any, information about the *nature* of the saponins present. Thus whilst chemical analysis and isolation reveal alfalfa seeds to contain soyasaponins, analyses based upon haemolytic or mould inhibition properties indicate the presence of little or no saponin, since soyasaponins are inactive in these assays.[1] The presence of small amounts of individual saponins possessing strong biological activity may also confuse the issue. As has been demonstrated by Nonaka[43] and Oleszek *et al.*[16] responses in particular bioassay systems are very dependent upon chemical structure. In the chronological development of saponin analysis, gravimetry, haemolytic and anti-fungal bioassays, spectrophotometry, and chromatographic techniques have all been employed. Examples, and limitations, of earlier methodologies have been fully described by Price *et al.*[1] and Oakenfull and Sidhu.[3]

From the standpoint of determination of saponins in crops used as animal feedingstuffs and human foods, the work of the late D. E. Fenwick and Oakenfull carried out a decade ago at the CSIRO Food Research Institute in Australia should be seen as particularly significant. This is not because of the reliability of the data—indeed there are excellent reasons for believing that the thin layer chromatographic procedures used introduced considerable overestimates (see below), rather it is because these workers approached the investigation in a systematic manner and demonstrated for the first time the possible extent of saponin occurrence in such plant species. These studies undoubtedly provided a spur to other research groups, who have subsequently extended and improved saponin analysis. The techniques adopted by Fenwick and Oakenfull[44,45] employed thin layer chromatography, the crude saponin-containing extracts being eluted with n-butanol:ethanol:concentrated ammonia (7:2:5). Despite elaborate procedures to identify non-saponin material which might interfere with the quantitation, it is now clear that the levels measured by these workers is considerably in excess of that found by other workers using alternative methods.[1] The latter include thin layer chromatography,[46] gas chromatography,[47,48] and high performance liquid chromatography[49,50] (Table 4). However at the present time there has still been no direct comparison between the methods and the possibility that genetic or environmental factors may be responsible for the discrepancies, although remote, cannot be entirely excluded.

Current attention has focused on the use of gas–liquid chromatography and high performance liquid chromatography. Whilst the former has the limitation that it can only be used for separation and quantitation of the aglycone portion of the saponin (after hydrolysis and suitable derivatization) it has been used for the analysis of saponins in soyabean, pea, quinoa, and liquorice. The former

---

[43] M. Nonaka, *Phytochemistry*, 1986, **25**, 73.
[44] D. E. Fenwick and D. G. Oakenfull, *J. Sci. Food Agric.*, 1981, **32**, 273.
[45] D. E. Fenwick and D. G. Oakenfull, *J. Sci. Food Agric.*, 1983, **34**, 186.
[46] C. L. Curl, K. R. Price, and G. R. Fenwick, *Food Chem.*, 1985, **18**, 241.
[47] I. Kitagawa, M. Yoshikawa, T. Hayashi, and T. Taniyama, *Yakugaku Zasshi*, 1984, **104**, 162.
[48] K. R. Price, C. L. Curl, and G. R. Fenwick, *J. Sci. Food Agric.*, 1986, **37**, 1185.
[49] I. Kitagawa, M. Yoshikawa, T. Hayashi, and T. Taniyama, *Yakugaku Zasshi*, 1984, **104**, 275.
[50] P. A. Ireland, S. Z. Dziedzic, and M. W. Karsley, *J. Sci. Food Agric.*, 1986, **37**, 694.

Table 4 *Saponin contents (%) of soya and its products*

Full fat soyabean	0.22–0.33[a], 0.22–0.35[b], 0.47[c]
Defatted soyabean	0.67[c]
Roasted, defatted soyabean flour	0.67[a]
Protein concentrate	not detected[c]
Miso (fermented bean paste)	0.15[a], 0.21[b]
Tofu (bean curd)	0.30[a], 0.33[b]
Kori-tofu (textured bean curd)	0.33[a], 0.35[b]
Natto (fermented soyabean)	0.25[a], 0.24[b]
Yuba (dried bean curd)	0.41[a], 0.44[b]
Tonyu (soyabean milk)	0.39[a], 0.41[b]

[a] HPLC method: I. Kitagawa *et al.*, *Yakugaku Zasshi*, 1984, **104**, 275.
[b] GC method: I. Kitagawa *et al.*, *Yakugaku Zasshi*, 1984, **104**, 162.
[c] HPLC method: P. A. Ireland *et al.*, *J. Sci. Food Agric.*, 1986, **37**, 694.

study,[47] required the separation and quantitation of soyasapogenols A and B after trimethylsilylation with N,O-bis(trimethylsilyl)trifluoroacetamide. Since in this case (and in the other examples mentioned) the ratio of sapogenol/saponin is known, one can readily convert the results of such analysis to saponin content. Significantly the data agreed very well with independent analysis of the intact saponins using HPLC.[49] Curl *et al.*[46] have, likewise, reported close correlation of saponin levels in the pea, determined by gas liquid chromatography, and a modified thin layer chromatographic procedure. The advent of fast atom bombardment mass spectrometry, and similar newer ionization techniques, has resulted in the molecular weights of many saponins being readily obtainable, even from complex mixtures. The use of FAB–MS and gas chromatography has allowed the levels of saponins containing oleanolic acid, hederagenin, and phytolaccagenic acid to be determined in quinoa.[23] The levels measured, 1.19 and 1.03 % were in general agreement with data in the earlier literature (Table 4); but the earlier data did not, of course, provide any information on the relative proportions of the individual saponin species.

High performance liquid chromatography has been employed for the separation and quantitation of both aglycones and intact saponins. Soyabean sapogenins have been so determined by Ireland and Dziedzic,[51] the authors then assuming a sapogenin/carbohydrate ration of 1:1 in order to calculate *saponin* content (Table 5). The same group has separated intact soyasaponins by HPLC but quantitation was not achieved.[50] Damon *et al.*[52] have employed small gradient changes in eluting solvent (methanol:water, acetonitrile:water) on reversed-phase C_8 or silica columns to separate oleanane saponins, while Kitagawa *et al.*[49] have converted soyasapogenols to their coumarin derivative prior to separation and quantitation on reversed phase C_{18} columns, with fluorescence detection. This technique, although very effective for soyasaponins and, by extension, other molecules possessing ether-linked sugars would not, however, appear to have general applicability. Recently Oleszek *et al.*[53] have separated

[51] P. A. Ireland and S. Z. Dziedzic, *J. Chromatogr.*, 1986, **361**, 410.
[52] B. Damon, A-C. Dorsaz, and K. Hostettman, *J. Chromatogr.*, 1984, **314**, 441.
[53] W. Oleszek, M. Jurzysta, K. R. Price, and G. R. Fenwick, *J. Chromatogr.*, 1990, **519**, 109.

Table 5 *Saponin contents (%) of peas and beans*

Dried pea	0.18[a]
Mung bean	0.05[a]
Runner bean	0.34[a]
Butter bean	0.10[a]
Kidney bean	0.35[a], 0.2[b]
Haricot bean	0.41[a]
Field bean	0.10[a]
Broad bean	0.35[a]
Lentil	0.11[a]
Yellow split pea	0.11[a]
Chickpea	0.23[a]

[a] GLC method: K. R. Price *et al.*, *J. Sci. Food Agric.*, 1986, **37**, 1185.
[b] HPLC method: P. A. Ireland and S. Z. Dziedzic, *Food Chem.*, 1987, **23**, 105.

the hederagenin and medicagenic acid-containing saponins from alfalfa leaf by HPLC, after initial reaction with 4-bromophenacyl bromide; currently the method is being put onto a quantitative basis and comparisons with antifungal assays conducted. The extraction of and isolation of twelve ginseng saponins has been carried out by Choi and co-workers[54] using a combination of preparative HPLC on silica gel carbohydrate and reversed-phase columns, and refractive index detection. Soldati[55] has separated ginsenosides and monitored the eluate at 203 nm, for ginsenosides Rb_1, Rb_2, Rc, Rd, Rf, and Rg_2. A detection limit of 300 ng was claimed. Quality control of ginseng preparations by the use of HPLC has been advocated by Ruckert *et al.*[56]

As mentioned earlier, glycyrrhizin possesses a conjugated dienone moiety and this has been used as the basis of quality control and screening methods.[26] The hydrolysis of glycyrrhizin to glycyrrhetinic acid, derivatization, and estimation by GLC has been used by various workers and such a procedure has been validated by the Association of Official Analytical Chemists, AOAC. Increasingly, however, such methods have been replaced by those based upon HPLC, since this does not require a hydrolytic step and can measure both glycyrrhizin and glycyrrhetinic acid (which may be formed by hydrolysis during processing). One such method[57] has been subjected to collaborative testing, has been included by the AOAC for 'final action', and is currently acceptable to the US Food and Drug Administration in relation to the GRAS status of liquorice products. As an indication of the greater specificity of HPLC, Bell[58] has concluded that such analysis yield figures generally ~ 30 % of those determined by gravimetric methods, which were regarded as standard until relatively recently. The levels of glycyrrhizin in liquorice roots, crude and finished products are shown in Table 8 (see page 305).

[54] J. H. Choi, W-J. Kim, H. W. Bae, S. K. Oh, and H. Oura, *J. Korean Agric. Chem. Soc.*, 1980, **23**, 199.
[55] F. Soldati, Proceedings of the Third International Ginseng Symposium, 1980, p. 59.
[56] K. W. Ruckert, Proceedings of the Third International Ginseng Symposium, 1980, p. 217.
[57] P. S. Vora, *J. Assoc. Off. Anal. Chem.*, 1982, **65**, 572.
[58] J. H. Bell, *Tob. Sci.*, 1980, **24**, 126.

5 Factors Affecting Saponin Content

A number of factors have been shown to affect saponin content, although in many cases the analytical methods employed were crude and further studies are warranted. Thus in addition to plant species, the genetic origin, the part of the plant examined, its physiological age, and state are all important, as are the environmental and agronomic factors associated with growth of the plant, and post harvest treatments, including storage and processing (cooking).[1] Kitagawa and co-workers[47,49] have thus found a variety of soyabean from China to be richer in soyasaponins (~ 0.3 %) than those from Japan, Canada, or the US. In a particularly detailed study, 457 varieties of soyabeans were classified into seven types according to the complement of group A saponins (Table 3, Aa–Af) present in their hypocotyls. Three types were predominant, exhibiting only soyasaponin Aa (Aa-type, 16.6 %), only soyasaponin Ab (Ab-type, 76.1 %), and both soyasaponins Aa and Ab (Aa–Ab type, 5.5 %), respectively. Further study indicated that these two saponins were under the control of co-dominant alleles at a single locus. An investigation or 18 wild relatives of soybean revealed seven in which group A saponins were absent from the hypocotyls—*G. wightii* Soja Perene, *G. latifolia*, *G. tomentella* Mopitou, *G. tomentella* Lindeman, *G. clandestina* P.I.233138, *G. canescens* White clifb, and *G. tabacina* Miyakojimaturu-mame. The lines possessing group A saponins were found to be exclusively either Aa type or Aa–Ab type. This suggests that the cultivated varieties within these two types are perhaps closer to wild relatives and at an initial stage of cultivation. It was also suggested by the authors that the Ab-type of cultivated soybean was more suitable for crossing in the early stage of breeding and that this was the reason for its predominance amongst the plant material examined. Recently the authors have found a mutant in *Glycine soja* which does not contain known group A saponins.[59] This, and other subsequent findings, will undoubtedly assist plant breeders in making selections and crosses with the aim of improving the quality characteristics of soybean by compositional manipulation.

Evidence of variation in saponin content has also been found within lentils, moth bean, alfalfa, and quinoa. Fermentation has been shown to reduce saponin levels in soyabeans[47,49] and alfalfa.[60] Soyasaponins possess undesirable bitterness and astringency characteristics which adversely affect the palatability of pea[61] and soyabean[62] products; according to Iijima *et al.*[62] the undesirable taste characteristics of soyabean saponins decreased significantly after acidic hydrolysis. Kudou *et al.*[8] examined 158 strains of *Aspergillus* (16 species) and selected *A. oryzae* KO-2 as possessing maximal soyasaponin hydrolysing potential. The enzyme was purified, shown to possess K_m and V_{max} of 0.48 mM and

---

[59] C. Tsukamoto, M. Shimoyamada, K. Harada, and K. Okubo, *Abstr. Ann. Meeting Jap. Soc. Biosci., Biotechnol., Agrochem.*, 1991, abstract 2Qa4.
[60] G. Y. Szakacs and E. Madas, *Biotechnol. Bioeng.*, 1979, **21**, 721.
[61] K. R. Price, N. M. Griffiths, C. L. Curl, and G. R. Fenwick, *Food Chem.*, 1985, **17**, 105.
[62] M. Iijima, K. Okubo, F. Yamauchi, H. Hirono, and M. Yoshikoshi, Proceedings of the International Symposium on New Technology in Vegetable Proteins, Oils, and Starch Processing (Part II), Beijing, 1987, pp. 2–109.

0.51 μmol h^{-1}, respectively, for deacetylsoyasaponin Aa, and demonstrated to be induced by triterpenoid saponins containing a β-glucuronide moiety. The enzyme was stable at temperatures below 40 °C and had a pH range of 5–8; a saponin hydrolase isolated from *A. niger* 7122 (the most active of 54 strains of this species examined) had similar properties.[63] These, and other[64] investigations suggest that improvement of the flavours of soyabean foods is possible via enzymic hydrolysis of soyasaponins. Another option which has been adopted commercially for the manufacture of tofu by the Taishi Foods Company involves the removal of the hypocotyl and seed coat, the removal of insoluble material from the soyamilk at temperatures below 50 °C and final defoaming. These three stages remove the acetylated saponin-rich hypocotyls (the main origin of bitter and astringent characteristics), and further reduce the levels of the remaining glycosides (saponins plus isoflavones) which are concentrated in the foam layer.[65,66] What is interesting about this process is that it duplicates what has been traditionally performed domestically by Chinese people living in Malaysia and Indonesia. Studies have shown that tofu prepared in this manner has improved acceptance when compared with products produced conventionally. Cooking and canning have been claimed to have little effect on saponin contents of broad beans and navy beans, but the soaking of moth beans did result in significant losses.[1] Cooking was claimed to have a significant affect on the saponin contents of chickpea and black gram (*Phaseolus mungo*).[67] However, the non-specific nature of the analytical methods employed suggest that these investigations would repay repetition, using more controlled conditions and GC/HPLC analysis.

The washing of quinoa grain has been shown to reduce saponin levels significantly, emphasizing the importance of the practice employed by the Incas[68]; recent investigations have shown the saponins to be concentrated in the outer layers of the grain, and to be significantly reduced by abrasive techniques.[23] Microscopic investigation reveals the presence of fine particles adhering to the rubbed grain by electrostatic attraction, these make a significant contribution to the saponin content of the processed product; this can, however, be further reduced by washing. Efficient abrasive-type dehullers have been described by Reichert *et al.*[69] and applied to the dehulling of a wide range of quinoa cultures. Removal of hulls (2–15 % total weight basis) reduced the saponin levels to those found in sweet quinoa varieties. In general the higher the initial saponin content of the seed the greater the loss of bran required to reduce the concentration in the product to this level.

[63] S. Kudou, T. Uchida, and K. Okubo, *Nippon Shokuhin Kogyo Gakkaishi*, (submitted).
[64] S. Kudou, S. Ojima, K. Okubo, F. Yamauchi, H. Fujinami, and H. Ebine, Proceedings of the International Symposium on New Technology in Vegetable Proteins, Oils, and Starch Processing (Part II), Beijing, 1987, pp. 2–148.
[65] M. Asano, K. Okubo, M. Igarashi, and F. Yamauchi, *Nippon Shokuhin Kogyo Gakkai-Shi*, 1987, **34**, 298.
[66] M. Asano, K. Okubo, and F. Yamauchi, *Nippon Shokuhin Kogyo Gakkai-Shi*, 1989, **36**, 318.
[67] S. Jood, B. M. Chauhan, and A. C. Kapoor, *J. Sci. Food Agric.*, 1986, **37**, 1121.
[68] N. W. Galwey, C. L. A. Leakey, K. R. Price, and G. R. Fenwick, *Hum. Nutr., Food Sci., Nutr.*, 1990, **42F**, 245.
[69] R. D. Reichert, J. T. Tatarynovitch, and R. T. Tyler, *Cereal Chem.*, 1986, **63**, 471.

Sprouting, once again, is claimed to have varying results on saponin content. Whilst saponins are found in both sprouted alfalfa and mung bean, they are reduced when moth bean is germinated.[1] According to Livingstone et al.[70] sprouted alfalfa contains 7 % saponins; these workers however used an antifungal bioassay and subsequent chemical/chromatographic investigations have not confirmed these high levels. Light was found[71] to have a profound effect on the saponin content of germinated soya bean; saponin levels increased under light irradiated germination, but decreased when light was excluded. Gorski et al.[72] have investigated the effect of plant age and plant part on saponin content of *Medicago lupulina*; saponins in seeds germinated for 6 days had increased significantly, whilst seeds and leaves were found to contain higher levels than stems and flowers. The amount of glycyrrhizin in roots and underground stems of liquorice is markedly affected by variety, source, and climatic conditions.[26] Függersberger-Heinz and Franz[73] have shown the concentrations to be higher in the main roots than in the lateral stems. These authors also found no detectable glycyrrhizin in the green parts of the plant. The economic importance of ginseng has meant that numerous studies have been conducted on the distribution of ginsenosides in the developing and mature plant.[1] Koizumi et al.[74] found lowest levels in the stalk and stem, intermediate levels in the main and lateral roots, with highest concentrations appearing in the root hairs and leaves. The effect of processing of ginseng root to produce, for example, red and white ginseng preparations has been studied.[75] Significant differences were noted as a result of hydrolysis of malonylginsenosides, deglycosylation, and C-20 isomerization.

Detailed histochemical investigations have revealed ginsenosides to be localized in the periderm and cortex rather than root xylem and pith.[76,77] In *Bupleurum falcatum* root, used as an oriental drug, Tani et al.[78] have determined saikosaponins a, c, and d to be localized in the outermost peripheral layer (comprising cork layer and pericycle); the physiological significance of this finding and also that silica, aluminium, and manganese also occur exclusively in this tissue (which contains minimal amounts of phosphorus) has yet to be determined. The majority of soyasaponins in soyabean seed are located in the plumule, hypocotyl and radicle,[79] those parts which develop into the mature plant. According to Taniyama et al.[6] soybeans from Japan, N. America, and China contained seed coats which were free of saponins whilst the cotyledons

[70] A. L. Livingstone, B. E. Knuckles, R. H. Edwards, D. deFremery, R. E. Miller, and G. O. Kohler, *J. Agric. Food Chem.*, 1979, **27**, 362.
[71] M. Shimoyamada and K. Okubo, *Agric. Biol. Chem.*, 1991, **55**, 577.
[72] P. M. Gorski, M. Jurzysta, S. Burda, W. A. Oleszek, and M. Ploczynski, *Acta Soc. Bot. Pol.*, 1984, **53**, 543.
[73] R. Függersberger-Heinz and G. Franz, *Planta Med.*, 1984, **50**, 409.
[74] H. Koizumi, S. Sanada, T. Ida, and J. Shoji, *Chem. Pharm. Bull.*, 1982, **30**, 2293.
[75] I. Kitagawa, T. Taniyama, H. Shibuya, T. Noda, and M. Yoshikawa, *Yakugaku Zasshi*, 1987, **107**, 495.
[76] M. Kubo, T. Tani, T. Katsuki, K. Ishizaki, and S. Arichi, *J. Nat. Prod.*, 1980, **43**, 278.
[77] T. Tani, M. Kubo, T. Katsuki, M. Higashino, T. Hayashi, and S. Arichi, *J. Nat. Prod.*, 1981, **44**, 401.
[78] T. Tani, T. Katsuki, Y. Okazaki, and S. Arichi, *Chem. Pharm. Bull.*, 1987, **35**, 3323.
[79] T. Tani, T. Katsuki, M. Kubo, M. Arichi, and I. Kitagawa, *Chem. Pharm. Bull.*, 1985, **33**, 3829.

contained lower levels of group A and B saponins than the hypocotyl (A: 0.07–0.09 % compared with 1.25–1.46 %, B: 0.14–0.18 % compared with 0.42–0.52 %). In a recent study Shimoyamada et al.[7] confirmed these findings and found aerial parts of the plant to be richer in saponins than seed hypocotyl > stem, branch, and petiole. In the underground parts the soyasaponin Bb level decreased in the order root hairs > lateral > main root. Preliminary studies have revealed that in quinoa the oleanolic acid-hederagenin- and phytolaccagenic acid-containing saponins are, at least partially, located at differential cellular sites.[23] The relevance of these findings for plant growth and protection has yet to be assessed.

6 Levels

Given the comments made above on the reliability and specificity of many of the earlier methods of quantitative analysis it will come as no surprise that many of the figures for 'saponin content' quoted in the literature are now regarded as wildly inaccurate.[1,3] Thus for example whilst Appelbaum et al.[80] reported figures ranging from 1.6 to 6.6 % for a variety of legumes as a result of gravimetric analyses, Livingstone et al.[70] described saponin levels in a number of peas and beans to be between 0.01 (detection limit) and 0.05 %, based upon the *Trichoderma* screening assay. If a reliable estimate of saponin content is to be made, an appropriate analytical method will, to a considerable degree, depend upon the chemical and biological properties of the saponin(s) present. The saponin contents of soyabeans, and its products, peas and beans, alfalfa fractions, quinoa and liquorice products are summarized in Tables 4–8. The figures quoted are intended to be indicative of the general levels, rather than being an exhaustive compilation of the published data. Such data will be found in references 1, 3, 23, 26, and 81.

On the basis of an analysis of various edible beans Price et al.[1] concluded that soyabean contained the highest levels, although detailed examination of the effects of genetic and agronomic variation had, at that time, not been conducted. Haricot, runner, and kidney beans appeared, after soyabeans, to contain most saponins, and although other grain legume have also been reported to be rich in saponins these claims must now be treated with scepticism.[1,3]

Alfalfa possesses saponin levels of < 0.3 % (for so-called sweet varieties) a much higher (typically > 1.5 %) in the case of the resistant, high saponin lines. The varieties Buffalo, Ranger, Lahontan, Vernal, and DuPuits were studied at eight different sites within the US. Saponin contents ranged from 2 % (Lahontan) to above 3 % (DuPuits), and those two varieties have been used subsequently in a number of feeding trials designed to establish the nature and extent of the anti-nutritional effects of alfalfa saponins[1] (see elsewhere). Other results showed that the first cutting on average exhibited a lower saponin content than the second or third. Over three successive harvests, saponin content was positively correlated with protein, ash, fat, and nitrogen-free extract and negatively correlated with crude fibre and hay yield. The average

[80] S. W. Appelbaum, S. Marco, and Y. Birk, *J. Agric. Food Chem.*, 1968, **17**, 618.

Table 6 Saponin content (%) of alfalfa fractions[a]

Products		
Whole alfalfa	0.14	1.71
Pressed alfalfa	0.13 (80)[c]	0.66 (76.3)[c]
Whole LCP[b] (Pro-Xan)	0.10	1.90
Alfalfa solubles[c]	0.06	0.65
Green-LPC (Pro-Xan II)[d]	0.08 (8.7)	1.54 (11)
White-LPC[c]	0.21 (1.5)	2.88 (1.6)
Alfalfa solubles[c]	0.06 (9.5)	0.47 (11.3)
White-LPC[d]	0.12	0.97 (1.9)
Alfalfa solubles[d]	0.20	1.17 (10.5)

[a] *Trichoderma* bioassay: A. L. Livingstone et al., in 'Nutritional and Toxicological Aspects of Food Safety', ed. M. Friedman, Plenum Press, New York, 1985, p. 253.
[b] Leaf protein concentrate.
[c] From coagulation and washing at pH 6.0.
[d] Coagulation and washing at pH 8.5.
[e] Dry matter, % yield of original alfalfa.

Table 7 Saponins content (%) of quinoa grain

Accessions	
17 cvs from Peru–Bolivia	0.14–0.73[a]
Real	0.44*[b]
Kancolla	2.1[c]
Kancolla	2.3[d]
UK-grown selections of lowland Chilean ecotypes	1.0, 1.2[e]

[a] Haemolytic assay: R. D. Reichert et al., *Cereal Chem.*, 1986, **63**, 471.
[b] HPLC method: M. Burnouf-Radosevich and N. E. Delfel, *J. Chrom.*, 1984, **292**, 403 (* expressed as aglycone).
[c] GC method: C. C. P. Elias and L. V. Diaz, *Arch. Latinoam. Nutr.*, 1988, **37**, 113.
[d] Spectrophotometry—GC method: C. C. P. Elias and L. V. Diaz, *Arch. Latinoam. Nutr.*, 1988, **37**, 113.
[e] G. C. method: C. L. Ridout et al., *J. Sci. Food Agric.*, 1991, **54**, 165.

Table 8 Glycyrrhizin content (mg g^{-1}) of liquorice products

UK confectionery	0.4–7.9[a]
US confectionery	0.1–1.7[b]
Belgian confectionery	> 2.2[c]
UK 'health' products	0.3–47.1[a]
US chewing tobacco	1.5–4.1[d]
Liquorice root	22.2–32.3[a]
Liquorice block	44.4–98.2[a]
Liquorice powder	79.4–112.8[a], 29–87[d]

[a] HPLC method: E. A. Spinks and G. R. Fenwick, *Food Add. Contam.*, 1990, **7**, 769.
[b] HPLC method: W. J. Hurst et al., *J. Agric. Food Chem.*, 1983, **31**, 387.
[c] HPLC method: B. Hermesse et al., *Arch. Belg. Med. Soc., Hyg., Med. Trav. Med. Leg.*, 1986, **44**, 60.
[d] J. H. Bell, *Tob. Sci.*, 1980, **24**, 126.

saponin contents of leaves were found to be twice that in stems, and a significant reduction occurred in the saponin contents of older plants.

Livingstone et al.[70] have reported the (medicagenic acid and hederagenin) saponin content of varieties of alfalfa grown commercially in the US. Levels ranged from 0.1–0.2 % (Lahontan, Ranger-low saponin) to 1.7 % (Ranger-high saponin). Despite the downward trend in saponin contents, the authors emphasized the need for saponin analysis to be conducted on alfalfa material intended for processing, with a view to using only those possessing low levels. Concern over the presence of anti-nutrients in alfalfa leaf proteins led these workers to investigate the effect of the nature of the raw material and the processing conditions on the saponin content in the final product. Saponins were determined by the *Trichoderma* bioassay and are thus indicative only of those containing medicagenic acid and hederagenin aglycones. Two processes were investigated: the Pro-Xan process, in which soluble and insoluble proteins are precipitated together, and the Pro-Xan II process, in which a chlorophyll-pigmented (green-LPC) and a white (white-LPC) concentrate are produced separately. The saponin contents of the alfalfa products prepared from high- and low-saponin cultivars are shown in Table 6. For fractions coagulated and washed at pH 6.0, the saponin contents (for both the high- and low-saponin alfalfa varieties) of the individual fractions were in the order white-LPC > whole-LPC > green-LPC. White-LPC, prepared by coagulation and washing at pH 8.5, contained lower levels of saponins, a finding that is consistent with the acidic character of the saponins being analysed. Such alkaline treatment also favours the removal of phenolic toxicants, such as isoflavone and coumestrol oestrogens.

Until very recently the limitations of the analytical methods employed meant that saponin data for quinoa could only be interpreted in a semi-quantitative manner. It is, however, clear that 'bitter', 'semi-sweet', and 'sweet' varieties occur and exhibit a significant gradation in total saponin content (Table 7). When sweet, saponin-free, plants were collected in Bolivia and crossed with bitter, saponin-rich, lines a 3:1 ratio of bitter to sweet character was found in the F_2 generation,[68] implying that bitterness (and hence, high saponin content) was determined by a single dominant gene. The occurrence of semi-sweet varieties of quinoa however, suggests that the amount of saponins is polygenically controlled.

Amongst thirteen samples of beer examined by Hermesse et al.[82] ten were found to contain less than 0.6 p.p.m. glycyrrhizin, the remainder containing below 6 p.p.m. A range of non-alcoholic drinks contained between 214–812 p.p.m. whilst alcohol drinks were generally rather lower (74 p.p.m.). In a recent survey of Japanese foods,[83] glycyrrhizin was measured in sixteen out of fifty-six products, but with the exception of takuan-zuke (a pickle), smoked squid, and saki-ika (a fish product), containing 52–224, 135, and 110 p.p.m., respectively, the amounts found were very small. Amongst the liquorice confec-

---

[81] E. A. Spinks and G. R. Fenwick, *Food Add. Contam.*, 1990, **7**, 769.
[82] B. Hermesse, A. Collinge, and A. Noirfalise, *Arch. Belg. Med. Soc., Hyg., Med. Trav. Med. Leg.*, 1986, **44**, 60.

tionery examined by Spinks and Fenwick,[81] (Table 8) liquorice allsorts, which comprise almost half the current UK liquorice confectionery sector (1988 value, £49 million) were found to contain less than 1 mg g^{-1} (1000 p.p.m.).

7 Metabolism and Human Exposure

As with many other classes of secondary metabolites, detailed information on the fate of saponins within the gut of mono-gastric animals is generally lacking. Gestetner et al.[84] fed soyabean meal to chicks, rats, and mice and observed that whilst saponins were present in the small intestine only sapogenols could be found in the caecum and large intestine. In vitro incubation with slices of small intestine from these species was without effect on soyasaponins whereas breakdown was observed when these components were incubated with caecal and large intestinal tissue. Other studies have been centred on material used in oriental drug preparations; thus, amongst others, the metabolism of saikosaponins from Bupleuri radix in the rat has been investigated by Shimizu et al.[85] and Odani et al.[86] have studied the fate of ginsenosides Rg_1 and Rb_1 in the same species. When administered orally Rg_1 was broken down in the stomach and small intestine to six products, identical to the behaviour of the compound with weak acid. Glycyrrhizin was converted by human intestinal flora (Ruminococcus and Clostridium spp.) to the aglycone, 18β-glycyrrhetinic acid, which was subsequently transformed to the 3-epimer via the 3,9-dione, 3-dehydroglycyrrhetinic acid.[87] Only a trace (1–2 %) of glycyrrhizin administered orally or by intravenous injection was excreted intact or as the aglycone.[88,89] The appearance of the latter was rather more rapid following oral administration; this finding is consistent with the hydrolysis of the saponin via gastric juice; intestinal bacteria, and intestinal enzymes. The aglycone is able to react readily with serum albumin and is hence rapidly excreted. Following intravenous administration of glycyrrhizin to the rat most of the saponin was found in the plasma and blood after 6 minutes; no glycyrrhizin could be detected in the brain whilst the lung, heart, stomach, and small intestine all contained intermediate amounts.[90] The highest concentration of glycyrrhetinic acid was found in the plasma and blood, smaller amounts being present in the brain and other organs. Kanaoka et al.[91] have isolated the monoglucuronide of glycyrrhetinic acid from the serum of a patient who had been treated with glycyrrhizin (200 mg daily by intravenous infusion for 7 weeks) for liver disease. Nearing the

---

[83] K. Fujinuma, K. Saito, M. Nakazata, Y. Kikuchi, A. Abe, and T. Nishima, Chem. Abstr., 1988, **109**, 228867t.
[84] B. Gestetner, T. Birk, and T. Tencer, J. Agric. Food Chem., 1968, **16**, 1031.
[85] K. Shimizu, S. Amagaya, and Y. Ogihara, J. Pharmacol., 1985, **8**, 718.
[86] T. Odani, H. Tanizawa, and Y. Takino, Chem. Pharm. Bull., 1983, **31**, 1059.
[87] M. Hattori, T. Sakamoto, T. Yamagishi, K. Sakamoto, K. Konishi, K. Hobashi, and T. Namba, Chem. Pharm. Bull., 1985, **33**, 210.
[88] N. Nakano, H. Kato, H. Suzuki, K. Nakao, S. Yano, and M. Kanaoka, Proc. Symp. WAKAN-YAKU, 1981, **14**, 97.
[89] K. Terasawa, M. Bandoh, H. Tosa, and J. Hirate, J. Pharmacobio-Dyn., 1986, **9**, 95.
[90] T. Ichikawa, S. Ishida, Y. Sakiya, and Y. Akada, Chem. Pharm. Bull., 1984, **32**, 3734.
[91] M. Kanaoka, S. Yano, H. Kato, and T. Nakada, Chem. Pharm. Bull., 1986, **34**, 4978.

Table 9 *Calculated UK mean daily intakes* (mg person^{-1}) *of saponins*

Sub-group	
Family	15
Children	13
Vegetarian	110
Males only:	
Caucasian	10
W. Indian	47
Asian	170
Asian (vegetarian)	214

[a] C. L. Ridout et al., Hum. Nutr. Food Sci. Nutr., 1988, **42F**, 111.

end of the treatment the patient, who was diagnosed to be mildly hypertensive, exhibited serum levels of glycyrrhetinic acid, its monoglucuronide, and glycyrrhizin of 0.6, 11.5, and 64 μg ml^{-1}, respectively.

Information on saponin intakes is very limited. Ridout et al.[92] have calculated mean daily intakes on the basis of limited screening of food products and information on the UK total diet. These intakes (shown in Table 9) are thus only approximate; however, they do serve to illustrate the large variation due to the composition of the diet and to the levels of saponin-containing foods actually consumed. Whereas the mean daily intake of saponins by vegetarians of East African origin was predominantly due to consumption of kidney beans and guar beans, the UK vegetarians were primarily exposed to saponins via soyabean products, lentil, and the ubiquitous 'baked bean'. This dietary difference is also responsible for significant differences in the chemical structures (and therefore biological activities) of the ingested saponins. Most of the plant foods consumed by Caucasian and West Indian groups contain primarily soyasaponins, the high consumption of guar (cluster) beans within Asian communities means that almost a third of their intake will be of non-soyasaponin components which—in very general terms—are likely to be more biologically active.

The significant biological activity of glycyrrhizin has resulted in several calculations of mean daily exposure, via liquorice-containing beverages, herbal products, and, primarily, confectionery. Spinks and Fenwick[81] quote mean intakes of 3 and 5 mg glycyrrhizin person^{-1} day^{-1} for the US and Belgium, respectively and have calculated a figure of approximately 2 mg p^{-1} for the UK. Kaas Ibsen[93] has reported a figure of 5–10 mg p^{-1} for Denmark. However given the variation in liquorice content in confectionery, and other, products it is certain that many individuals have a much higher—and continuous—exposure.

8 Role in Plants

Most crop species contain lines that are resistant to existing fungal, bacterial, and viral diseases. The plant breeder, however, encounters a number of problems in combining such resistance into commercial varieties. Resistance genes

[92] C. L. Ridout, S. G. Wharf, K. R. Price, I. T. Johnson, and G. R. Fenwick, *Hum. Nutr. Food Sci. Nutr.*, 1988, **42F**, 111.
[93] K. Kaas Ibsen, *Dan. Med. Bull.*, 1981, **28**, 124.

are often found in wild progenitors or primitive cultivars which lack the agronomic requirements of high-yielding varieties. Incorporating genes from such backgrounds requires backcrossing and selection programmes which are time consuming and hence costly. In addition, the process would be magnified by the number of resistance genes required for incorporation. This number may be high, because pathogenic organisms are themselves variable and often contain different races/pathotypes. Many of the problems associated with breeding for disease resistance could be overcome if there was a better understanding of the mechanism responsible for disease and for resistance to disease. It is anticipated that progress in understanding these processes will be accelerated through the application of modern molecular genetic techniques. In general, these techniques depend on the ability to excise a gene from an organism, manipulate it in some way, and reintroduce it into the genome by transformation. In this way, the structure expression and function of genes can be analysed.

Although some secondary metabolites are often associated with disease resistance, to date no product of any resistance gene has been isolated. Therefore, resistance genes are still defined only in terms of their phenotypic effect and by means of genetic analysis. It is obviously important to establish that the synthesis of particular compounds in response to infection, following the incorporation of a resistance gene, is essential for conferring disease resistance and not due to genetic linkage or gene interaction. It may also be possible to engineer novel forms of resistance similar to that reported for plant viruses. In this case, protection against the viral disease can be engineered by introducing, via a transform mechanism, a stable 'interfering' molecule, which then effectively 'immunizes' the plant against a narrow range of closely related viruses. Knowledge of the products of resistance genes may help to explain how resistance is conferred and perhaps allow genes to be modified to confer more widespread resistance, but it is also essential in order to identify any possible adverse consequences of introducing new and possibly toxic natural chemicals into the food chain.

The molecular basis of disease resistance is, of course, highly complex. Many plants intended for human consumption (including legumes) contain secondary metabolites, which offer protection against fungal infection and pest infestation, or deter herbivores. Other species, when exposed to micro-organisms, nematodes, or insects, synthesize and accumulate biologically active, low molecular weight compounds (phytoalexins). In many cases, plant protection is a consequence of the presence of both pre-formed secondary metabolites and phytoalexins. Thus, such compounds can contribute to the harvest of healthy plants for human and animal consumption.

Whilst it is evident (see below) that saponin-containing plants, crude extracts, and isolated, purified saponins have wide-ranging biological activities, the exact biological and physiological role of these secondary metabolites in plants remains obscure. The interaction between saponin-containing plants and fungi, micro-organisms, and insects has been discussed by Appelbaum and Birk,[94]

---

[94] S. W. Applebaum and Y. Birk, in 'Herbivores, their Interaction with Secondary Plant Metabolites', ed. G. A. Rosenthal and D. H. Janzen, Academic Press, New York, 1979, p. 539.

Birk and Peri,[95] Fenwick et al.[1] and Oakenfull and Sidhu.[3] The first two reviews in particular were focused on the detailed structure/function studies carried out on legume saponins, especially those from alfalfa, by Israeli workers who were particularly active in the period 1960 to 1980. In retrospect, their efforts may be seen as particularly noteworthy given the absence of reliable methods for separating, quantifying, and isolating purified saponins. However, as Appelbaum and Birk[94] concluded in 1980, if the role of saponins in resistance is to be fully resolved, there is a pressing need for studies which measure the nature and amount of the individual saponins present in the plant tissue eaten and which determine the response of the pest to changes in this composition and content. It is, for example, crucial to distinguish between a direct effect of the saponins on the pests, and an indirect effect based upon taste distortion (as has been claimed for the gymnemic acid saponins[96]) or antifeedant activity (for example reference 97).

As a result of detailed studies,[7,98] it has been suggested that soyasaponin Bb is associated with plant protection and disease resistance because of its distribution in hypocotyl, nodule, leaf, and root hairs. The use of mutant species, containing well defined saponin compositions, will certainly facilitate an understanding of the role of such compounds in soyabean growth and development. Amongst the findings reported by Appelbaum and Birk[94] and Birk and Peri[95] were several which suggested that 'alfalfa' saponins were more active than 'soyabean' saponins against fungi and insects. This finding is in agreement with the general (but by no means total) observation[1,3] that biological activity decreases with aglycone structure in the order medicagenic acid > hederagenin > soyasapogenol (A or B). There is a general consensus[68] that increased saponin content in quinoa is associated with increased resistance to insect and bird damage, and visual observation of selections of UK grown quinoa supports this, but the exact nature of the resistance is as yet unknown. Amongst recent reports of the effects of saponins on insects, Harmantha et al.[99] have described the inhibition of growth of leek-moth (*Acroleprosis assectella* Z.) larvae by the steroidal saponin, aginosid, found in leek flowers. As has been found to be the case elsewhere,[94] the toxicity of this compound is mediated by the addition of cholesterol or sitosterol. Oleszek et al.[17] and Oleszek and Jurzysta[100] have examined the allelopathic effect of *Medicago lupulina*, *M. media* (alfalfa), and red clover; the effect of the latter was much lower than that of alfalfa, a findling attributed to the difference in saponin content. Soil textures had a significant influence on the inhibitory effect of alfalfa root towards winter wheat seedling growth, with plants grown on light soils showing significantly greater effects—presumably a reflection of the greater sorption of the inhibitors on the heavy

[95] Y. Birk and I. Peri, in 'Toxic Constituents of Plant Foodstuffs', 2nd ed., ed. I. E. Liener, Academic Press, New York, 1980, p. 161.
[96] T. Eisner and B. P. Halpern, *Science*, 1971, **172**, 1362.
[97] E. Horber, in 'Insect and Mite Nutrition', ed. J. G. Rodriquez, North-Holland, Amsterdam, 1972, p. 611.
[98] M. Shimoyamada, Doctoral thesis, Faculty of Agriculture, Tohoku Univ., Japan, 1991.
[99] J. Harmatha, B. Mauchamp, C. Arnault, and K. Slama, *Biochem. System. Ecol.*, 1987, **15**, 113.
[100] W. Oleszek and M. Jurzysta, *Plant Soil*, 1987, **98**, 67.

soils and their readier breakdown and detoxification by soil micro-organisms. Tarikov et al.[101] have reported germination of cotton and growth of wheat coleoptiles to be much affected by the concentration of the alfalfa root saponins present.

According to recent studies[102,103] the inhibitory activity of alfalfa saponins towards *Sclerotium rolfsii*, a plant pathogenic fungus, was dependent upon the presence of a potentially free C-3 hydroxyl group in the saponin aglycone. The most active compound was the 3-*O*-β-D-glucopyranoside of medicagenic acid, presumably because of its ready transport to sites of action, its improved incorporation into fungal membranes, and its potential breakdown by fungal β-glucosidase to produce the fungitoxic aglycone.

Crombie and co-workers[104] have carried out elegant chemical studies on the avenacins, novel saponins present in oat roots, and their rearranged aglycone products, the avenestergenins. The 'take-all' fungus of wheat, which rejoices in the name *Gaeumannomyces graminis* var. *tritici* does not attack oat roots in contrast to the var. *avenae*. It has been shown that whereas the former is unable to detoxify the fungicidal avenacins, the latter can degrade the branched trisaccharide chain of avenacin A-1 (the most active of these saponins) to yield the less toxic and less soluble mono- and bis-deglucoavenacin A-1. In this way the presence of these sugar-degrading enzymes in var. *avenae* is considered fundamental to its ability to attack the roots of both oats and wheat. Finally although saponins occur overwhelmingly in plants, they are found in starfish such as *Acanthaster planci* and *Patiriella calcar*. Lucas et al.[105] have found evidence that saponins in the eggs and larvae of *A. planci* serve as chemical defences against planktivorous fish.

Selective breeding for disease resistance undoubtedly offers great potential benefits, both for agriculture and for the environment. Nevertheless, such strategies should not be undertaken without a proper awareness of the biochemical mechanisms that underlie the practical results. Such biochemical insight should be combined with toxicological studies designed to achieve an understanding of the effects of the relevant 'resistance factors' in mammalian metabolism. Since conventional toxicological methods are probably inadequate for the assessment of risks due to chronic low level exposure, such studies should preferably be undertaken at a cellular level and will probably require the establishment of appropriate multidisciplinary centres or networks of scientific co-operation. It is hoped that, given the recent advances in methodologies for saponin isolation and analysis, this area of research at the interface of the chemical/plant sciences will develop with renewed impetus.

[101] S. Tarikov, A. E. Timbekova, N. K. Abubakirov, and R. K. Koblov, *Chem. Abstr.*, 1989, **110**, 209429f.
[102] M. Levy, U. Zehavi, M. Naim, I. Polachek, and R. Evron, *J. Phytopath.*, 1989, **125**, 209.
[103] M. Levy, U. Zahavi, and M. Naim, *Carbohydr. Res.*, 1989, **193**, 15.
[104] L. Crombie, W. M. L. Crombie, and D. A. Whiting, *J. Chem. Soc., Perkin Trans. 1*, 1986, 1917.
[105] V. S. Lucas, R. J. Hart, M. E. H. Howden, and R. Salathe, *J. Exp. Mar. Biol. Ecol.*, 1979, **40**, 155.

9 Physical Properties

The potential for the use of certain saponins as 'natural additives' is considerable given their physical and physicochemical properties and the prevailing 'green' attitudes amongst consumers which favour the use of such products rather than 'synthetic additives' in food. Saponins have been shown to be excellent emulsifiers,[3] and Gohtani and co-workers[106-110] have reported in detail on the foaming power, emulsification properties, and surface activity of commercial soyasaponin and soyasaponins Ab, deacetyl Ab and Bb. The foamability, foam stability, and creaming ability of soyasaponin Ab were all greater than that of its deacetylated analogue. The bark of the soaptree, *Quillaja saponaria*, contains approximately 10 % saponin and has been used as a foaming agent in confectionery, baked goods, and beverages; quillaja saponins have been classified as generally recognized as safe (GRAS) for permitted food use in the US, as have saponins from *Yucca mohavensis*. *Saponaria officinalis* saponins are traditionally used in Greece and the Near East as emulsifiers for the preparation of halweh, a confectionery product.

The self-aggregation of saponins has been described by Oakenfull and Sidhu[3]; the extent of aggregation depends upon the saponin structure. Whereas soyasaponins yield only dimeric structures, those from *Gypsophila* and *Saponaria* form larger aggregate, perhaps ten and fifty molecules, respectively. The formation of mixed micelles between saponins and bile acids is a property having far reaching consequences; this mechanism having been proposed for the anti-fungal properties of many saponins, and for many of the effects observed in insects, animals, and man. Whereas the size of self-aggregating saponin polymers is limited by the size of the bulky oligosaccharide moieties, and those of bile salts alone are limited by electrostatic repulsions of the acid groups, in the mixed micelles these destabilizing factors are relieved and greatly extended stacks, with saponin and bile acid molecules alternating, are produced.[3] Oakenfull[111] has examined the sizes and structures of these micelles in detail; difference in size and shape are observed and current opinion suggests the micelles contain a loose internal structure with the sugar moieties being both on the surface and within the interior. There is currently considerable interest in such structures as adjuvants in the veterinary and pharmacological areas (see below).

---

[106] S. Gohtani, Y. Ohsuka, and Y. Yamano, *Nippon Nogei Kagaku Kaishi*, 1985, **59**, 25.
[107] S. Gohtani and Y. Yamano, *Nippon Nogei Kagaku Kaishi*, 1985, **57**, 153.
[108] S. Gohtani and Y. Yamano, *Nippon Nogei Kagaku Kaishi*, 1987, **61**, 1113.
[109] S. Gohtani and Y. Yamano, *Nippon Nogei Kagaku Kaishi*, 1990, **64**, 139.
[110] S. Gohtani, K. Shinomoto, Y. Honda, K. Okubo, and Y. Yamano, *Nippon Nogei Kagaku Kaishi*, 1990, **64**, 901.
[111] D. G. Oakenfull, *Aust. J. Chem.*, 1986, **39**, 1671.

10 Physiological and Pharmacological Properties

Interactions with Biological Membranes

Saponins have long been known to cause lysis of erythrocytes *in vitro*, indeed the quantifiable release of haemoglobin, coupled with the inhibition of this effect by cholesterol has been extensively used as a means of detecting and 'quantifying' saponins in plant material.[1,3] The primary action of membranolytic saponins upon cells is to increase the permeability of the plasma membrane. This leads to loss of essential electrochemical concentration gradients and to the inevitable destruction of the cell. The suggestion that the haemolytical activity of saponins resulted from their affinity for membrane sterols was advanced by Fischer[112] many years ago and whilst there is considerable evidence to support the irreversible binding of membranolytic saponins to discrete sites within the plasma membranes, precise details of the interactions involved are still sought. There is very clear variation both in the haemolytic activity of individual saponins, and in the order of susceptibility within the animal kingdom, generally the guinea pig and horse are most susceptible.[3]

It has been suggested that saponins combine permanently with membrane cholesterol to form permeable, micelle-like aggregates within the plane of the membrane.[113,114] The precise molecular structure of these lesions remains unclear, although in one model[115] the saponin molecules are arranged in a ring with their hydrophobic moieties combined with cholesterol around the outer perimeter. It has been suggested[116] that, rather than being a specific site for saponin binding, cholesterol exerts its effect by altering the structure—and hence susceptibility to membranolytic effects—of the membrane. At present it is impossible to predict with any certainty the effect of individual saponins on biological membranes; it is considered from recent work[117,118] that the nature of the aglycone is important, with increasing numbers of polar (carboxylic acid) groupings enhancing permeabilizing activity. According to Namba *et al.*[119] ginseng saponins possessing 20-*S*-protopanaxtriol as the aglycone are strongly haemolytic whereas those containing 20-*S*-protopanaxdiol have no such activity, and may even be protective. Glycyrrhizin once again appears to be anomalous, showing an ability to protect membranes against saponin-induced damage.[118] This is not to suggest that the oligosaccharide moiety is unimportant, indeed current work[120] indicates that branching of the sugars may be a particularly significant structural feature.

[112] R. Fischer, *Biochem. Z.*, 1929, **209**, 319.
[113] A. D. Bangham and R. W. Horne, *Nature (London)*, 1962, **194**, 952.
[114] A. M. Glauert, A. T. Dingle, and J. A. Lucy, *Nature (London)*, 1962, **194**, 953.
[115] P. Seeman, *Fed. Proc.*, 1974, **33**, 2116.
[116] R. Segal and I. Milo-Goldzweig, *Biochim. Biophys. Acta*, 1978, **512**, 223.
[117] I. T. Johnson, G. M. Gee, K. R. Price, C. L. Curl, and G. R. Fenwick, *J. Nutr.*, 1986, **116**, 2270.
[118] J. M. Gee, K. R. Price, C. L. Ridout, I. T. Johnson, and G. R. Fenwick, *Toxicol. in vitro*, 1989, **3**, 85.
[119] T. Namba, M. Yoshizaki, T. Tomimori, K. Kobashi, K. Mitsui, and J. Hase, *Chem. Pharm. Bull.*, 1973, **21**, 459.
[120] R. Bomford, K. R. Price, and G. R. Fenwick, unpublished observations.

Although it has been claimed[121] that only the aglycone is capable of combining with receptor cells in the membrane (and implicit in this hypothesis is the presence of β-glycosidase activity in the cell membrane), this appears to be at variance with results[122] showing that the free aglycone of the membranolytic glycoalkaloid tomatine, (which may be seen in this context as a basic saponin), combines with neither cholesterol nor sitosterol. Other saponins may interact with, and affect the permeability of, artificial membranes which totally lack such hydrolytic activity.[123] It is clear that this is a very complex area, but the study of the mechanisms involved, and their consequences for plants, animals, and humans would seem currently to be facilitated by the availability of well-defined and purified natural products and by the various techniques which may be employed for their chemical or enzymic modification. One such investigation has been recently reported.[124] The haemolytic and antifungal properties of α-hederin (obtained from *Hedera rhombea*) and its derivatives were examined; the data obtained indicated that the terminal rhamnose unit of the disaccharide moiety was more important for antifungal activity than for haemolytic activity, whilst the later was more affected by the presence of a free carboxylic acid group in the hederagenin aglycone.

Effects on Fungi, Micro-organisms, and Viruses

Many saponins exhibit significant activity against fungi and micro-organisms under experimental conditions,[1,125-129] and inhibition of the growth of *Trichoderma viride* has been used as a screen for alfalfa saponins.[70] When assayed against *Saccharomyces carlsbergensis*, triterpenoid saponins possessing oleanolic acid or hederagenin aglycones were more active than those possessing lupanine, macedonic acid, or meristropic acid. Pathogenic fungi of tomato tend to be more resistant to α-tomatine than do non-pathogenic fungi.[123] The toxicity of oat roots, root extracts, and isolated saponins to *Pythium* and other fungi has been reported.[130]

Saponins from, amongst other species, *Aesculus hippocastum* (horse chestnut), *Bupleurum falcatum*, and *Thea sinensis* have been found to possess activity against influenza A2 virus,[131] whilst glycyrrhizin is active against several DNA

---

[121] R. Segal, I. Milo-Goldzweig, H. Schupper, and D. V. Zaitschek, *Biochem. Pharmacol.*, 1970, **19**, 2501.
[122] J. G. Roddick, *Phytochemistry*, 1979, **18**, 1467.
[123] J. G. Roddick and R. B. Drysdale, *Phytochemistry*, 1984, **23**, 543.
[124] M. Takechi and T. Tanaka, *Phytochemistry*, 1990, **29**, 451.
[125] B. Wolters, *Planta*, 1968, **79**, 77.
[126] B. Gestetner, Y. Assa, Y. Henis, Y. Birk, and A. Bondi, *J. Sci. Food Agric.*, 1971, **22**, 168.
[127] M. M. Anisimov, V. V. Shcheglov, and S. N. Dzizenko, *Prikl. Biokhim. Mikrobiol.*, 1978, **14**, 573.
[128] H. Betz and E. Schlösser, *Tagungsber., Akad. Landwirtschaftwiss. DDR*, 1984, **222**, 179.
[129] T. Nagata, T. Tsushida, E. Hamaya, E. Enoki, S. Manabe, and C. Nishino, *Agric. Biol. Chem.*, 1985, **49**, 1181.
[130] J. W. Deason and R. T. Mitchell, *Trans. Br. Mycol. Soc.*, 1985, **84**, 479.
[131] G. S. Rao, J. E. Sinsheimer, and K. W. Cochran, *J. Pharm. Sci.*, 1974, **63**, 471.

and RNA viruses and varicellazoster virus.[132,133] Whilst mechanisms are unclear it is possible that saponins might interfere with the attachment of viruses to the cell surface, hence preventing or seriously inhibiting penetration and replication. An interesting observation regarding the behaviours of group A and group B saponins from soyabeans has been made by Nakashima *et al.*[134] Whilst both the group A and deacetylated group A soyasaponins (Table 3) showed inhibitory effects towards HIV infection *in vitro*, these were small as compared with the behaviour of the group B saponins. Since the latter are found widely distributed throughout the soyabean plant, as well as amongst members of the Leguminosae generally (whereas the group A saponins are confined to the hypocotyl of soyabean) it is tempting to speculate that the group B saponins have important physiological roles to play in leguminous plants.

Saponins have been utilized as adjuvants in an anti-malarial vaccine (*Plasmodium yoelii*) and in parenterally adminstered foot-and-mouth and rabies vaccines.[135] These reports prompted Campbell and co-workers to examine the effect of quillaja saponin, administered orally, on the humoral response of mice fed inactivated rabies vaccine. The heterogeneous 'saponin' fraction examined potentiated the immune response to the inactivated vaccine and led the authors[135] to suggest that many saponins might be further exploited for use in oral immunization procedures. Subsequent work[136,137] with crude *Quillaja* saponin preparations has reinforced this conclusion. Morein and co-workers[138] have described the formation of ISCOMs (Immuno Stimulating COMplexes) and have applied these in commercial vaccines. ISCOMs, cage-like micellar structures, are readily formed with a *Quillaja* saponin preparation and a detergent extract of the viral membrane are mixed and the excess detergent removed by dialysis, ultrafiltration, or density gradient centrifugation. The first ISCOM-based vaccine, against equine influenza virus, was marketed in 1987. Currently the active saponin species within the crude mixture is under investigation and other plant sources are being screened as alternatives to *Quillaja*.

Effects on Insects and Fish

The effects of saponins on insects, molluscs, and aquatic vertebrates have been recently summarized.[3] Thus, for example, alfalfa saponins are active against the larvae of the grass grub (*Costelytra zealandica*)[139,140] whilst soyasaponin is reported to be highly toxic to the rice weevil (*Sitophilus oryzae*).[141] As would be

[132] R. Pompei, O. Flore, M. A. Marccialis, A. Pani, and B. Loddo, *Nature (London)*, 1979, **281**, 689.
[133] M. Baba and S. Shigeta, *Antiviral Res.*, 1987, **7**, 99.
[134] H. Nakashima, K. Okubo, Y. Honda, T. Tamura, S. Matsuda, and N. Yamamoto, *AIDS*, 1989, **3**, 655.
[135] I. Maharaj, K. J. Froh, and J. B. Campbell, *Can. J. Microbiol.*, 1986, **32**, 414.
[136] S. R. Chavali and J. B. Campbell, *Immunobiology*, 1987, **174**, 347.
[137] S. Rao Chavali and J. B. Campbell, *Int. Arch. Allergy Appl. Immunol.*, 1987, **84**, 129.
[138] B. Morein, B. Sundquist, S. Höglund, K. Dalsgaard, and A. Osterhaus, *Nature (London)*, 1984, **308**, 457.
[139] B. Morein and K. Simon, *Vaccine*, 1985, **3**, 83.
[140] J. Krzmanska and D. Waligoria, *Chem. Abstr.*, 1985, **102**, 128998.
[141] H. C. F. Su, R. D. Spiers, and P. G. Mahany, *J. Econ. Entomol.*, 1972, **65**, 844.

expected from earlier comments there is evidence for structure-related activity. Thus alfalfa root saponins (rich in medicagenic acid derivatives) are markedly more toxic to the flour beetle (*Tribolium castraneum*) than soyasaponins; whilst saponin-containing extracts of chick pea and lentil were much less active against the azuki beetle (*Collobrachus chinensis*) than were similar extracts from alfalfa or soyabeans.[142] Pracros[143,144] has used the response of the yellow mealworm larvae (*Tenebrio molitor*) toward alfalfa fractions as a basis for the development of a bioassay for alfalfa saponins in support of plant breeding programmes.

When saponins are added to water, fish may be rapidly paralysed and die. This toxic effect of saponins to fish is a property which has been exploited by man since ancient times, and in more recent times has been the basis for saponin bioassays based upon the response of snails, tadpoles, and other small fish. Hostettman and co-workers in Lausanne have been actively engaged in the search for new and natural molluscides which may be of value in controlling water snails such as *Biomphalaria glabrata*, vectors of schistosomiasis, which is endemic in parts of Africa. Saponins in a number of African plant species have been shown to be active in this respect,[145–147] the starting point of many investigations being the activities of local populations in areas where the disease is endemic.[148]

Effects on Birds and Mammals

Despite the high toxicity of many saponins when given intravenously to higher animals,[1,3] their toxic effects are greatly reduced when administered orally and the most common saponins of foods and feedingstuffs do seem to be free of significant oral toxicity. Whilst haemolytic saponins will interact with the small intestinal mucosal cells of mammals[117,118] it is probable that the permeabilized cells will be readily eliminated by the normal processes of epithelial replacement. Whilst this would explain the failure of haemolytic saponins to cross the gut mucosa the physiological consequences of increased permeability and mucosal cell tumours require further investigation. It may be hypothesized that the low toxicity of orally-ingested saponins to man and other large mammals is due primarily to the large surface area of the gastrointestinal tract in relation to the concentration of saponins to which it is exposed.

No adverse effects were observed when high concentrations of soyabean saponins were fed to chicks, mice, and rats[149] and although it has been suggested that the goitrogenic activity of soybean products observed in rats is partly due to soyasaponins,[150] Liener[151] has proposed that saponins should no longer be

[142] S. W. Appelbaum and Y. Birk, in 'Insect and Mite Nutrition', ed. J. G. Rodriguez, North-Holland, Amsterdam, 1972, p. 629.
[143] P. Pracros, *Agronomie*, 1988, **8**, 257.
[144] P. Pracros, *Agronomie*, 1988, **8**, 793.
[145] K. Hostettmann, *Schweiz. Apoth. Ztg*, 1985, **123**, 233.
[146] F. Gafner, J. D. Msonthi, and K. Hostettmann, *Helv. Chim. Acta*, 1985, **68**, 555.
[147] A-C. Dorsaz, M. Hostettmann, and K. Hostettmann, *Planta Med.*, 1988, 191.
[148] A. Tekle, *Ethiopian Med. J.*, 1977, **15**, 131.
[149] I. Ishaaya, Y. Birk, A. Bondi, and Y. Tencer, *J. Sci. Food Agric.*, 1969, **20**, 433.
[150] J. Suwa and S. Kimura, *Ann. Rep. Fac. Educ. Gunma Univ. Art Technol. Ser.*, 1981, **17**, 71.

regarded as significant anti-nutritional factors in soyabeans. Whilst their activity *per se* may justify this conclusion their membrane-permeabilizing effect may result in an increase in the absorption, and hence toxicity, of other dietary components or xenobiotics.

Intraruminal administration of alfalfa saponins to sheep resulted in a reduction in nutrient degradation and microbial fermentation in the rumen. Fractional digestive coefficients of organic matter, hemicellulose, cellulose, and nitrogen were reduced in the stomach, but increased in the small intestine, in the presence of saponins.[152] The effects of a steroidal saponin preparation isolated from *Yucca schidigera* in improving ruminal organic matter digestion and feedlot performance of cattle and stimulating anaerobic fermentation have resulted in the commercial production of this material, trademark 'Sarsaponin', which is sold in the United States as Sevarin (cattle), and Microaid (pigs).

The safety of alfalfa saponins for human consumption has been extensively investigated by Malinow and co-workers.[153] Rats fed alfalfa saponins at levels of 1 % in the diet for up to 26 weeks showed no ill effects although a potentially beneficial reduction in serum cholesterol and triglycerides was observed. No adverse effects were observed in the primate species *Macaca fascicularis*,[154] following consumption of an undefined mixture of alfalfa-top saponins for up to 78 weeks.

Prolonged administration of disodium glycyrrhizinate to mice produced no evidence of toxicity, nor of any change in the incidence or distribution of tumours.[155] However, glycyrrhizin is known to mimic the action of the mineralocorticoid hormone, aldosterone, and hence to cause retention of water and sodium.[156] There are various reports of patients who developed hypertension as a result of consuming large, or regular, quantities of liquorice-based confectionery.[1,157]

The triterpenoid saponins obtained from *Quillaja* are widely used as a food additive and have, consequently, been subjected to intense toxicological scrutiny. No significant toxic effects resulted from short term feeding trials with rats nor following the feeding of up to 1.5 % *Quillaja* saponin to mice for prolonged periods.[158,159] Steroidal-saponin-containing extracts of *Yucca mohavensis* were found to be less haemolytic than soyasaponins and the feeding of the material (0.5 %) to rats for 12 weeks produced no significant effects in terms of growth,

[151] I. E. Liener, *J. Am. Oil Chem. Soc.*, 1981, **58**, 406.
[152] C. D. Lu and N. A. Jorgensen, *J. Nutr.*, 1987, **117**, 919.
[153] M. R. Malinow, W. P. McNulty, P. McLaughlin, C. Stafford, A. K. Burns, A. I. Livingston, and G. O. Kohler, *Food Cosmet. Toxicol.*, 1981, **19**, 443.
[154] M. R. Malinow, P. McLaughlin, W. P. McNulty, D. C. Houghton, S. Kessler, P. Stenzi, S. H. Goodnight jnr., E. J. Bardana, and J. C. Paltay, *J. Med. Primatol.*, 1982, **11**, 106.
[155] T. Kobake, S. Inai, K. Namba, K. Takemoto, K. Matsuki, H. Nishina, I.-B. Huang, and S. Tokuoka, *Food Chem. Toxicol.*, 1985, **23**, 979.
[156] P. M. Stewart, M. A. Wallace, R. Valentino, D. Burt, C. H. L. Shackleton, and R. W. Edwards, *Lancet*, 1987, 821.
[157] 'Martindale, The Extra Pharmacopoeia', 28th Ed., ed. J. E. F. Reynolds, The Pharmaceutical Press, London, 1982, pp. 691–692.
[158] I. F. Gaunt, P. Grasso, and S. D. Gangolli, *Food Cosmet. Toxicol.*, 1974, **12**, 641.
[159] J. C. Phillips, K. R. Butterworth, I. F. Gaunt, J. G. Evans, and P. Grasso, *Food Cosmet. Toxicol.*, 1979, **17**, 23.

food utilization, blood counts, blood glucose, or non-protein nitrogen or in gross or histological findings postmortem.[160]

The growth-retarding effect of a variety of saponin-containing feedingstuffs has been commented on by numerous workers and isolated extracts rich in saponins have in some cases been shown to produce similar effects.[1,3] The underlying mechanisms for these effects are not clear. Given the bitterness of the vast majority of saponins it is tempting to suggest that the primary effect of saponins is to reduce the palatability of diets. In support of this it has been noted that many species are sensitive to flavours perceived by humans as bitter, and that foods rich in saponins are avoided by such species.[161,162] Unfortunately for this hypothesis, the addition of cholesterol to the diet at levels which abolish the growth inhibiting effects of saponins does not reduce bitterness or palatability. Saponins may, however, inhibit growth by reducing the availability of essential nutrients. Thus whilst a range of saponins had no effect on vitamin E, A, and D_3, there was evidence of binding between ammoniated glycyrrhizin, alfalfa saponins, and zinc.[163,164] Southon et al.[165] have shown Gypsophilla and soyabean saponins to produce similar effects; reduced zinc availability is associated with anorexia and growth inhibition in experimental animals. Ginseng saponins, fed to rats at levels between 0.02 and 1.5% have been found to increase the absorption of iron.[166]

Differences in the effects of saponins observed in ruminants and non-ruminant animals[1,3] has been attributed to alterations in metabolism brought about by ruminal micro-organisms. Sarsaponin apparently alters protein metabolism by micro-organisms *in vitro*[167] and modified ruminal digestion and flow rate in intact animals.[168] However such effects have not significantly improved the performance of lactating dairy cows.[169] A potential beneficial interaction between saponins and tannin has been noted by Freeland et al.[170] The authors speculate that saponins and tannins interact within the gut to form non-toxic complexes and that herbivores may have evolved the capability to select these constituents in the critical proportions necessary for detoxification to occur in natural diets.

Effects on Cholesterol

Although there are few published studies to support or refute the suggestion that saponins seriously limit the availability of essential nutrients, there is

[160] B. L. Oser, *Food Cosmet. Toxicol.*, 1966, **4**, 57.
[161] B. R. Leamaster and P. R. Cheeke, *Can. J. Anim. Sci.*, 1979, **59**, 467.
[162] P. R. Cheeke, J. S. Powley, H. S. Nakaue, and G. H. Arscott, *Can. J. Anim. Sci.*, 1983, **63**, 707.
[163] L. G. West and J. C. Greger, *J. Food Sci.*, 1978, **43**, 1340.
[164] L. G. West, J. L. Greger, A. White, and B. J. Nornamaker, *J. Food Sci.*, 1978, **43**, 1342.
[165] S. Southon, I. T. Johnson, and K. R. Price, *Br. J. Nutr.*, 1987, **59**, 49.
[166] S. J. Lee and B. R. Yang, *Korean Biochem. J.*, 1984, **17**, 456.
[167] M. A. Grobner, D. E. Johnson, S. R. Goodall, and D. A. Benz, *J. Anim. Sci. West Sect. Proc.*, 1982, 64.
[168] A. L. Goetsch and F. N. Owens, *J. Dairy Sci.*, 1985, **68**, 2377.
[169] F. R. Valdez, L. J. Bash, A. L. Goetsch, and F. N. Owens, *J. Dairy Sci.*, 1986, **69**, 1568.
[170] W. J. Freeland, P. H. Calcott, and L. R. Anderson, *Biochem. Syst. Ecol.*, 1985, **13**, 189.

abundant evidence that saponins do interact with sterols in the gastrointestinal tract in a way which might prove beneficial to humans. It is generally agreed that elevated plasma cholesterol levels are a significant risk factor in the etiology of cardiovascular disease. The hypocholesterolemic effect of some saponins has therefore provoked considerable clinical interest, and prompted some workers to propose that their consumption may provide a useful means of dietary management of plasma cholesterol in man. To investigate the mechanism underlying the hypocholesterolemic effect of alfalfa, Malinow et al.[171] investigated the effects of isolated alfalfa-top saponins on cholesterol absorption in rats. Five, 10 or 20 mg doses of saponins were given to each animal by intragastric administration, together with radioactively labelled cholesterol. The extent of cholesterol absorption was determined by the analysis of plasma radioactivity and the subsequent appearance of faecal sterols. A progressive inhibition of cholesterol absorption with increasing saponin dosage was observed. Later work confirmed that the saponin content rather than dietary fibre was responsible for the inhibition of cholesterol absorption by alfalfa meal.[172]

In several later studies with experimental animals, Malinow and co-workers have shown that a reduction in experimentally-induced hypocholesterolemia, brought about with saponins, leads to changes in systemic cholesterol metabolism which can beneficially alter the progress of cardiovascular disease.[173] The possibility that such an effect may be of benefit to man is strengthened by recent work with primates. The consumption of a semi-purified diet containing 1.0 % isolated alfalfa-root saponins or 0.6 % alfalfa-top saponins by cynomolgas macaques (*Macaca fuscicularis*),[174] led to reduced cholesterol absorption and a reduction in the plasma response to a cholesterol-supplemented diet.

Alfalfa saponins appear not to be toxic at the levels necessary to produce a useful physiological response.[153,154] Bingham et al.[175] gave a group of 174 arthritic patients tablets of a saponin-rich extract from *Yucca schidigera* in a treatment which also included controlled diets, exercise, and physiotherapy. Substantial reductions in plasma cholesterol were observed in response to the saponin, particularly in individuals with higher levels initially. Improved plasma cholesterol levels have recently been measured in a group of hyperlipidemic human subjects fed alfalfa seeds.[176] Interestingly, an improvement in plasma cholesterol levels in hyperlipidemic patients given a preparation of *Panax ginseng* root, which is high in saponins, has also been reported recently.[177]

[171] M. R. Malinow, P. McLaughlin, L. Papworth, C. Stafford, G. O. Kohler, L. Livingston, and P. R. Cheeke, *Am. J. Clin. Nutr.*, 1977, **30**, 2061.
[172] M. R. Malinow, P. McLaughlin, C. Stafford, L. Livingston, G. O. Kohler, and P. R. Cheeke, *Am. J. Clin. Nutr.*, 1979, **32**, 1810.
[173] M. R. Malinow, P. McLaughlin, C. Stafford, A. L. Livingston, and G. O. Kohler, *Atherosclerosis*, 1980, **37**, 433.
[174] M. R. Malinow, W. E. Connor, P. McLaughlin, C. Stafford, D. S. Lin, A. L. Livingston, G. O. Kohler, and W. P. McNulty, *J. Clin. Invest.*, 1981, **67**, 156.
[175] R. Bingham, D. H. Harris, and T. Laga, *J. Appl. Nutr.*, 1978, **30**, 127.
[176] J. Molghaard, H. von Schenk, and A. G. Olsson, *Atherosclerosis*, 1987, **65**, 173.
[177] M. Yamamoto, T. Uemura, S. Nakama, A. Uemiya, and A. Kumagai, *Am. J. Clin. Med.*, 1983, **11**, 96.

The mechanism by which alfalfa saponins interfere with cholesterol absorption has not been determined, although there is a long standing view that they form insoluble complexes with cholesterol in the gut lumen[178] as they do *in vitro*.[179] Studies by Story *et al.*[180] using chemically undefined saponins extracted from alfalfa plants and sprouts offer some support for this proposed mechanism.

Apart from binding, and hence limiting the absorption of dietary cholesterol, certain substances have the capacity to alter cholesterol metabolism by interfering with the enterohepatic bile salt circulation. Bile salts are synthesized from cholesterol in the liver, and pass via the bile duct into the small intestine where they form mixed micelles with fatty acids, monoglycerides, and cholesterol. Lipid absorption occurs in the upper small bowel, but the liberated bile salts are eventually reabsorbed by an active transport mechanism in the distal small intestine and conveyed back to the liver to begin the cycle again. However, non-absorbable materials which bind bile salts can render them unavailable for reabsorption and increase their faecal excretion. This is compensated for by increased synthesis from endogenous cholesterol, resulting in a reduction in plasma cholesterol levels. It has frequently been proposed that various constituents of dietary fibre, particularly lignin, may bind bile salts and hence reduce plasma cholesterol levels in man.

Topping *et al.*[181] showed that the addition of saponins from *Saponaria officinalis* to the drinking water of pigs led to an increase in faecal bile acid excretion. Previously Oakenfull and Topping[182] had suggested that only those sources of dietary fibre that are also rich in saponins have the capacity to bind bile salts to any significant degree *in vitro*, and they reported that saponins facilitate the adsorption of bile salts to non-digestible macromolecular substrates *in vitro*. This view was challenged by Calvert and Yates[183] who, using a different *in vitro* technique, were unable to show any effect of saponins upon the binding of bile salts by various constituents of dietary fibre. However, these authors appear to have made no attempt to purify their fibre preparations. They simply measured binding of bile acids to soyabean flour, for example, where any effect of saponins would have been swamped by the strong binding capacity of the protein present. This finding appears to be consistent with another report, showing that the removal of saponins from alfalfa plant material actually enhances its ability to bind bile salts *in vitro*.[184] In a further study, Rotenberg and Eggum[185] investigated the possible role of saponins in the hypocholesterolemic activity of pectin. The addition of 0.2 % *Gypsophilla* saponins to a diet

[178] C. B. Coulson and R. A. Evans, *Br. J. Nutr.*, 1960, **14**, 121.
[179] B. Gestetner, Y. Assa, Y. Henis, Y. Tencer, M. Royman, Y. Birk, and A. Bondi, *Biochem. Biophys. Acta*, 1972, **270**, 181.
[180] J. A. Story, S. L. LePage, M. S. Petro, L. G. West, M. M. Cassidy, F. G. Lightfoot, and G. V. Vahouny, *Am. J. Clin. Nutr.*, 1984, **39**, 917.
[181] D. L. Topping, G. D. Storer, G. D. Calvert, R. J. Illman, D. G. Oakenfull, and R. A. Weller, *Am. J. Clin. Nutr.*, 1980, **33**, 783.
[182] D. G. Oakenfull and D. L. Topping, *Br. J. Nutr.*, 1978, **40**, 299.
[183] G. D. Calvert and R. A. Yates, *Br. J. Nutr.*, 1982, **47**, 45.
[184] J. A. Story, A. White, and L. G. West, *J. Food Sci.*, 1982, **47**, 1276.
[185] S. Rotenberg and B. O. Eggum, *Acta Agric. Scand.*, 1986, **36**, 211.

containing pectins that had been washed with ethanol, had no effect on the ability of this diet to reduce liver cholesterol concentrations. The authors concluded that saponins play no part in the effect of pectin on cholesterol metabolism.

In other studies, Oakenfull[111] demonstrated that the purified saponins obtained from soapwort (*Saponaria officinalis*), soyabean, and *Quillaja saponaria* form large micellar aggregates with bile acids. Furthermore, the formation of these structures significantly reduces the absorption of bile salts from the perfused small intestine of the rat.[186,187] Isolated soyabean and *Quillaja* saponins were shown to reduce diet-induced hypercholesterolemia in rats and to lead to significant increases in the faecal excretion of neutral sterols and bile acids, respectively.[188] The authors suggested that particular saponins reduce hypercholesterolemia by different mechanisms. It is proposed that *Quillaja* and alfalfa saponins reduce primarily the absorption of dietary cholesterol, while saponins derived from *Saponaria*, and, more importantly, from a nutritional point of view, soyabeans, increase the faecal excretion of bile salts and/or cholesterol.[189] *Quillaja* saponin appears to operate by both mechanisms. A recent brief report by Rao and Kendall[190] describes a significant hypocholesterolemic effect even in rats not fed a high cholesterol diet, which would suggest a reduction in bile salt absorption in addition to a direct interaction with cholesterol.

The importance of saponins in the hypocholesterolemic effect of alfalfa is thrown into question by the finding of Story *et al.*[180] that the binding of bile salts by alfalfa plant material *in vitro* is not reduced by the extraction of saponins, and that the ability of alfalfa to reduce the accumulation of liver cholesterol in cholesterol-supplemented rats was significantly increased by the removal of saponins. Clearly, this is a topic of considerable complexity.

Undoubtedly some of the apparently contradictory findings reflect the fact that the saponins in particular samples and preparations of alfalfa are only poorly characterized and their concentration may only be approximately known.[191] The difference in activity of different saponins has been clearly indicated in a recent study in which soyasaponins (85 % pure) and tigogenin β-cellobioside[192] reduced cholesterol uptake by 50 % in the rat at concentrations of 2.0 g l^{-1} and 0.27 g l^{-1} respectively.[189]

There is further controversy in relation to the significance of the hypocholesterolemic effect of soyabean saponins. In several studies,[193-195] a significant

---

[186] D. G. Oakenfull and G. S. Sidhu, *Nutr. Rep. Int.*, 1983, **27**, 1253.
[187] D. G. Oakenfull and G. S. Sidhu, *Br. J. Nutr.*, 1986, **55**, 643.
[188] D. G. Oakenfull, D. L. Topping, R. J. Illman, and D. E. Fenwick, *Nutr. Rep. Int.*, 1984, **29**, 1039.
[189] G. S. Sidhu, B. Upson, and M. R. Malinow, *Nutr. Rep. Int.*, 1987, **35**, 615.
[190] A. V. Rao and C. W. Kendall, *Food Chem. Toxicol.*, 1986, **24**, 441.
[191] M. R. Malinow, *Atherosclerosis*, 1984, **50**, 117.
[192] M. R. Malinow, *Ann. N.Y. Acad. Sci.*, 1985, **454**, 23.
[193] R. E. Hodges, W. A. Krehl, D. B. Stone, and A. Lopez, *Am. J. Clin. Nutr.*, 1967, **20**, 198.
[194] C. R. Sirtori, E. Agradi, E. Conti, O. Mantero, and E. Gatti, *Lancet*, 1977, 275.
[195] C. R. Sirtori, E. Gatti, O. Mantero, E. Conte, E. Agradi, E. Tremoli, M. Sirtori, L. Fratterigo, L. Tarrazzi, and D. Kritchevsky, *Am. J. Clin. Nutr.*, 1979, **32**, 1645.

reduction in plasma cholesterol has been achieved in human subjects by substituting textured vegetable protein for animal protein in their diets and it has been suggested[196,197] that this is due to the relatively high levels of saponins present in this material. A subsequent study with hypocholesterolemic human volunteers failed to confirm this hypothesis,[198] but Oakenfull and Topping have challenged the significance of these findings on the grounds that the compliance of the subjects with the experimental diet was not adequately controlled.[199] A well-controlled study with a group of normal human volunteers showed that the saponins present in soyabean flour (fed at 50 g day^{-1}) significantly increased faecal excretion of bile acids.[200] The hypocholesterolemic effect of soyabean protein fed to rabbits was not influenced by supplementation with an undefined commercial saponin, and this led the authors to suggest that the composition of the protein itself may be of primary importance in the modification of cholesterol metabolism.[201]

One possibility, which although occasionally mentioned in passing[174,179] has not been investigated by those working on hypocholesterolemic saponins, is that they may interact directly with sterols in the brush border membranes of the intestinal mucosal cells. Recent studies[117,118] have established that hemolytic saponins significantly increase the permeability of isolated intestine *in vitro*, and it would be surprising if a similar effect did not occur in the intact animal. Strong evidence for this is provided by recent reports describing the use of saponins as oral adjuvants for vaccines (see above).

One consequence of this may be that saponins increase the exfoliation, and hence the replacement, of intestinal mucosal cells, while simultaneously rendering their membrane cholesterol constituents unavailable for reabsorption. Such an effect might increase cholesterol excretion and thereby make a useful contribution to the hypocholesterolemic activity of saponins. However, Ewart[202] drew attention to the irritation of the bowel by saponins in the absence of adverse systematic effects over fifty years ago. Changes in the permeability of the gastrointestinal mucosa might thus have undesirable consequences, which will need to be evaluated before substantial increases in saponin consumption by human beings can be recommended.

Much of the previous uncertainty as to the usefulness and importance of saponins in the control of cholesterol metabolism in man appears to be due to the fact that saponins have often tended to be treated as a single, well-defined substance of known concentration. In fact, as Malinow[191] has recently pointed out, in nutritional studies the term 'saponin' is often only an operational definition applied to mixtures of substances whose purity and chemical struc-

---

[196] J. D. Potter, D. L. Topping, and D. G. Oakenfull, *Lancet*, 1979, 223.
[197] D. G. Oakenfull, *Food Tech. Austr.*, 1981, **33**, 432.
[198] G. D. Calvert, L. Blight, R. J. Illman, D. L. Topping, and J. D. Potter, *Br. J. Nutr.*, 1981, **45**, 277.
[199] D. G. Oakenfull and D. L. Topping, *Atherosclerosis*, 1983, **48**, 301.
[200] J. D. Porter, R. J. Illman, G. D. Calvert, D. G. Oakenfull, and D. L. Topping, *Nutr. Rep. Int.*, 1980, **22**, 521.
[201] C. Pathirana, M. J. Gibney, and T. G. Taylor, *Br. J. Nutr.*, 1981, **46**, 421.
[202] A. J. Ewart, *Commonw. Austr. Counc. Sci. Ind. Res. Bull.*, 1931, **20**, 1.

Table 10 *Saponin-containing plants used as flavourings, herbs, health foods, and tonics*

Achillea millefolium	Yarrow
Aletris farinosa	Colic root
Aloe baradensis	Aloe
Calendula officinalis	Marigold
Cassia occidentalis	Coffee senna
Chenopodium ambrosioides	Wormseed
Clematis vitalba	Traveller's joy
Cymbopogon citratus	Lemongrass
Eleutherococcus senticosus	Wujia ginseng
Glycyrrhiza species	Liquorice
Lathyrus hirsutus	Rough pea
Ornithogalum umbellatum	Star of Bethlehem
Pachyrhizus species	Yam bean
Panax species	Ginseng
Phytolacca americana	Pokeweed
Salvia officinalis	Sage
Smilax aristolochiifolia	Mexican sarsaparilla
Symphytum peregrinum	Comfrey
Taraxacum officinale	Dandelion
Trigonella foenum-graecum	Fenugreek

ture are unknown. Furthermore, the various physiological indicators of cholesterol metabolism which have been measured by workers in the field have often differed, so that realistic comparisons of studies are difficult.

11 Herbs and 'Health' Products

Many plants being used as flavourings, herbs, 'health' products, and tonics contain saponins.[1,203] Table 10 summarizes some, but by no means all, of these. Both yarrow and aloe have found use as bitterants in beverages and confectionery; according to Duke[203] the roots of wormseed, which may be used as a flavouring for soup, contain 2.5 % saponins, although the analytical method used to obtain this figure was not given. Bulbs of the Star of Bethlehem have been ground and mixed with flour to make bread; this practice is to be discouraged since the plant contains toxic sarmentogenin glycosides. Young shoots of pokeweed are used as greens or potherbs, and a number of saponins have been isolated from this source.[1] The presence of these originally precluded using the juice of pokeberry fruit as a source of betanin natural colouring, but Forni *et al.*[204] have described a method to remove the undesirable saponin fraction. Extracts of sarsaparilla root are used for flavouring purposes in root beers and confectionery. This plant is interesting in that it contains two novel, highly branched saponins, parillin and sarsaparilloside.[1] Marigold leaves may be used, like saffron, as a seasoning or colouring and the fresh plant may be

---
[203] J. A. Duke, 'Handbook of Medicinal Herbs', CRC Press, Boca Raton, 1985.
[204] E. Forni, A. Frifilo, and A. Polesello, *Food Chem.*, 1983, **10**, 35.

brewed to a herbal tea. The plant contains a number of biologically-active saponins.[205] The oriental herb, *Dianthus superbus* contains saponins exhibiting analgesic activity, whilst a number of oriental species—including *Luffa*, *Gynostemma*, *Panax*, and *Glycyrrhiza* are extensively used in herbal medicine and, more recently, in so-called 'functional' foods.[1] Steroidal saponins isolated from the aerial parts of *Luffa cylindrica* have been patented as anti-obesity agents,[206] while Japanese patents include those for the production of honey containing *Gynostemma* saponins[1] and a refreshing carbonated beverage containing liquorice root saponins.[207] Saponins from *Buffa operculata*, primrose, and tea seeds have expectorant properties and also promote the secretion of mucus from the respiratory tract.[208] Fenugreek is an important cash crop in many parts of the world and it is also used as a food, spice, and medicine. Extensive studies[1] have revealed the presence of steroidal saponins and the plant has attracted considerable interest as a source of diosgenin and yamogenin, raw materials for steroid production.[209] It is probable that other saponins also occur in fenugreek, since Oleszek *et al.*[210] have found both soyasapogenol and medicagenic acid glycosides in *Trigonella monspeliaca* tops.

Undoubtedly ginseng and liquorice have been most widely studied amongst the oriental saponin-containing drugs. Detailed information on the former has been published by Hu,[211,212] Shibata,[213] and Sonnenborn.[214] Amongst the pharmacologically-confirmed effects reported by Shibata[213] are CNS stimulating (and depressing) action, papaverin-like activity, serotonin-like activity, ganglion-stimulating activity, analgesic, antipyretic, and anti-inflammatory activities, and antihistamine-like activity. Undoubtedly there are many more effects,[1] some of which will be linked to the presence of dammarane saponins. Such saponins are not confined to *Panax* species, but are found in other medicinally-important rhamnaceous plants, most notably *Zizyphus jujuba*, and in the cucurbit, *Gynostemma pentaphyllum*.

Lutomski[215] has described the biological properties of liquorice, many of which can be traced to the presence of glycyrrhizin and, thereby, to glycyrrhetinic acid. It was the Dutch physician Revers[216] who discovered that liquorice extract contains a substance with spasmolytic power and with a beneficial influence on the healing process of gastric ulcers. Since then, numerous investigators have substantiated the curative effect of liquorice on peptic ulcers; however, the identity of the antipeptic factor remain obscure. Studies in rats

[205] R. Chemli, A. Toumi, S. Oueslati, B. Zouaghi, K. Boukef, and G. Balansard, *J. Pharm. Belg.*, 1990, **45**, 12.
[206] Osaka Yakuhin Kenkyusho K K patent, *Chem. Abstr.*, 1985, **102**, 165552u.
[207] K. Tanaka, *Chem. Abstr.*, 1986, **105**, 962298p.
[208] E. S. Lower, *East. Pharm.*, 1985, **28**, 55.
[209] R. Saunders, P. S. J. Cheetham, and R. Hardman, *Enzyme Microb. Technol.*, 1986, **9**, 549.
[210] W. Oleszek, M. Jurzysta, S. Burda, and M. Ploszynski, *Acta Soc. Bot. Polon.*, 1987, **56**, 281.
[211] S-Y. Hu, *Econ. Bot.*, 1976, **30**, 11.
[212] S-Y. Hu, *Am. J. Chin. Med.*, 1977, **5**, 1.
[213] S. Shibata, in 'Advances in Medicinal Phytochemistry', ed. D. Barton and W. D. Ollis, John Libbey, London, 1986, 159.
[214] U. Sonnenborn, *Dtsch. Apoth. Ztg*, 1987, **127**, 433.
[215] J. Lutomski, *Pharm. Unserer Zeit*, 1983, **12**, 49.
[216] F. E. Revers, *Ned. Tijdschr. Geneeskd.*, 1946, **90**, 135.

and humans with peptic ulcers conducted by Nishiyama[217] showed that the therapeutic effect on ulcers is not due to glycyrrhizin. This is also the opinion of Revers[218]; however, other authors claim that glycyrrhizin or rather glycyrrhetinic acid is the active factor that cures peptic ulcer. This controversy has even penetrated the patent courts because patents have been granted to glycyrrhetinic acid derivatives[219] as well as deglycyrrhizinized liquorice[220] preparations which allegedly cure peptic ulcers.

Glycyrrhetinic acid has also been shown to possess anti-inflammatory activity which has led to its use in ointments to cure dermatological disorders. The uncoupling of oxidative phosphorylation by glycyrrhetinic acid, as demonstrated *in vitro* by Whitehouse *et al.*,[221] may be a factor in the anti-inflammatory mechanism. Several Japanese papers on biomedical parameters of glycyrrhizin and clinical studies showing its use in liver diseases were collected by Maruzen Pharmaceutical Corporation. From the same source stems a brochure reporting on other biological activities for glycyrrhizin acid or its derivatives, running the gamut from antitussive to antibacterial and even anticancer effects. Furthermore, Tanaka *et al.* recently reported that liquorice extract as well as glycyrrhizin inhibited the activity of several mutagenic agents.[23,215] The effect of glycyrrhizin on AIDS virus has also recently been studied.[222]

Glycyrrhizin may also show promise in combating dental caries. Segal *et al.*[223] have reported that the compound in a 1% solution markedly inhibits *in vitro* plaque formation by *Streptococcus mutans*, the oral micro-organism that transforms sucrose into caries-producing lactic acid.

The use of liquorice extract in the treatment of peptic ulcer sometimes appeared to invoke oedema and other side effects which could be ascribed to a deoxycorticosteron-mimetic effect of the glycyrrhizin present. Many investigations followed and it was shown that glycyrrhizin and glycyrrhetinic acid decrease the output of ACTH, reduce urinary excretion of sodium and chloride and increase potassium excretion, elevate blood pressure, and induce metabolic alkalosis with severe hypokalemia and hypernatremia.[215] It would seem that there is a complex interaction with both the renin–angiotensin–aldosterone axis and the adrenal–pituitary axis of man and experimental animals, but further study is needed. It is now well established that the consumption of liquorice or glycyrrhizin in excessive amounts and over long periods of time causes a so-called pseudo-aldosteronism consisting of sodium and water retention (even oedema) and hypokalaemic alkalosis, hypertension, and suppressed plasma renin activity and aldosterone secretion (see references in references 26 and 81). These symptoms resemble primary aldosteronism except that in the latter

[217] T. Nishiyama, *Igaku Kenkyu*, 1955, **25**, 366.
[218] F. E. Revers, *Acta Med. Scand.*, 1956, **154**, 749.
[219] D. V. Parke, in 'Topics in Gastroenterology', ed. S. C. Truelove and J. A. Ritchie, Blackwell, Oxford, 1976, p. 329.
[220] W. Larkworthy and P. F. L. Holgate, *Practitioner*, 1975, **215**, 787.
[221] M. W. Whitehouse, P. D. G. Dean, and T. G. Halsall, *J. Pharm. Pharmacol.*, 1967, **19**, 533.
[222] H. Nakashima, T. Matsui, O. Yoshida, Y. Isowa, Y. Kido, Y. Motoki, M. Iro, S. Shigeta, T. Mori, and N. Yamamoto, *Gann*, 1987, **78**, 767.
[223] R. Segal, S. Pisanty, R. Wormser, E. Azaz, and M. N. Sela, *J. Pharm. Sci.*, 1985, **74**, 79.

situation aldosterone secretion is increased. The pseudo-aldosteronism induced by glycyrrhizin may be accompanied by myopathy (muscular weakness) when the serum potassium level is decreased below a critical value.[224] Nevertheless, a 60-year-old athlete who had suffered liquorice intoxication won the Italian senior marathon championship with a still reduced potassium level.[225]

12 Requirements for Future Research

It is clear that saponins occur in a wide range of plants which are consumed by animals and man. Many of the reports about the occurrence must, however, be treated with caution, since the methods employed are now known to be open to question. At the natural product level there remains considerable scope for further research into the nature of the saponins present in forage crops, vegetables, grain legumes, and oilseeds. There is an obvious desire on the part of many Western consumers to replace *additives*, which are perceived as man-made, generally harmful and—specifically—cancer producing, by naturally-occurring compounds, which are believed to be harmless and safe. This perception has been challenged in the present volume and elsewhere[226,227] but it remains probable that the market for natural flavourings, antibacterial agents and anti-oxidants will substantially increase within the current decade; many saponins are attracting attention in this respect, especially in the Orient. It is probable that more will be made of biotechnological methods for increasing the yield and reducing product variability as compared to that obtained from conventional agricultural practices.[228]

In assessing the risk/benefit of saponins (or any other dietary constituent)— irrespective of whether they are naturally present in, or added to, the diet—it is necessary to have information on both the extent of dietary exposure, and on the potency of such compounds in man. Whilst analytical methods have improved significantly in the last decade there is still a pressing need for 'base-line' data on the levels of saponins present in foods and feedingstuffs, and on the effects of plant breeding, environmental and agronomic variables, processing, and domestic cooking. As will be seen from the earlier section, this information is not readily available at present. Since such data as we have at present clearly point to a significant link between chemical structure and biological activity it is necessary that such analytical methods take account of *individual* saponin content. Information on dietary exposures should reflect not only the overall population, but should also consider particular sub-groups judged likely to be exposed to particularly high levels, for examples vegetarians and 'health' product devotees. Information on exposure to particular saponins (*e.g.* glycyrrhizin) may also be important.

Aside from the obvious examples of ginseng and liquorice there is very little data available on the metabolism and biological effects of plant saponins. The

[224] S. Ishikawa, T. Saito, and K. Okada, *Endocrinol. Jap.*, 1985, **32**, 793.
[225] S. Roti, E. Violi, and D. Manari, *Med. Sport (Turin)*, 1982, **35**, 431.
[226] P. R. Cheeke, 'Toxicants of Plant Origin', CRC Press Inc., Boca Raton, 1989.
[227] M. R. A. Morgan and G. R. Fenwick, *Lancet*, 1990, **336**, 1492.
[228] T. Yoshikawa and T. Furuya, *Plant Cell Rep.*, 1987, **6**, 449.

availability of purified saponins in amounts sufficient to facilitate toxicological and, perhaps, clinical investigations should improve this situation. It should, however, be borne in mind that we do not eat saponins as such and so any studies should realistically address the fate and activity of saponins as part of the complex medium of 'food'. Thus, there is at least one indication that saponins and haemagglutinins (both contained in legumes) exhibit synergistic activity.[229] The role of saponins in lowering plasma cholesterol and the effect of these compounds on gut wall integrity are but two of the areas which will undoubtedly receive further attention in the coming years. Evidence for the beneficial or deleterious effects of saponins in plant foods and feedingstuffs will lead to pressures on plant breeders to 'improve' crops in the light of these findings. In addition, adverse consumer attitudes towards the widespread use of agrochemicals has already led breeders and plant pathologists to screen for improved 'natural' plant resistance characteristics. Such activities will result in more being learned about the cellular location of individual saponins, their effects on fungi, insect pests, and herbivores, and the opportunities that exist for manipulation of saponin levels by conventional breeding, or through applications of the newer technologies. Such researches, however, must take account of the long term effects of saponins in animals and man. Thus the future development of saponin research is seen as necessitating the fullest interaction between plant and food scientists, chemists, animal and human nutritionists, and clinicians. As Oakenfull and Sidhu[3] concluded 'in saponins, nature has provided a vast stock of pharmacologically-active and individually useful compounds'—it is now up to scientists to more effectively exploit their occurrence.

Acknowledgement

The authors are grateful to Mrs. Rosemary Williams and Mrs Shirley Newman for their assistance, patience, and forebearing during the preparation of this chapter.

---

[229] J. R. Alvarez and R. Torres-Pinedo, *Pediatr. Res.*, 1982, **16**, 728.

Subject Index

Absorption, 4, 7
Acacia, 209
    trypsin inhibitor, 77, 78
Achillea millefolium, 323
Adzuki bean proteinase inhibitor, 86
Aesculus hippocastum, 314
Aflatoxins, 227–232, 240, 247–251, 253–257
Aglycone(s), 204, 209, 214, 216, 220
    of saponins, 287
Alcohols, 17
Aletris farinosa, 323
Aleurone, 9
Alfalfa, 286
    definition, 289
    levels of saponins in, 305
Alfalfa saponins,
    effects on cholesterol absorption, 319
    effects on pathogens and pests, 310, 311
    structures, 290
    toxicity to insects, 316
Alimentary toxic aleukia (ATA), 227, 235, 236
Alkali treatment, 122, 199
Alkaloids, 14, 148–179
Allelophathy, 171–172
Allergenic factors, 9
Allergens, 13
Aloe baradensis, 323
Amaryllidaceae, 14, 149
Amelanchier alnifolia (service berry), 209
Amino acids,
    analogues, 27–30
    antagonists, 27
    dispensable, 8, 21, 22
    indispensable, 21, 22
    non-protein,
        analytical methodology, 23, 24

        applications, 41, 42
        detoxification, 46–48
        distribution in plants, 23–26
        lathyrogenic, 24–26, 36–40
        mechanisms of toxicity, 42–46, 48
        structural features, 26–28
        toxicity, 28–41, 48
β-Aminopropionitrile, 24, 38, 41
Amygdalin, 208, 214, 216, 220, 222
Anabasine, 160
Anatabine, 160
Angustifoline, 156, 157, 170
Anticoagulant factors, 9
Arachis hypogaea lectin, 60, 67
Arginine,
    analogues, 27–33
    antagonists, 27
Anthocyanidins, 182, 186
Apioglycyrrhizin, 294
Apple, 183
Araboglycyrrhizin, 294
Armyworm, 188
Ascorbic acid, 185
Asparagosides, 293
Asparasaponins, 287
    structures, 292
Aspergillus, 227–229, 242, 248–252
Astringency, 183
    in soyabeans, 290, 302
Atriplex, 18
Atropine, 148, 149, 164, 172
Autumn crocus (*Colchicum autumnale*), 152, 165
Avenacins, 311
Avenestergenins, 311
Azukisaponins, 287

Balkan endemic nephropathy, 244
Baking, 9

Banana, 183
Bamboo, 208
Barley, 1, 4, 183, 188, 226, 236, 237,
    244–247, 252, 253
  trypsin inhibitor (BTI), 89
  trypsin inhibitor family, 89
  bitter gourd trypsin inhibitor, 90
Beans, 13, 183, 191, 192, 194, 196, 198,
    199, 200
  saponin contents, 300
Beverage crops, 152, 159, 160
Bile salts, effects of saponins on,
    318–323
Bilharzia, effect of saponins on vectors
    of, 316
Biological role (alkaloids), 171, 172
Biological value, 8, 191
Biosynthesis,
  beverage crop alkaloids, 159, 160
  cocaine, 164, 165
  lupin alkaloids, 156–159
  opium alkaloids, 163, 164
  other crop plant alkaloids, 165,
    166
  Solanaceae alkaloids, 153–155
  tobacco alkaloids, 160–163
Birdsfoot trefoil, 183, 196
Bitter, 214, 223
Bitter gourd trypsin inhibitor, 90
Bitterness,
  in asparagus, 293
  in quinoa, 306
  in soyabean products, 290, 302
Bloat, 196
Bovine serum albumin, 187, 193
Bracken, 189
Bracken carcinogens, 19
Bran, 6, 7, 9
Brassica, 2, 127
  napus, 127
Bread composition, 6
  Naan, 6
  unleavened, 9
  white, 6, 10
  wholemeal, 10
Broad bean, 19
Browse, 209, 223
Buckwheat, 20
Butan-1-ol, 186
Buffa operculata, 324
Bupleurum falcatum,
  antiviral effects of saponins, 314
  tissue distribution of saponins, 303
Bupleuri radix, 307
Buxaceae, 14

Cadaverine, 148, 156–158
Caffeine, 148, 149, 159, 174–177
Cake, fruit, 6
Calcium oxide, 199
Calendula officinalis, 323
Caley pea, 12
Canaline, 28, 32, 33, 43
Canavalia ensiformis, 2, 60, 62
Canavanine,
  biosynthesis in plants, 26, 28
  detoxification, 46
  mechanisms of toxicity, 42–45
  metabolism, 32, 33
  occurrence in plants, 23, 24
  structural features, 27
  toxicity, 31–33
Cancer, 222, 230, 232, 239, 240, 257
Carbohydrates, 5, 6, 195
Cardioactive glycosides, 18
Casein, 187, 195
Carica papaya, 208
Cardiospermin-p-hydroxybenzoate, 208
Cardiospermin-p-hydroxycinnamate, 208
Cardiovascular disease, 319
Carrageenan and colitis, 273
Cassava, 203, 208, 214, 217, 220,
    223–225
Cassia occidentalis, 323
Castor bean, 13
Catharanthus roseus, 165, 170, 175–178
Cattle, 189, 196
Celery, 20
Cellulose, 7
Cereal grains, 2, 4, 8
Cereals, 1, 2, 4, 7, 14, 226, 227, 232, 237,
    239, 242–244, 247–255, 257
α-Chaconine, 153, 154, 169
Chlamydosporol, 242
Chenopodiaceae chick peas, 12, 13
Chenopodium ambrosioides, 323
Chenopodium quinoa, 286
Chickpea, saponin content, 300
Cholesterol, effects of saponins on,
    318–323
Chromatography, 211–213
Cinchona spp., 166, 175, 178, 179
Cinchonidine, 166
Cinchonine, 166
Citrinin, 229, 243, 244, 247
Claviceps, 226, 244–246
Clematis vitalba, 323
Cloves, 184
Cobalamin (vitamin B_{12}), 220, 224
Coca (*Erythroxylon coca*), 152, 160, 175,
    179

Cocaine, 164, 170
Codeine, 162, 163, 164, 176, 177
Coffee (*Coffea arabica*), 152, 159, 171, 172, 174, 175, 177, 179
Clover, 2, 16
Coeliac disease, 107, 109, 112, 113, 116, 119–121, 124
Coeliac sprue, 13
Colchicine, 152, 153, 165, 166, 175
Collagen, 186
Colon function, 269, 270
Colonic bacterial enzymes, 276, 277, 280
Colonic cancer, 258, 283, 284
Colonic carcinogenesis, 281, 282
Compositae, 14, 149
Composition,
    bread, 6
    food, 5, 6
    wheat flour, 6
β-Conglycinin, 108, 121, 122, 124
Conium maculatum, 14
Cooking, 8, 9, 15
Corn, 195
    trypsin inhibitor, 89
Cotinine, 161, 176
Cotton, 232, 238, 247, 250
Cottonseeds, 9, 227, 229, 250, 253
Cowpea proteinase inhibitors (CPI), 74, 78, 85
Cretinism, 223
Crown vetch, 196
Cruciferae, 14, 127
β-Cyanoalanine,
    biosynthesis and distribution in plants, 24–26, 215, 216
    mechanism of toxicity, 45
    toxicity, 37, 38
Cyanide, 145, 202–225
Cyanide hydratase, 215
Cynodon spp. (stargrass), 223
Cyanogens, 15, 202–225
Cyanoglucoside, 202–225
Cyanoglycoside, 202–225
Cyanohydrin, 202, 204, 210, 213, 220
Cyanolipids, 202, 214, 216, 217
Cycadaceae, 202
Cyclodextrins, 195
Cyclopentenyl glycine, 204, 206, 215
Cymbopogon citratus, 323

Deer, 189
Defence mechanisms in the gut, 109, 110
Denaturation, protein, 10
Deoxycorticosterone-mimetic activity of glycyrrhizin, 325

Deoxynivalenol (DON), 233–236, 239, 247, 248, 250, 251, 253, 257
Detoxification, 4
    alkaloids, 175, 176
    glucosinolates, 143–147
    digestive metabolism, 143, 144
    glutathione, role in, 145
    gut microflora, effects of, 143
    of amino acids, 46–48
    of crop plants and their seeds, 47
    systemic metabolism, 144–147
α,γ-Diaminobutyric acid,
    mechanisms of toxicity, 45
    occurrence in plants, 24–26
    toxicity, 38, 40
Dhurrin, 208, 209, 222
Diabetes, 225
Diacetoxyscirpenol (DAS), 233–236, 241, 247, 248, 253
Dianthus superbus, 324
Diarrhoea, 112, 121
Diet,
    availability, 5
    energy, 1, 2
    fibre, 5, 6, 7, 9
    human, 5
    lipid, 9
    phytate, 9
    roughage, 7
Dietary fibre,
    antitoxic effects, 278, 279
    caecal fermentation, 265, 282, 283
    cellulose, 259, 261
    chemical analysis, 264, 265
    definition, 258
    and drug absorption, 276
    hemicellulose, 259, 261
    and intestinal obstruction, 272, 273
    and nutrient absorption, 266, 267
    pectins, 260, 261
    physical properties, 263, 264
    physicochemical properties, 283, 284
    and short chain fatty acids, 282
    and sterol metabolism, 268, 269
    and stool weight, 271
    structure, 259–263
    and toxicology, 271
Dietary proteinase inhibitors,
    and CCK, 100, 101
    and chymotrypsin synthesis, 100
    fate in GIT, 99, 100
    general effects of, 97, 98
    growth effects, 97
    and human digestive enzymes, 98, 99
    metabolic effects, 97

Dietary proteinase inhibitors *cont.*
   and pancreatic carcinogenesis, 101–103
   and pancreatic hyperplasia, 100, 101
   and pancreatic hypertrophy, 100, 101
   physiological responses, 100–103
   and protein digestibility, 97
   and trypsin synthesis, 100
Digestibility, 1, 5
   determination, 5
   dehydrochalcones, 186
   organic matter, 196
   protein, 191, 192, 195, 199, 200
Digitalis purpurea, 18
3, 4-Dihydroxyphenylalanine, 29, 41
Dispensable amino acids, 8
Djenkolic acid, 24, 26, 30, 41

Egg production, 191
Eleutherococcus senticosus, 323
Ellagic acid, 181
Emulsifiers, saponins as, 312
Endosperm, 7, 9
Energy content, 5
   dietary, 1, 2
   digestible, 7
   food, 7
   gross, 7
   metabolizable, 7
   net, 7
   supply, 7
   value, 5, 6, 7
Enzyme,
   inhibition, 187, 190, 193, 194, 196
   kinetics, 194
   tannin complexes, 193–195
Enzymes,
   alkaline phosphatase, 194
   α-amylase, 187, 194
   cellulase, 193, 196
   glucosidase, 187, 194, 195
   lipase, 193, 194
   phosphodiesterase, 194
   takadiastase, 187
   trypsin, 193, 194
Enzyme-linked immunosorbent assay (ELISA), 122, 169, 170
Ephedrine, 148, 149
Epidemic spastic parapesis, 225
Equine leukoencephalomalacia, 240, 241
Ergot alkaloids, 229, 244–246
Ergotism, 227, 245, 246
Erythrina spp. trypsin inhibitor, 77, 78
Euphorbiaceae, 14

Factors,
   allergenic, 9
   anticoagulant, 9
   goitrogenic, 9
   safety, 11
Faeces, 5, 7, 270
Fagopyrum esculentum, 20
Favism factors, 19
Feed,
   efficiency, 191, 192, 198
   intake, 189, 191, 198
Fenugreek, 323
Fermentation, 9
Ferns, 182, 185, 207, 218
Ferric chloride, 185–187
Fibre diet, 5, 7, 9
Fibrous polysaccharides, 15
Flash chromatography, separation, and purification of saponins, 295
Flavan-3-ol, 181
Flavan-3,4-ol, 186
Flour,
   extraction rate, 9
   wheat, 6
   white, 9
   wholemeal, 9
Food,
   composition, 5, 6
   energy, 7
   nutritive value, 9
   preparation, 5
   processing, 5
Forage, 1, 14
Forage legumes, 16
Foxglove, 18
Fruit cake, 6
Fumonisins, 240, 241
Furocoumarins, 19, 20
Fusarin C, 239, 240, 256
Fusarium, 227, 232–242, 246, 247, 250, 251, 253, 257
Fusarochromanones, 237, 238

Gallic acid, 181, 184, 193
Garden bean proteinase inhibitor (GBI), 78, 84
Gas chromatography–mass spectrometry (GC–MS), 168, 170
Gas chromatography, saponin analysis, 298, 299
Gelatin, 184, 185
Geraniaceae, 17
Ginseng, 319, 324
Gliadin, 109, 122
β-Glycosidase, 280

Subject Index

Glycosidase, 211, 220, 221, 280
Glucosinolates, 13, 126–147, 208, 209
 analysis, 131–133
 biosynthesis, 128
 digestive metabolism, 143
 enzymic hydrolysis, 133
 natural functions, 128
 structure, 126, 127
 systemic metabolism, 144
 toxicity, 136–143
Glucosinolate concentrations, 127
 environmental influences, 127
 genetic influences, 127
Glucosinolate hydrolysis, 133–136
 iron concentration, effects of, 133, 135
 pH, effects of, 133–135
 protein co-factors, effects of, 133, 135
β-Glucuronidase, 280
Gluten induced enteropathy, 13
Glycine max, 286
 lectin, 52–54, 64, 66
Glycine soja, 290
Glycinin, 108, 121, 122, 124
 isotypes, 108
Glycoalkaloids, 8
 (Solanaceae) (steroidal alkaloids), 150, 152–155, 167–173
Glycoprotein, 189
Glycyrrhetinic acid, 293, 307, 325
Glycyrrhiza glabra, 286
 aglycone, 293
 levels of saponins in, 305
 saponins, 293
Glycyrrhiza inflata, 294
Glycyrrhizin, 287, 325
 human metabolism, 307
 physiological effects, 325
 structure, 293
 sweet taste, 293
Goats, 189
Goitre, 223
Goitrogenic factors, 9
Goitrogens, 13
Gorillas, 189
Gramineae, 15
Grass, 1, 2, 208
 silage, 4
Groundnuts, 9
 proteinase inhibitor (GI), 78, 84, 85
Gums, 261, 262
Gut motility, 120, 121
Gymnosperms, 207
Gynostemma saponins, 324

Gypsophila saponins, 320
 effect on mineral availability, 318

Haemoglobin, 187
Hairy vetchling, 12
Hamsters, 191, 198
Heat treatment, 199
Hedera rhombea, 314
Hederagenin, 291, 292
Hederin, 3
Heterodendrin, 208
Hevea brasiliensis (rubber), 208, 215, 216
Hexahydroxydiphenic acid, 181
High performance liquid chromatography, 187
Homoarginine,
 distribution in plants, 23, 24
 in saponin research, 294–300
 structural features, 27
 toxicity, 33
Hydroxyamino acid, 209, 210
Hydroxybenzoic acid, 180
27-Hydroxycholesterol, 155
Hydroxyhydroquinone, 184
Hydroxynitrile, 202
 lyase, 220, 221
4-Hydroxylupanine, 155, 156
13-Hydroxylupanine, 155–157, 170
Hyoscyamine, 164, 165
Hypericum perforatum, 20
Hypersensitivity,
 in avian species, 118
 in the calf, 112, 114, 117
 in humans, 112, 118
 in piglets, 112, 117
Hypocholesterolaemic effects of saponins, 318–323
Hypokalaemic effect of liquorice, 325
Hyponatraemic effect of liquorice, 325

Immune response,
 cell-mediated, 109, 110, 112, 117, 119
 humoral, 109, 110, 112, 119
2-Iminothiazolidine carboxylic acid, 220
Immunoglobulins,
 IgA, 109, 112, 118, 119
 IgE, 109, 112, 117
 IgG, 109, 112, 117, 118, 119
 IgM, 109, 112, 119
Immunology,
 and gliadin, 119
 of the gut, 109, 110
 and soyabean antigens, 114, 117, 118
Immunostimulating complexes, role of saponins in, 315

Indigofera spicata, 2, 24, 32
Indospicine,
  mechanisms of toxicity, 44, 45
  occurrence in plants, 24
  structural features, 27
  toxicity, 32
Inositol hexaphosphate, 7
Intestinal absorption, 266, 267
Intestinal lesions
  in coeliacs, 114, 116
  and soyabean antigens, 111–113, 115, 116
Iron absorption, 195
Isoleucine, 204
Isothiocyanates, 141, 142
  enzymes, inactivation of, 141, 142
  excretion, 144, 145
  thyroid function, effects, on, 141

Kale, 17
Kashin–Beck disease, 42, 237
Kwashiorkor, 231

Laetrile, 222
Lathyrogenic amino acids, 24–26, 36–40
Lathyrogens, 12
Lathyrus species, 12
  Lathyrus hirsutus, 2, 323
  Lathyrus odoratus, 12
  Lathyrus sativus, 12
  Lathyrus sylvestris, 12
Lead acetate, 184
Leather, 180, 184, 186
Leber's hereditary optic atrophy, 224
Leucine, 204
Lectins, 13, 49–67
Leek, effect of saponins in pests, 310
Leguminosae, 2, 12, 14, 15
Lens culinaris lectin, 52
Lentils, saponin content, 300
Leucaena leucocephala, 2, 23, 24, 34–36
Leuco-anthocyanidin, 186
Lignin, 5, 180
Liliaceae, 14
Lima bean proteinase inhibitor (LBPI), 78, 83, 84
Linamarin, 208, 215, 216–218, 220
Linum usitatissimum (flax, linseed), 208, 216, 217
Linustatin, 208, 216
Lotaustralin, 208, 216, 217
Lotus corniculatus (birds-foot trefoil), 208, 215, 217, 223
Liquorice, 286, 325
Lucerne, 2, 16

Luffa saponins, 324
Lupanine, 155–157, 170
Lupins, 13
  (*Lupinus* species), 150, 156–159, 167, 170, 179
  Lupinus albus, 150, 151, 156, 170
  Lupinus angustifolius, 150, 170, 171
  Lupinus luteus, 150, 157
  Lupinus mutabilis, 150, 151, 156
  Lupinus polyphyllus, 157, 158, 170
Lycopersicon esculentum lectin, 52, 55, 60
Lymphocytes, 109, 111, 112, 114
Lysine toxicity, 28, 31, 42

Macrotyloma axillare proteinase inhibitor, 85
Maintenance, 7
Maize, 1, 227, 229, 231, 233, 235, 236, 238–241, 244, 246–253, 257
Malabsorption,
  in coeliacs, 120
  and soyabean antigens, 120
Mangels, 17, 18
Manihot esculenta (cassava, manioc, tapioca, yucca), 203, 208, 214, 217, 220, 223–225
Marasmus, 231
Mass spectrometry,
  fast atom bombardment, 296, 297
  for saponin identification, 296, 297
Medicagenic acid, 291
Medicago lupulina, 303
Medicago sativa, 2, 16, 286
  saponins, 290
Methods of analysis, 166–171
  Chromatographic, 167–169
  Colorimetric, 166, 167
  Fast atom bombardment mass spectrometry, 171
  Gravimetric, 171
  Immunoassay, 169, 170
  Laser microprobe, 170
  Liquid chromatography, 169
  Near infrared reflectance spectroscopy, 170
  Radioisotope dilution, 170
  Titrimetric, 168
β-N-Methyl-α,β-diaminopropionic acid, 45
S-Methylcysteine sulphoxide,
  detoxification, 46
  occurrence, 26
  metabolism and toxicity, 40, 41
  structural features, 28, 30

Methylxanthines (purine alkaloids), 159, 170, 172, 174, 177
Mice, 191, 198
Microbial protein, 8
Micro-organisms, 217, 221, 222
Milling, 9
Mimosine,
  detoxification, 46
  metabolism, 34–36
  occurrence in plants, 23, 24
  structural features, 27, 29
  toxicity, 33–36
  and wool growth, 34
Momordica repens proteinase inhibitor, 90
Moniliformin, 238, 239, 242, 247, 248
Monkeys, 189
Monodesmoside, definition, 289
Monosaccharides, 5
Morphine, 150, 162–164, 175–177, 178
Moths, 190
Mouldy corn toxicosis, 236
Mucilages, 262, 263
Mycetism, 226
Mycotoxicosis, 226
Mycotoxins, 8, 9, 15, 20
  bioassay for, 254, 255
  carcinogenicity of, 227, 230, 240, 243, 252
  detection/quantification of, 255–257
  mutagenicity of, 230, 239, 240, 243
  natural occurrence of, 253
  phytotoxity of, 246, 247
  production of, 248–254
  sampling for, 253, 254
  synergy between, 242, 244, 257
  myelin, 224
Myosmine, 161

Naan bread, 6
Narthecium ossifragum, 20
Near infrared reflectance (NIR), 188
Neolinustatin, 208, 216
Neurolathyrism, 12
Neurotoxins, 37–40
Nicotine, 148, 149, 160–163, 170, 172, 176
Nicotine-N-oxide, 161, 176
Nicotinic acid, 204
Ninhydrin, 187
Nitrate reductase, 277
Nitrates, 17
Nitriles, 139–141, 202, 209, 210
  cellular respiration, effects on, 139, 140
  excretion, 145
  kidney, effects on, 139
  liver, effects on, 139–141
Nitrites, 17
Nitro compounds, 209, 210
Norlaudanosoline, 163, 164
Nornicotine, 160, 161
Noscapine, 150, 163
Notification of New Substances Regulations, 10
Nuclear magnetic resonance (NMR), 183
  for saponin identification, 295, 296
Nuclear magnetic resonance spectroscopy, 152, 157
Nutrition of farm animals, 4, 5

Oak, 180, 184, 190
Oat saponins, effects on disease organisms, 311
Oatmeal, 6
Ochratoxin A (OA), 229, 242–244, 252–257
Officinalisnins, 287
Oilseeds, 2
  meals and cakes, 9
Oleanolic acid, 292
Opium, 14
  (morphinan) alkaloids, 150, 162–164, 170, 175, 176, 179
Ornithogalum umbellatum, 323
Osteolathyrism, 12
β-N-Oxalyl-α,β-diaminopropionic acid
  biosynthesis and distribution in plants, 24–26
  mechanisms of toxicity, 45
  toxicity, 37–39

Pachyrhizus species, 323
Palatability, 189
Panaz ginseng, 319, 324
Papaver species, 162, 172, 178
Papaveraceae, 14
Papaverine, 162, 163
Patulin, 229, 247, 248, 252, 254
Pea, 237, 247
Peaches, 183
Peanut (groundnut), 227, 229, 250, 251, 253, 254, 257
Pears, 183
Peas, 183
  saponin contents, 300
Penicillium, 242, 244, 249, 252
Periwinkle, 152, 165, 179
  Madagascan (*Catharanthus roseus*), 165

Persimmon, 183
Pesticides, 9
  residues, 16
Peyer's patches, 109
Phaseollosides, 287
Phaseolus lunatus (butter or lima bean), 208, 215, 218
Phaseolus vulgaris lectins, 52–54, 59, 61, 62, 64, 66
Phenylalanine, 204
Phlobaphenes, 182
Phloroglucinol, 184
Photosensitive agents, 19
Photosensitization, 19
Phylloerythrin, 20
Phytase, 9
Phytate, 195
Phytic acid, 7, 9, 10
Phytohaemagglutinins, 13
Phytolacca americana, 323
Phytolaccagenic acid, 292
Pigs, 2, 4, 7, 191
Pisum sativum lectin, 52, 60, 295
Plant breeding, 199, 200
Pokeweed, 323
Polygonaceae, 17
Polymorphism, 214, 217
Polyphenolics, 14
Polysaccharide composition, 260–263
Polysaccharides, non-starch, 7
Polyvinyl pyrrolidone, 194
Potassium carbonate, 199
Potato (*Solanum tuberosum*), 150, 153–155, 171–173
  carboxypeptidase inhibitors, 88
  chymotrypsin inhibitor, 87
  polypeptide inhibitors, 88, 89
  proteinase inhibitor I (YI, 1), 87
  proteinase inhibitor II (PI, II), 87, 88
  tuber, 8
Potato protoalkaloids, 148
Poultry, 2, 5, 191, 192, 198
Poultry proanthocyanidin, 182, 186
Primary hepatocellular carcinoma (PHC), 232
Processing, 1, 5, 8, 9
Protease inhibitors, See Proteinase inhibitors
Protein, 1, 2, 4–7
  absorption, 199
  availability, 192, 193
  denaturation, 10
  fraction, 1, 197
  microbial, 8
  mucosal, 183

  nutritive value, 10
  proline-rich, 193, 198
  protection, 196, 197
  quality, 9
  radioactive, 187, 195
  rumen degradable, 196
  salivary, 192
  soluble, 196
  tannin interactions, 186, 190, 192–197
  utilization, 5
Proteinase inhibitors, 13
  amino acid sequences, 74, 75
  anti-carcinogenic effects, 103
  anti-inflammatory effects, 103, 104
  assay procedures, 91, 92
  biosynthesis, 92–94
  characteristics, 73–91
  classification, 73–91
  defence mechanism, 104, 105
  definitions, 69
  distribution, 72
  double-headed, 70
  endogenous proteinase regulation, 104
  evolutionary aspects, 86, 87
  families, 71–91
  fate in intestinal tract, 99, 100
  functions in plant, 104, 105
  human chymotrypsin sensitivity, 98, 99
  human trypsin sensitivity, 98, 99
  inactivation, 94, 97
  mechanism of action, 69
  metabolic effects, 97–103
  molecular biology, 92–104
  nutritional effects, 97–103
  occurrence, 72
  pancreatitis treatment, 104
  physico-chemical properties, 73–91
  physiological effects, 97–103
  physiological responses, 100–103
  proteinase complexes, 67, 70, 71
  protein reserves, 104
  reactive site, 69
  reactive site sequence, 78, 79
  single headed, 70
  standard mechanism, 70
  sub-cellular location, 72
Prunasin, 207–209, 216, 220, 222
Prunus spp. (almond, apricot, peach), 208, 209
Pseudoalkaloids, 148
Pseudocyanoglycosides, 202
Psophocarpus tetragonolobus lectin, 52, 54, 64

Pteridium aquilinum (bracken), 218
Pulses, 1, 2, 14

Quillaja saponaria, 312
    micelle formation, 321
    saponins as ISCOMs, 314
    use in foods, 317
Quinidine, 166
Quinine, 166, 167, 172, 177
Quinoa, 286
    levels of saponins in, 305
    saponins, 291
Quinolizidine, 155, 156
Quinolizidine (lupin) alkaloids, 150, 152, 156–159, 167, 170–172, 174, 177
Quinosides, 292
    Saikosaponins, 307
    Salvia officinalis, 323

Ranunculaceae, 14
Rape, 17
Rats, 191, 195, 197, 198
Red mould disease, 236
Redox reactions, 185
Reye's syndrome, 231
Rhodanese, 218–220
Rhubarb, 18
Rice, 1, 229, 231, 238, 245, 247, 252, 253, 257
Rice bran trypsin inhibitor, 74, 78, 86
Ricinus communis lectins, 63, 65
Rosaceae, 15, 207, 209, 216
Roughage, 7
Rumen, 7, 196, 197
Rye, 227, 245

Safety factors, 11
Sainfoin, 183, 197
Sapindaceae, 202
Sapogenols, 289
Saponaria officinalis, 312
    effect of saponins on bile acid excretion, 320
    micelle formation, 321
Saponins, 16, 285–327
    from alfalfa, 290
    analysis, 297–300
    from asparagus, 292
    dietary intakes of, 307, 308
    effect on cholesterol levels, 318–323
    effects of processing, 302, 303
    effects on birds, 316
    effects on fish, 315, 316
    effects on fungi, 314, 315
    effects on insects, 315, 316
    effects on mammals, 316–318
    effects on microorganisms, 314, 315
    effects on viruses, 314, 315
    as emulsifiers, 312
    factors affecting levels, 301–304
    from ginseng, 319, 324
    interaction with bile salts and cholesterol, 318–323
    interaction with brush border sterols, 322
    interactions with biological membranes, 313, 314
    isolation, 294, 295
    levels in foodstuffs, 299, 300
    from liquorice, 293, 325
    nomenclature, 288
    occurrence in herbal products, 323–326
    physical effects, 312
    from *Quinoa*, 291
    role in plants, 308–311
    separation, 294, 295
    from soyabean, 289
    structural determination, 294–297
    sugars, 287
    synergy with haemagglutinins, 327
Sarsaparilla, 323
Sarsaponin, use in animal feeds, 317
Schistosomiasis, effects of saponins, 316
Selenocystine, 30
Selenomethionine,
    mechanisms of toxicity, 45, 46
    metabolism and toxicity, 39, 40
    occurrence in plants, 26
    structural features, 28, 30
Sequestration of minerals and vitamins, 273–275
Sericea, 183, 188, 189
Sheep, 196, 197
Silk tree trypsin inhibitor, 77
Smilax aristolochiifolia, 323
Solanaceae, 14, 149, 152, 153
Solanidine, 148, 149, 153–155, 169
α-Solanine, 153, 154, 167, 169
Solvent extraction, 9, 10
Sorghum, 183, 185, 191, 194, 195, 196, 198, 199, 208, 214, 222, 247, 252
Soyabean(s), 9, 13, 239, 247, 286
    antigens, 108, 111
    denaturation, 122, 123
    distribution of saponins in seed, 304
    flour, 6
    meal, 4
    proteinase inhibitor (Bowman-Birk) (BBI), 80–83

Soyabean(s) *cont.*
   proteinase inhibitor family, 80–87
   saponin contents, 299
   trypsin inhibitor (Kunitz) (STI), 73–80
   trypsin inhibitor family, 80–87
Soyasapogenols, 289
Soyasaponin,
   group A, 288, 289
   group B, 288, 289
   group E, 288, 289
   hypocholesterolaemic effects, 321
   nomenclature, 288
   structures, 289
Sterigmatocystin, 240, 252, 255, 256
Squash,
   trypsin inhibitor, 90
   trypsin inhibitor family, 90
Sparteine, 155, 156, 157, 167, 170, 174, 176
Spruce, 182
St John's wort, 20
Starch, 5–7, 194, 195, 261
Stems, 1
Storage, 8, 9
Storage proteins,
   of cereals, 108
   of legumes, 108
Strawberry, 182
Sugar, 2, 5, 6, 8
Sugar beet, 18
Sulphur, 219, 220
Supply, energy, 7
Swayback, 224
Swedes, 17
Sweet pea, 12
Symphytum peregrinium, 323

Take-all disease of oats, effects of saponins, 311
Tannic acid, 195
Tannins, 14, 182
   ellagi, 181
   gallo, 181
   hydrolysable, 181, 193, 195
   oxy, 182
   polymerization, 182, 183, 199
   vegetable, 180
Tannin analysis,
   chromatographic, 187, 188
   colorimetric, 185, 186
   enzymatic, 187
   gravimetric, 184
   volumetric, 184
Taraxacum officinale, 323

Taxiphyllin, 207, 208
Tea (*Camellia sinensis*), 152, 159, 171, 179, 185
Tetraphyllin B, 208
Thea sinensis, 314
Thebaine, 162, 163, 164, 176
Theobromine, 159, 177
Theophylline, 159, 177
Thin layer chromatography, assay for saponins, 298
Thiocyanate, 218–225
Thioglucosides, 13
Thyroid, 223, 224
Tibial dyschondroplasia, 237
Tigogenin cellobioside, effect on cholesterol, 3
Toasting, 9
Tobacco (*Nicotiana tabacum*), 152, 160–163, 172, 179
Tobacco amblyopia, 224
Tomato (*Lycopersicon esculentum*), 150, 153–155, 160, 179, 232, 247
   proteinase inhibitor I, 87
   proteinase inhibitor II, 87–88
Toxic amino acids, 12, 21–48
Toxicity, glucosinolates, 136–138
   rapeseed meal, 136
T-2 toxin, 233–236, 242, 247, 248, 253, 255–257
Trichoderma viride, assay for saponins, 297, 298, 304, 314
Trichothecenes, 227, 229, 233–236, 241, 242, 246–249, 253–257
Trifolium repens (white clover), 16, 208, 217, 223
Triglochin, 208
Triglochin spp. (arrowgrass), 208, 222
Trigonella foenum-graecum, 323
Trigonella monspeliaca, 324
Triticum vulgare lectin, 63
Tropane alkaloids, 164
Tropical ataxic neuropathy, 224, 225
Trypsin inhibitors, 13, 69, 73, 76, 78–83, 89
Tubers, 1, 2, 8
Turnips, 17
Tyrosine, 204

Ungnadia speciosa, 216
Umbelliferae, 20
Urea cycle, 28, 32, 33

Valine, 204
Vegetables, 1, 2
Vicia faba, 19

Vicia faba lectin, 52, 62
Vicianin, 207
Vinblastine (Vincaleukoblastine), 165, 166
Vincamine, 165
Vincristine (Leurocristine), 165, 166
5-Vinyl oxazolidinethione, 142, 143

Water absorption, 270
Wheat, 1, 226, 236, 238, 245, 247, 252, 253, 257
 bran, 6
 flour, 6
 wheat germ trypsin inhibitor, 74, 78, 86
White bread, 6, 10
White flour, 9
Wholemeal,
 bread, 10
 flour, 9
Wild pea, 12
Wilson's disease, 224

Winged bean trypsin inhibitor (WBI), 77–80
Wortmannin, 241–242

Xanthine, 152, 156, 159, 163
X-ray crystallography, 152, 156, 163

Yamogenin, 293
Yucca mohavensis, 312
 use of saponins in food, 317
Yucca schidigera,
 effects on plasma cholesterol, 319
 improvement of ruminal digestion, 317

Zahnic acid, 291
Zearalenol, 237
Zearalenone, 227–229, 235–237, 247–251, 253–256
Zizyphus jujuba, 324
Zucchini trypsin inhibitor, 90